Centrifugal Pump Reliability 101

This comprehensive guide addresses the critical elements required to achieve reliable centrifugal pump systems in process industries. Moving beyond isolated pump components, the book takes a holistic approach to pump system reliability, examining the entire ecosystem in which pumps operate. The author draws from decades of hands-on experience to provide a practical framework that bridges theoretical knowledge with real-world application, focusing specifically on roto-dynamic (centrifugal) pumps and the complex systems they inhabit.

Centrifugal Pump Reliability 101: Practical Solutions Through Fundamentals and Case Studies distinguishes itself from other technical resources; this book delivers exceptional value through its collection of 60 detailed case studies documenting actual pump failures and their resolutions. These real-world examples illuminate the often-hidden root causes of reliability issues that might otherwise take years to discover. The content follows a cradle-to-grave approach, covering essential aspects from proper sizing, selection, and installation through operation and maintenance. Readers benefit from accelerated learning as the author distills 10–20 years of professional insights into accessible knowledge, teaching them to identify the few critical factors among many variables that most significantly impact pump reliability.

The book serves professionals across the organizational spectrum in process industries – from novice maintenance staff seeking fundamental concepts to experienced reliability engineers hunting for root causes. Project engineers will find crucial guidance on design decisions, while operations managers will understand how their choices affect long-term pump reliability.

Reliability, Maintenance, and Safety Engineering: A Practical Field View on Getting Work Done Effectively

Series Editor: Robert J. Latino, Reliability Center, Inc., VA

This series will focus on the "been there, done that" concept in order to provide readers with experiences and related trade-off decisions that those in the field have to make daily, between production processes and costs no matter what the policy or procedure states. The books in this new series will offer tips and tricks from the field to help others navigate their work in the areas of Reliability, Maintenance and Safety. The concept of "Work as Imagined" and "Work as Done", as coined by Dr. Erik Hollnagel (an author of ours), this series will bridge the gap between the two perspectives to focus on books written by authors who work on the front-lines and provide trade-off decisions that those in the field have to make daily, between production pressures and costs – no matter what the policy or procedure states. The topics covered will include Root Cause Analysis (RCA), Reliability, Maintenance, Safety, Digital Transformation, Asset Management, Asset Performance Management, Predictive Analytics, Artificial Intelligent (AI), Industrial Internet of Things (IIoT), and Machine Learning (ML).

Lubrication Degradation: Getting into the Root Causes
Sanya Mathura and Robert J. Latino

Practical Root Cause Failure Analysis: Key Elements, Case Studies, and Common Equipment Failures
Randy Riddell

Predictive Safety Analytics: Reducing Risk through Modeling and Machine Learning
Robert Stevens

Centrifugal Pump Reliability 101: Practical Solutions Through Fundamentals and Case Studies
Randy Riddell

For more information on this series, please visit: https://www.routledge.com/Reliability-Maintenance-and-Safety-Engineering-A-Practical-Field-View-on-Getting-Work-Done-Effectively/book-series/CRCRMSEGWDE

Centrifugal Pump Reliability 101

Practical Solutions Through Fundamentals and Case Studies

Randy Riddell

CRC Press
Taylor & Francis Group
Boca Raton London New York

CRC Press is an imprint of the
Taylor & Francis Group, an **informa** business

Designed cover image: Shutterstock

First edition published 2027
by CRC Press
2385 NW Executive Center Drive, Suite 320, Boca Raton FL 33431

and by CRC Press
4 Park Square, Milton Park, Abingdon, Oxon, OX14 4RN

CRC Press is an imprint of Taylor & Francis Group, LLC

ISBN: 978-1-041-21886-9 (hbk)
ISBN: 978-1-041-21701-5 (pbk)
ISBN: 978-1-003-72952-5 (ebk)

DOI: 10.1201/9781003729525

Typeset in Times
by SPi Technologies India Pvt Ltd (Straive)

Contents

Foreword

I've had the pleasure to work with Randy for over 20 years as a trusted supplier and he's one of the best problem solvers I've ever been around. Being in the pump industry myself for 22+ years. . .I highly recommend *Centrifugal Pump Reliability 101* to everyone – young professionals and seasoned professionals.

Randy's 35-year career in reliability pays dividends in this book. He shares the basics as well as lessons learned along the way. With 60 case studies, there are real-world examples for everyone in the pump world.

The book is centered on centrifugal pump reliability and is broken down into an easy-to-read format. As we all know, a centrifugal pump is only as good as its foundation, piping, operation, maintenance, etc. There are several factors that affect the life of a pump. Working for an OEM, the pump often takes the blame, but in reality, there are so many people and processes that contribute to the total life of the pump. *Centrifugal Pump Reliability 101* walks us through each part of the process in a relatable way.

Randy and I have worked together on creative solutions when the original pump was not properly chosen or the application changed over the years. Due to cost or available real estate, a new pump was not an option so we had to adapt and make modifications that would work within budget constraints.

A great example of this is in Case Study 43. The application had a lot of trash and stringy plastics/wires and the pump would often plug multiple times per day. This would typically be handled better by a recessed impeller pump, but a standard paper stock pump had been installed from the beginning and the mill had just been living with the problems and loss of production for years. A few different options were tried, but the solution ended up being fairly simple and low cost. Please read the case study for full details!

When it comes to reliability, funding can play a crucial part. I call on quite a few reliability engineers in different industries and they often find the root cause of the failure or reason for poor performance, but the solution may be cost-prohibitive. This can be very frustrating and that's why it's important to be creative in your solutions. Randy has tackled problems both large and small at each of his stops and his maintenance physics triangle is very accurate:

- Good and Fast – This combination will not be cheap.
- Good and Cheap – This combination will not be fast.
- Fast and Cheap – This combination will not be of good quality.

After reading this book, you should be able to find some value in a case study that relates to one of your real-world applications and be able to apply a similar solution that Randy has given to us all. Thank you Randy for sharing all of your valuable experience!

Brett Moore – *Sales Engineer, ITT Goulds Pumps*
Brett.Moore@itt.com

Preface

The primary scope of this book is roto-dynamic (centrifugal) pumps and pump systems, looking at the elements that will achieve pump reliability. If you work in maintenance, reliability, or engineering for a process industry that utilizes centrifugal pumps, one of the goals is to achieve optimum pump reliability. There are fundamentals that must be followed to achieve a reliable pump system. Pump reliability involves more than just the pump but also involves the pump system.

For the inexperienced professional, learning and applying fundamentals is a great jumpstart toward achieving pump system reliability. The fundamental elements of pump reliability on the front end include a broad view of pump sizing, selection, piping, and installation. When operating the pump system, it must be operated as designed. Maintenance also must be performed in a manner to restore the pump to original condition and performance. Pump system reliability like most systems encompasses the full life cycle – from cradle-to-grave approach, which also includes design, installation, operation, and maintenance.

As the next generations enter the industrial workforce, much of the reliability learnings have not been effectively passed on to that generation. The intent of this book is to document some pump reliability learnings for the next generation of professionals just coming into reliability. The term "corporate amnesia" is often applied where the learnings of yesterday are forgotten today, and the same type of failures are repeated. Part of failure analysis corrective actions is to transfer the learnings to other applications. This is one of the main elements of this book – to pass on key learnings given in the 60 case studies. The case studies highlight some key principles and learnings in real-life pump system failures, where root causes of pump failures were solved and solutions implemented. While there is an emphasis on pulp and paper pump systems throughout the pump reliability case studies, the lessons learned are still applicable to any pump system.

There are many good pump design and pump system books on the market which highlight fundamentals for pump professionals. Pump design and installation know-how has been clearly documented in pump standards, papers, and books. The case studies, where real reliability problems were solved, are what sets this book apart. As mistakes are made in pump design, operation, and maintenance, the pump reliability issues must be solved almost in reverse to find the root cause. These can seem hidden from reliability professionals trying to solve the riddle. The case studies in the book will highlight many of the basic principles that cause poor pump reliability.

Pump systems always have many things going on simultaneously, but usually only a few key issues that are contributing to most of the reliability problems. Pump professionals have to wade through the noise and determine what are the root causes effecting the reliability events. Waiting 20 years to figure out some keys to pump reliability is a long time to learn, which is how personal learnings happen. Many of the pump reliability learnings I've had took 10–20 years for me to understand and realize. Just looking at the case studies in this book, almost 60% occurred in the last

15 years of my 35-year career. The learning, understanding, and applying took time in the first 15 years before better analysis led to delivering pump reliability solutions. This book intends to speed up that process for the reliability professional.

With the advancement of artificial intelligence (AI) and machine learning (ML), there is still a great need for real intelligence (RI) from humans to interface with AI systems to solve fundamental pump reliability problems. AI needs good information input and proper interpretation of the output. That requires knowledgeable human action. The AI models may appear to reduce the human element but it is still required to understand, interpret, and apply the feedback from AI systems to achieve pump reliability.

Pump system reliability reaches all facets of an organization. For the project engineer and system designer, their role is critical in the design and installation phase. Much of a plant's reliability is designed in the front end. Pump system design can cement the reliability for years and even decades of plant operation. This book is also for the manager who decides to purchase the low-cost option or decides to cut out the engineering of the pump system to use the pump that the salesperson has chosen as a guess. This book is for the operations manager who decides to tie into an existing system without having engineering evaluate the impact to the pump system.

My learning on pump reliability never stops, and my hope is that the contents of this book can help every reliability professional improve their effectiveness in solving their pump reliability problems.

One final note of reading or referencing the content of this book. It may be read like any book beginning to end; however, for some, a different approach may be more effective. For the less experienced or beginner pump professional, reading straight through may be more beneficial. For a more experienced pump professional, just reading the 60 case studies may provide more efficient information as they already have pump and system fundamentals mastered. The book in that sense is almost a book within a book. However you choose to read, hoping some shared learnings improve your centrifugal pump reliability.

About the Author

Randy Riddell, a native of Kossuth, Mississippi, attended Mississippi State University where he received a Bachelor of Science in Mechanical Engineering. Randy currently resides in Killen, Alabama, with his wife Amy and their two children – Marlie and Lacee. Randy has over 35 years of industrial experience with a career focused on maintenance and reliability in the paper industry. His formal certifications include CMRP from SMRP, MLTI from ICML, CLS from STLE, Level 1 Vibration, and Pump System Assessment Professional, PSAP, from the Hydraulic Institute. Randy has published many articles and two books around industrial equipment reliability topics. He has also been a guest speaker at several industry and university engineering settings, and the Hydraulic Institute.

1 Introduction to Pump Reliability

PUMP BEGINNINGS

Pumps are used to move fluids from one location to another overcoming some piping system resistance. Pumps have been around since the beginning when God created the human heart which pumps 103,680 times a day pumping about 1.3 gpm at a blood pressure of 120/80 mm of Hg (mercury). Around 2000 BC, Egyptians used rotating buckets to water fields. Around 200 BC, the Greeks used a reciprocating pump supplying water to communities. Also near that same time, Archimedes invented the screw pump. In the 1400s, the first centrifugal pump was used in a mud lifting application. In the 1600s, Dennis Papin, a French physicist, invented the first straight vane centrifugal pump. In the 1800s, Henry Worthington invented the first direct acting steam pumping engine. Also in the 1800s, Seabury Gould purchases Edward Mynderse & HC Silsby Company and later makes the first all metal pump in Seneca Falls, New York. Later in the 1800s, British engineer, John Appold, designed the first curved vane impeller which greatly improved centrifugal pump efficiency.[1]

Pumps have become an integral part of society performing so many critical functions in almost every application such as industrial, agricultural, commercial, residential, and medical. As other advancements have been made, the demand for pumps and pump systems to perform at high levels of reliability have also increased. Much has been learned and forgotten in pump systems as the technology revolution has taken off. I say much has been forgotten as much of what has been learned has now been programmed into artificial intelligence models and deep within the information technology systems. The pump practitioner relies less on his own knowledge and wisdom of pump systems and more on pump AI; yet, we must never quit learning, understanding, and practicing sound pump system best practices as pump system professionals. Even relying on AI, pump application still requires fundamental knowledge of pump systems. It starts with pump safety.

PUMP SAFETY

Pump reliability and safety like other areas can have some connection or correlation at least on some hydraulic items. While there are many safety rules to follow with pumps and pump systems, below are a few safety guidelines around pump safety to always consider when designing, operating, or maintaining an industrial pump system.

1. Pumps are pressure vessels which have design limits so do not use a centrifugal pump in an application that exceeds its design pressure or temperature.

DOI: 10.1201/9781003729525-1

2. Don't ever use a suction valve to throttle a pump to control flow or pressure. More on suction issues in another chapter, but there is no benefit or reason to ever throttle a suction valve in operation.
3. Never force piping to connect to a pump suction or discharge flange or nozzle. This causes pipe strain on the pump affecting casing stress, vibration, and shaft misalignment. More on this subject in pump installation chapter.
4. Never operate a pump without proper guarding in place. Certain coupling gaps should be closed to prevent potential human interaction with rotating elements. It is worth noting that guards are meant to keep people away from rotating elements, not to contain a catastrophic failure of the pump or components.
5. Ensure the pump is assembled correctly and all fasteners tight per specifications. Pump components operating loose are not a safe condition.
6. Always isolate all energy sources when working on a pump. This seems obvious but there are many kinds of energy. Electrical energy converted to shaft power to the pump is most obvious but other hazardous energy such as potential energy from fluid pressure, flow, and chemical hazards must be isolated. Locking out a pump driver and isolating fluid from pump (also draining pump of fluid) are key steps. I know of many injuries and one fatality that occurred from not isolating the pumping fluid correctly as employees were hit with hot and dangerous fluid when the pump assembly was removed from the casing.
7. Never operate a pump in a dead head condition. This can be from operating with a discharge valve closed or a process dead head condition. There will be more on this topic in other chapters but below is a short case study of what can happen in a deadhead from a safety standpoint.

CASE STUDY 1 PUMP EXPLOSION

A paper machine starch system had a simple machine starch supply pump with header pressure control and two control valves as shown in Figure 1.1. The pump was a small ANSI $3 \times 2 \times 13$ end suction pump and had a 12.25″ diameter, four vane semi-open impeller. The pump had a design point of 325 gpm at 134 ft which was just left of best efficiency point (BEP). The pump was driven by a 30 HP, 1800 rpm fixed speed motor.

The machine control valve regulates flow to the machine size press and the header had a pressure control recirculation valve. When the machine had a break and interrupted starch flow, the machine flow valve would close. To keep the header pressure constant and starch in a holding pattern ready mode, the header pressure recirculation valve opened recirculating to the tank. This also kept the pump from a dead head condition.

The failure started when there was a functional failure in the pressure control loop. For short term, the loop was put in manual mode and the recirculation valve was closed so starch could flow to the machine. Operators were required

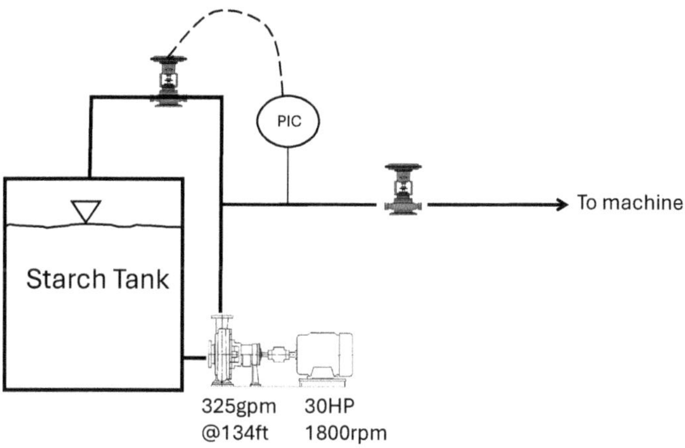

FIGURE 1.1 Starch system process flow sheet.

to remember on a machine upset condition to manually open the recirculation valve on the starch system. This worked for some time. Nothing could go wrong with this plan, right?

One day on an extended machine upset condition, no one opened the recirculation valve, and the pump ran dead head (no flow) until heat buildup over-pressurized the pump and it exploded. The starch was already at a 150°F normal operating temperature so the amount of time for starch temperatures to get extreme didn't take very long. Pump parts were found at a 25 ft radius in the area. Motor assembly was found about 20 ft away and motor wiring is what kept it from going further. This was a small unit around 25 HP but destructive none the less. Fortunately, no one was in the area and no injuries resulted but if they had been, there could have been serious injuries including a potential fatality. While this type of explosion failure is rare, it is possible under the right circumstances in pump systems.

This failure is one that had a root cause in maintenance and operations area. A failure occurred in the pressure control system, which is a maintenance-related root cause area. Operations also contributed to the failure. When the pressure control did not work in automatic mode anymore, the system was placed in manual and not operated correctly. Also, proper work order was not entered into the failed system and the corrective action was not given proper priority within the maintenance function.

The hazard and increased risk of a catastrophic failure was not recognized by anyone on the maintenance or operations team. At worst, the teams probably assumed that the primary impact would be reduced pump reliability and this was all that was considered. Also, most likely, the control valve would never be operated in manual mode. Most likely, the machine breaks were short, and the pump fluid never reached a critical temperature.

Case Study 1 Learning – Never operate pump in dead head condition. Understand pump controls and function within pump system so whenever anything stops functioning proper risk assessment is completed.

PUMP RELIABILITY FOUNDATION

Pump system reliability is just like any other equipment reliability in that when the process or plant calls for the pump system to provide a certain flow and head that the pump system can provide that. Any interruption in this function becomes a cost of unreliability and a functional failure. While some may think of reliability only being part of the job function of the maintenance department or reliability department, the fact is every part of the organization has a role to play in total reliability, from original engineering design, selection and procurement, storeroom, original installation to maintenance, operations management, and so on. Every person that has made decisions in an asset's life has impacted its total reliability up to that point as shown in the expression below. The reader will see this played out in many case studies throughout the book.

$$\begin{aligned}\text{Total System Reliability} = \sum &\text{Engineering} + \text{Purchasing}\\ &+ \text{Construction (Installation)} + \text{Operations}\\ &+ \text{Maintenance (Preventive, Predictive, Corrective)}\\ &+ \text{Safety} + \text{Environmental} + \text{Management} + \text{Other}\end{aligned}$$

The impact of all the input elements listed above might show up in a pump system by some of these past decisions by each. Engineering might oversize the pump for future needs initially which can create the pump to operate initially too far from the best efficiency point (BEP) resulting in poor pump system reliability. Purchasing may bid out the pumps using limited criteria that only require a pump meeting a simple design flow and head at the lowest price, with little or no consideration for many other pump parameters, such as pump efficiency, NPSH, cavitation potential, pump speed, shaft stiffness, and pump materials.

Construction and installation may also set up pump systems in the cheapest or fastest way to meet deadlines, without considering proper pump installation guidelines or best practices from pump manufacturer instructions or hydraulic institute standards. Installation may also ignore future maintenance access and needs for the pump system. Maintainability of the pump system is often overlooked during the initial stages of design and planning.

Operations may ignore tank levels to satisfy minimum submergence or operate pumps with suction valves partially closed or discharge valves closed too much. Automatic control systems may be run in manual mode, putting the pump system operating outside of its proper range. Other process conditions such as pressure, temperature, or consistency may be operated outside of design ranges for various reasons impacting pump system reliability.

Maintenance may not have good procedures and training for maintenance technicians to properly maintain pump impeller clearances. The storeroom may not have the right parts to rebuild a pump rotating assembly (powerend) resulting in reusing worn parts or using substitutes. The reliability group may decide not to include all the pumps on the vibration analysis route to identify pump system defects prior to failure. Lubrication routes may not have been completed for the installed pump systems.

The safety group may prohibit maintenance technicians from adjusting pump packing while the pumps are running, which may lead to either leakage or over-tightening when the pump is shut down. Unadjusted packing then leads to water in the pump and eventually bearing failure. The environmental group may decide that all pumps must have certain types of mechanical seals to prevent leakage; yet, some pumps are not suitable for mechanical seals, so the result is poor reliability of some pump systems and chronic seal failures.

Management has a much larger impact on plant reliability than many realize. Management decisions can last a lifetime or at least until the next manager change. They may make a decision to dispose of all storeroom pump parts that have not been used within three years, which can result in a significant shortage of spare parts over time. Management may decide to replace all pumps every three years based on the advertised L10 life. Management may decide that they want to operate both pumps in parallel when only one pump is needed to run the process. Management may instruct operators to keep the tank level low to prevent overflow during process upsets. Management may tell maintenance that they want every part changed on a pump regardless at every rebuild. Management may instruct pump repairs to be completed by shop A because that is who they like. They make this decision without any knowledge of the original equipment manufacturer (OEM) shop that is currently used with great success. I've seen all these and many more as just a sample.

Some of these decisions can be reversed and some can have far-reaching implications. For example, poor design at the original installation is a very costly piece of any fixed asset. These types of issues are baked in to, so to speak, the infrastructure and DNA of a facility which can be difficult to diagnose and sometimes difficult to correct with limited budgets. Many times, these issues are overlooked because people assume there are no problems with the design. It is also easy to blame the last people who worked on a system for any issues or to rely on the latest corporate program.

One question to consider is how long should my pump system last? What is good pump life? What is infant mortality? Infant mortality is where an asset fails in its infancy. Infant mortality is different for every type of asset and can be different for components within the same asset. I've seen pumps last three years or 10–15 years or maybe even only a few months before some component fails.

ISO/ANSI requirements have pumps designed for an overall life of 20–25 years based on a clean, cold water. This 25-year life is the entire pump assembly. This does not mean that all pump components will last that long. Many pump and bearing manufacturers will say bearings are chosen to operate for 40,000 hrs in centrifugal pumps. For a 24/7 operation, this would be around 4.5 years which is in line with much of this study. Look closer at an API 610 specification and the bearing L10 life is ~25,000 hrs (2.9 years). An ANSI B73.1 is ~17,000 hrs (1.9 years). Some mechanical seal manufacturers will give an average seal life of four years. Packing may be 12 months. These short life components are managed through proper maintenance including preventive maintenance.

There have been several published studies which show pump mean time between failure (MTBF) to be three to four years for ANSI pumps and six to seven years for

API pumps. Best in class pumps have been found to be near ten years. In a study I conducted in 2013 across ten papermills, I found that pumps were the third-highest maintenance cost equipment in a papermill. The average cost per year in 2013 for pump maintenance was $2.2 million annually, with an average cost per WO at $2,000. The mean time between work order (MTBWO > $500) was 3.5 years. I have also seen well-designed, installed, and operated pumps that last 10–15 years before any major maintenance. A low-reliability pump may require major maintenance every 9–12 months due to wear or failure of major components.

Obviously, there are lots of differences in the pump life results from plant to plant and industry to industry. These differences may be pump design (ANSI versus API or end suction versus double suction), process environment, applications, industries, regions, and the organizations themselves that design, operate, maintain, and manage pump systems. The key takeaway is if a pump does not last three to four years (regardless of design), then something is likely not correct. The good news is that the reliability issue can be fixed.

A reliable and well-maintained pump is also an energy efficient pump. Energy has become a major focus for most nations and for all industries. Pumps consume an enormous amount of energy not only in manufacturing but also in utilities and municipal applications. Pumps in industry can also consume as much as 25% of the energy in most process plants. The larger pump industries such as petrochem, pulp, and paper also consume as much as 6,000 Gwh/yr of energy.[2]

Pump reliability can also be evaluated by pump components. A study in one large facility revealed the pump failure by a component as shown in Figure 1.2. This is just one study in one facility so results can vary from plant to plant, industry to industry,

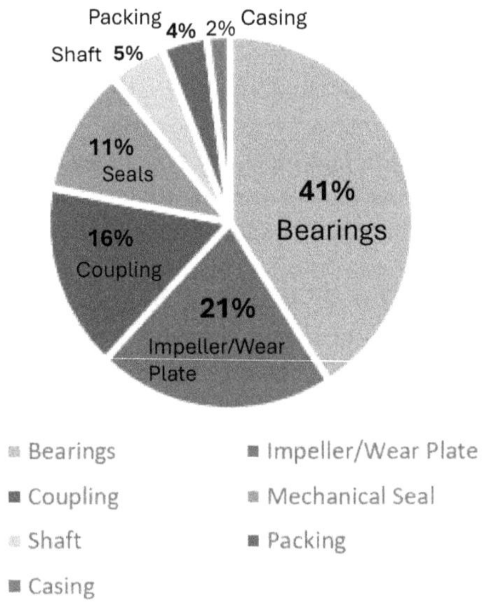

FIGURE 1.2 Pump failure components by percentage.

and pump manufacturer to manufacturer. I've seen some plants where pump wear is very high and casing failure is probably in the 10–15%, but pump bearings and pump hydraulic components are certainly typically high-failure components.

Now the component failure breakdown will depend on many other factors, but in general the list above is probably pretty typical and somewhat expected. Many of these failure components are closely linked. For example, bearings can fail after the seal has failed letting water into the pump bearing housing. If the bearings run long enough in a failing condition, the impeller may wear or rub on the pump casing, wear plate, or stuffing box. This is where root cause failure analysis becomes very important even at a high level of data recording to determine the primary root cause component failure and identify secondary failures. Secondary failures are purely a result of the primary failures that end up driving up repair cost.

Pump reliability can be affected during any part of the pump or pump system's life as shown in a DIPF curve in Figure 1.3. Many pump reliability outcomes are decided in the design and installation phase before pump system operation even starts. Design and some installation elements include the reliability DNA of the pump system. Design decisions may include pump selection (hydraulic fit and mechanical specifications), operating point with respect to AOR/POR, system design elements such as piping, controls, valves, and process considerations of submergence, air entrainment, and NPSH. Mechanical specifications will include all mechanical elements such as shaft design, couplings, driver, sealing, baseplate, foundation, and pipe stress.

Pump installation can also greatly affect pump reliability not only in the short term but also in the long term, especially for items that are not easily changed such as foundation, baseplate, and piping. Some elements of pump installation may be corrected later after initial installation, such as alignment, but some items like pipe strain can be difficult to address once structures are in place. Installation elements such as alignment can be corrected many times later, but the other design/installation elements are cemented into the plant reliability DNA once installed. After pump installation, startup and commissioning activities begin to impact the pump system reliability.

For pump operation many factors will influence pump system reliability. Is the pump being operating at its design point? Is it still within the design AOR and POR for the pump? Just some of the important questions that will factor into the P-F curve interval and determine how much life the pump will exhibit. The rest of the book will discuss many of these reliability elements with case studies showing the effects on pump reliability.

After the pump system is in operation, routine maintenance tasks begin to take effect. Many of these preventative maintenance tasks are setup during installation or startup. Some of these may include vibration analysis, oil analysis, and PM inspections. All these routine maintenance tasks continue until a defect is detected, indicated by point P, which requires intervention at some point.

All the items in design and installation actually help to prevent failure or another word extend the good running performance longer until a failure begins at point P. After an initial failure is identified later at point P, the pump system reliability can only be managed and not prevented. Many maintenance and reliability systems must be in place to not only detect a pump system defect but also to identify and help manage it to repair. Some of these include vibration analysis, oil analysis, ultrasound,

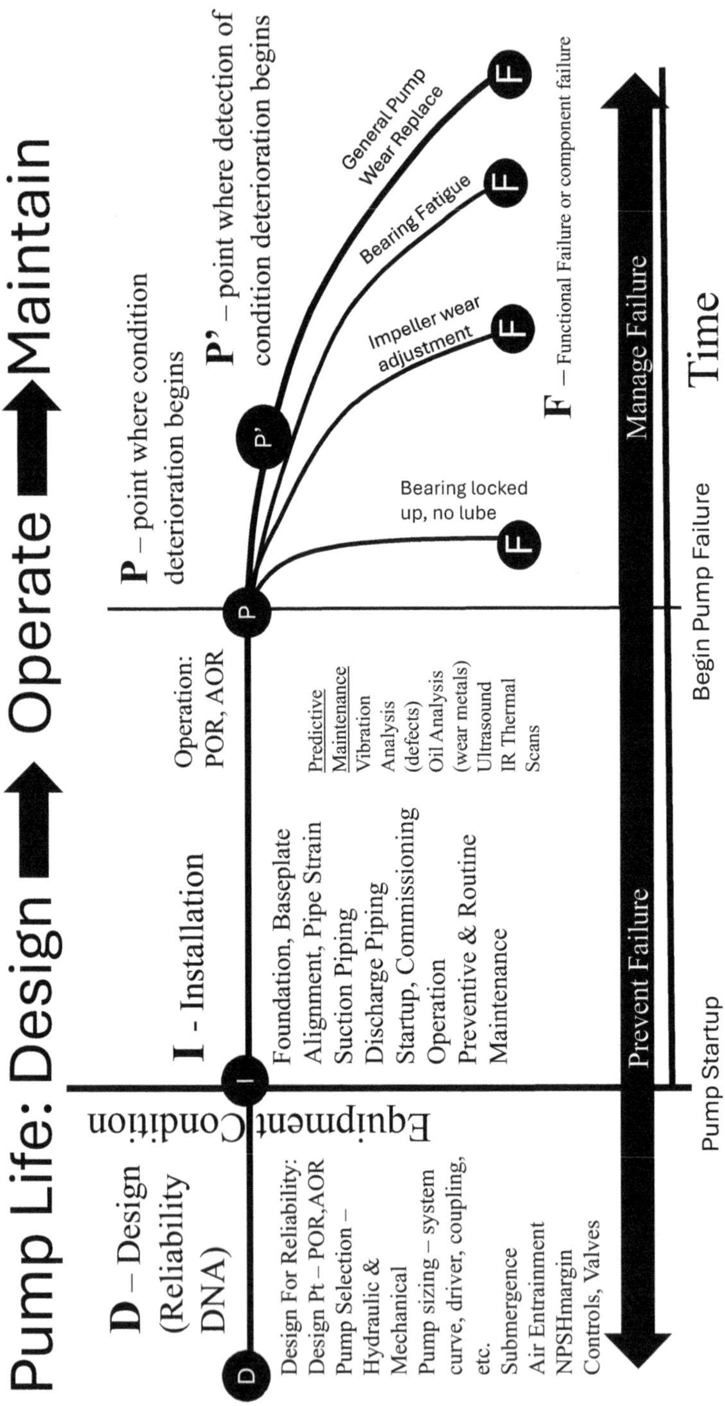

FIGURE 1.3 DIPF reliability curve pump considerations.

and thermal imaging. Once the defect has been identified, the condition is managed until it can be restored to a good or original standard condition when the DIPF cycle begins again at a new startup.

After the initial defect has begun at point P, the P-F interval at this point will be different as it approaches functional failure, F. Functional failure may be when the pump hydraulic performance fails to deliver enough fluid pressure and flow to support the process. The length of time or shape of the P-F region will vary by the type of failure mode. Another word, if the pump performance deteriorates over time due to general wear (impeller, wear plate, casing), then the P-F interval may be relatively large compared to other failure modes. A pump running without proper lubrication will have a short P-F interval – may be hours or days instead of years – before a functional failure occurs.

When the pump functionally fails, it will be in three major categories of which to have its reliability determined or altered. The pump system will have failure root cause in one of three areas – design, operation, or maintenance. Operation and maintenance can usually be corrected easily and are usually easier to find. Design issues can be harder to identify and difficult as well as costly to correct.

One key design element that will be highlighted in the book case studies is the hydraulic operating point and the impact on pump reliability. The hydraulic operating point, pump efficiency and pump reliability have a direct correlation to each other as shown in Figure 1.4. For best reliability, a pump operating at −10%/+5% of best efficiency point (BEP) is a great goal. This is a tight target so if the pump can be operating at the preferred operating region (POR) for low-specific speed pumps, then very good reliability can still be obtained. Any pump operating outside the allowable operating region (AOR) will be assured to have some level of poor reliability.

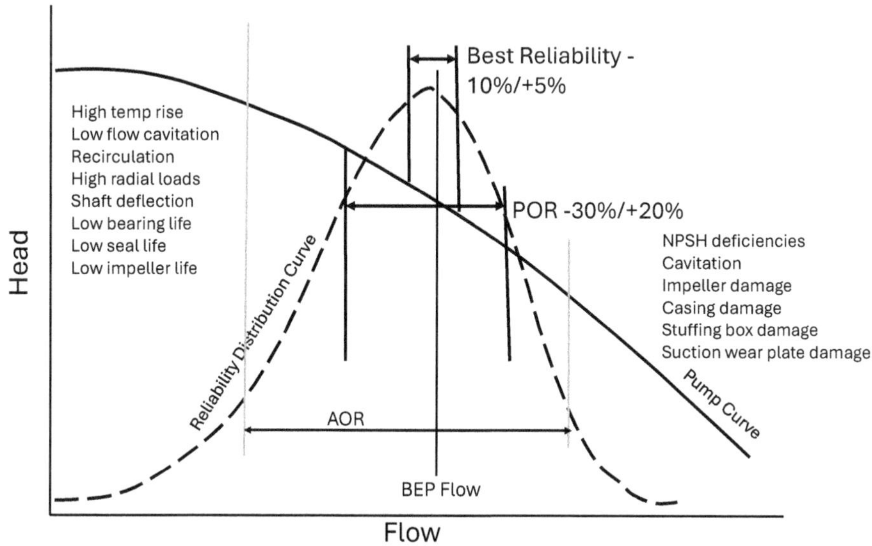

FIGURE 1.4 Pump operating point and reliability curves.

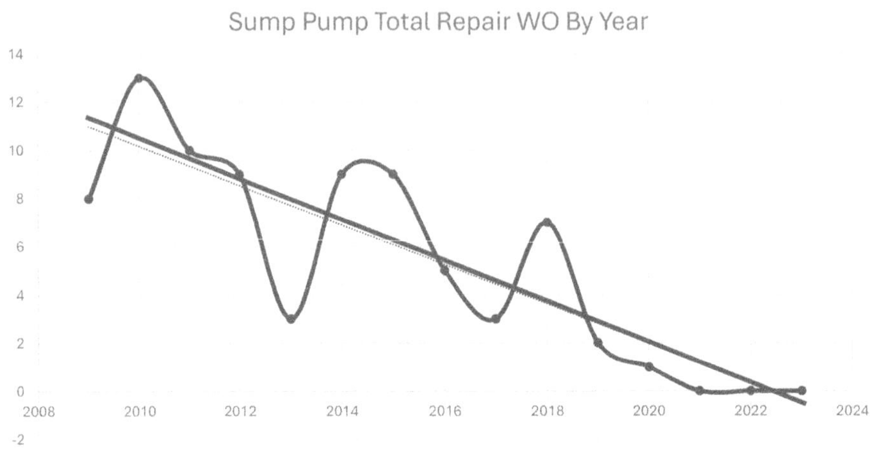

FIGURE 1.5 Maintenance work order trend showing pump reliability.

The AOR is determined by the pump manufacturer and not by some percentage of flow from BEP.

As an example, a sump pump system (Case Study 54) in a plant had operated for years with just an operating point outside the allowable operating region, AOR. The result was high pump vibration leading to damaged pump bearings, seals, shaft, impeller, and pump casing. Once the correction was made to the pump system, the pump ran for many years with very little pump maintenance. The repair work order trend of the pump is shown in Figure 1.5, which shows when correction was completed in 2019, that resulted in significant improvement in pump maintenance.

NOTES

1 Ref. Pumps & Systems Magazine January, 2012.

2 Source: U.S. Industrial Motor Systems, Market Opportunities Assessment, U.S. Department of Energy.

2 Pump Selection

INTRODUCTION

Pump selection can take on several approaches, but each requires the engineer to understand both the fluid being pumped and the system through which it flows. The end goal is to match the pump with the system so that the fluid transfer goals for the process can be achieved, and the pump also provides reliable service. If the pump selection is not right on the front end, it will be difficult to achieve pump reliability on the back end even when the process demands have been met.

Pump selection isn't the sales engineer coming in and looking at the plant and saying you need this pump. Too many times, the plant contacts a pump sales engineer requesting a pump for this process without any system design information or process baseline conditions. Pump selection involves several roles in the process plant. The end user (process engineer or operations manager) needs to set the production process needs and conditions (mainly flow and process flow sheet with all the system components). The pump system designer has to design the pump system and determine the sizing of piping, components, and calculate the system head curve. The sales engineer or supplier then will do their part to select the best pump for the specific system.

I think one simple question to answer at the very beginning is – do I even need a pump for the system? This may sound unnecessary, but if the fluid system already has a high static head and you are flowing downhill and there is enough static head to achieve the desired flow for the process then why install a pump? Most of the time the answer is no, and a pump is needed but every once in a while, the static head can provide the flow in the right system conditions. The original paper machine stuff box designs operate on this principle, as shown in Figure 2.1. A constant static head from the stuff box feeds the machine chest stock through a basis weight control valve into the fan pump suction. This all works off static head across a control valve to get the desired flow. The stuff box design is done to remove any head variations that might come from a pump. The overflow in the stuff box ensures a constant head with little variation.

I have had this happen a few other times. One occurrence involved a situation where a small flow of alum to a paper machine fan pump suction was needed. The alum tank was located on a mezzanine level while the fan pump was in the basement, providing about 55 feet of available head. Because the flow was small, all that was needed was a control valve to regulate the flow. The natural reaction is to install a pump whenever process flow is needed, but that is not always true. The process to determine if a pump is needed is a similar process to that used for sizing and selecting a pump: determine the total system head (friction and static) and size the pump.

DOI: 10.1201/9781003729525-2

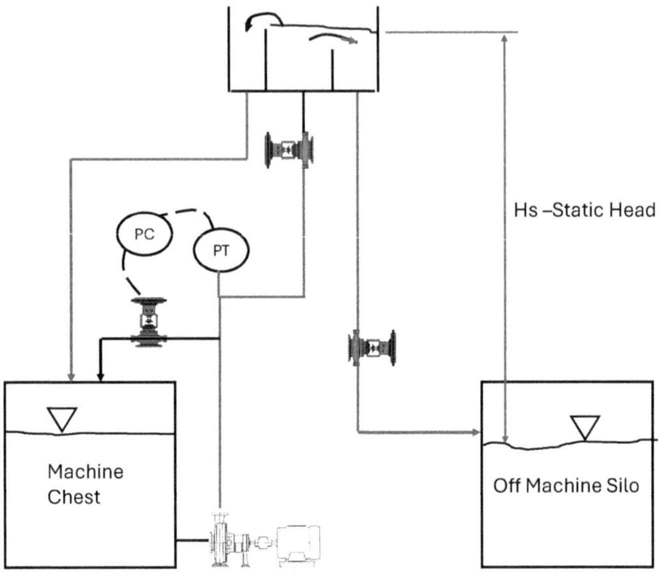

FIGURE 2.1 Machine chest process flow diagram.

In a static head system, the friction head will have to be lower than the total static head. The difference will be managed by the control valve. In the alum system example, there was around 25 ft of friction head and 55 ft of static head available. That left 30 ft of head for the flow control valve to be sized to manage the design flow. The control valve would have a 30 ft of head pressure drop across the valve at the design flowrate. The pump (static head in this case) and the system were in perfect harmony and the process was also satisfied. The result would be improved system reliability.

PUMP SYSTEM HYDRAULICS

There are several progressive steps to pump selection with each one adding more detail to refine the pump for final selection. The better this process is executed, the more likely the pump system will operate in a reliable manner to achieve its full life cycle. Mistakes on the front end of pump selection will lead to many years of unreliable performance, therefore, pump selection is an important piece of pump system reliability.

The type of fluid being pumped is very important to determine the type or model pump to use in an application. Water or waterlike fluids are the baseline for centrifugal pumps but if the fluid is a slurry, suspension, high viscosity, specialty chemical, has solids or trash, suction lift, high pressure, low flow, or high temperature, then other type or model pumps may need to be considered.

In the pre-computer days, a generic pump selection chart from a pump manufacturer showing the hydraulic coverage may be used to see several sizes of a certain

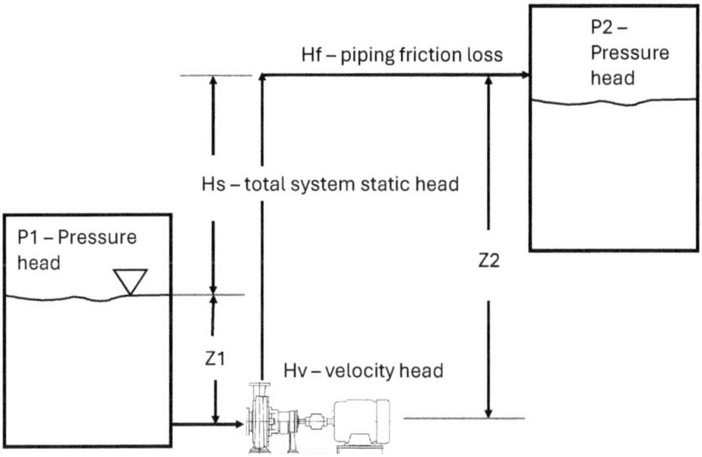

FIGURE 2.2 Typical pump system elements.

model that could be used. These are still available in most pump print catalogs. Today, pump selection software does much of the manual work to give a list of possible pumps. However, the first step to selecting a pump is to detail out the pump system. While the correct pump for a system must be selected, the system head for the pump must be determined based on the pump system makeup. The basic components for a pump system head consists of the total static head, friction head, velocity head, and pressure head as shown in Figure 2.2 and expressed by the Bernoulli equation.

A review of what is happening in a pump system is a good place to start. The pump system in Figure 2.3 shows the head (pressure) at each point as the fluid travels through the system. The pump system is really a balance of energy in various forms which are pressure head, velocity head, static (elevation) head, friction head, and pump head. This is expressed in the modified Bernoulli equation, which summarizes the total energy in a fluid at any two points in the system.

$$Hp = (Z1 - Z2) + \left(\frac{P1*2.31}{SG}\right) - \left(\frac{P2*2.31}{SG}\right) + \frac{V1^2}{2g} - \frac{V2^2}{2g} - Hf$$

where:

Hp = pump head, ft
Z = elevation above the reference (pump centerline), ft
$Hs = Z1 - Z2$, ft
P = pressure in the fluid, psi
SG = specific gravity
V = fluid velocity, ft/s
g = gravitational constant, 32.174 ft/s^2
Hf = friction head loss of system, ft

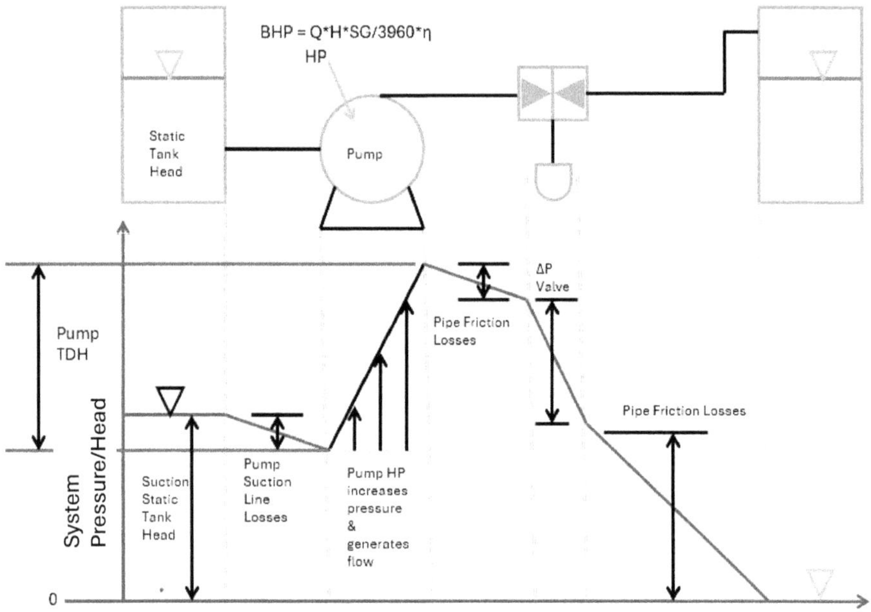

FIGURE 2.3 Pump system head and pressure effects through each component.

Apply the head energy balance for any pump system to determine the pump head required. For a system where the pressure at the supply tank and destination tank, such as in Figure 2.3, is the same, then the pressure head component becomes zero. The velocity head is typically very small and can most of the time be neglected. On many systems, the static head and friction head become the main components to determining pump head requirements.

Following the head in Figure 2.3, the system starts out with some static head in the tank. As fluid flows from the tank, friction losses occur in the suction piping, therefore, total head of the fluid decreases. The pump inputs energy into the fluid to increase head and produce flow through the system. The energy imparted to the system is fluid horsepower, also known as water horsepower (WHP). The brake horsepower, BHP, of the pump is simply the WHP divided by the pump efficiency. This is the energy transferred to the fluid as flow and head energy, and expressed by the equations below.

$$\mathrm{WHP} = \frac{Q \times H \times SG}{3960}$$

$$\mathrm{BHP} = \mathrm{WHP}/\eta$$

where:

WHP = water horsepower
Q = pump flow, gpm
H = pump head, feet

SG = specific gravity of fluid
BHP = brake horsepower
η = pump efficiency
3,960 = constant conversion to HP, ft-gal/min

The constant 3,960 in the WHP formula is the conversion to HP with units ft-gal/min. This comes from 1 HP/weight of one gallon of water. One HP = 33,000 ft-lbs/min divided by 8.34 lbs/gal of water, which is 3,960 ft-gal/min. In the WHP formula, all the units that are left is HP. This is for water, so this is why the BHP formula has SG in it to account for density of other fluids.

As the fluid leaves the pump, more system friction losses occur in piping, valves, and other process system components (filters, heat exchangers, etc.). If the friction or static head increases, then the required HP of the pump increases if the flow remains constant. For a centrifugal pump, the flow does not remain constant (more on that in pump and system section). As the fluid enters the destination tank, the elevation static head must be determined. These energy elements must balance as the pump accounts for the energy unbalance within the system. The system head and the pump head balance at a specific flowrate. This is where the pump curve and the system curve intersect.

One important design condition to consider is when the discharge piping has higher elevation head than the destination tank, as shown in Figure 2.4. This can occur as piping routing may be in a pipe rack or other route due to physical plant layouts. While the static head in the tank is lower, the system is typically designed for the highest head to overcome. Upon initial filling of the pipe on startup, the pump must overcome that head at the highest point of the piping. If the piping is in full

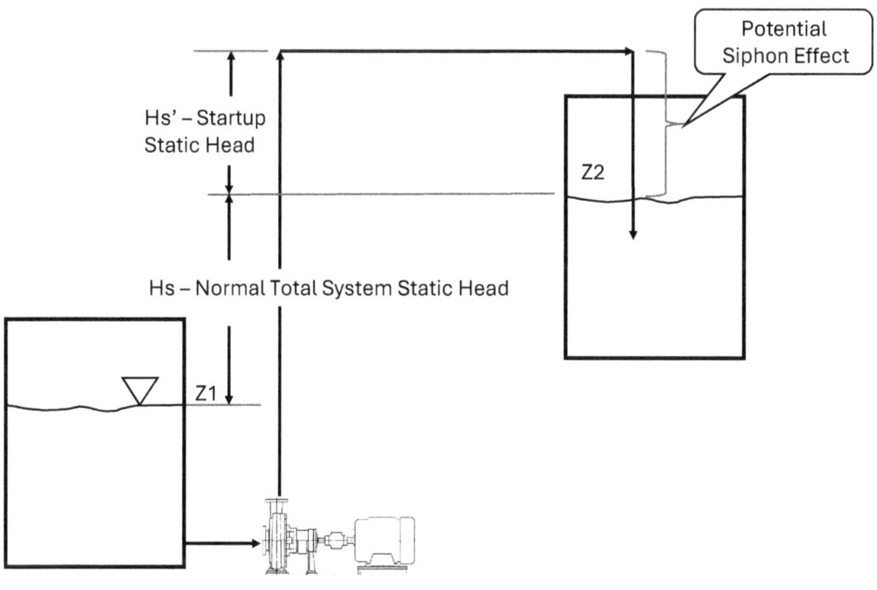

FIGURE 2.4 Static head considerations on piping.

flow, the head of the higher pipe can sometimes be offset by the siphon effect as the pipe descends. Therefore, the pump system head at steady operation may be lower than the initial startup head. On some installations, this may be enough to compensate a valve for both conditions to keep the pump operating in a reliable range.

The specific gravity cannot be overlooked on pump and driver selection. Small changes in SG could change motor sizing. Pump curves are typically figured with water at 68°F, therefore, handling hot or cold fluids will change pump HP. Pump curves are shown in feet of head so any density fluid can be used with one curve. Pumping a higher density fluid may mean needing a larger HP motor, so verify this upfront, especially when the design point is close to the motor HP selected.

The wire to water pump efficiency and HP defines the energy from the motor wiring into the pump system through to the pump fluid. The BHP in a pump system can be measured and evaluated at any point in this machine system, if the right parameters are known. The BHP from the motor or the pump direction is equal where the two shafts connect.

The electrical power into the motor is the highest power number and the water horsepower at the pump fluid end is the lowest. Energy is consumed or lost from the input of the motor wiring to the fluid on the pump end. The wire to water efficiency represents these losses through the pump and drive system. Figure 2.5 shows the pump system power from wire to pump fluid. Available power decreases from the wire to the fluid as losses add up. These losses not only cause a reduction in available power but also generate heat, vibration, and noise. The efficiencies can be expressed by the equations below.

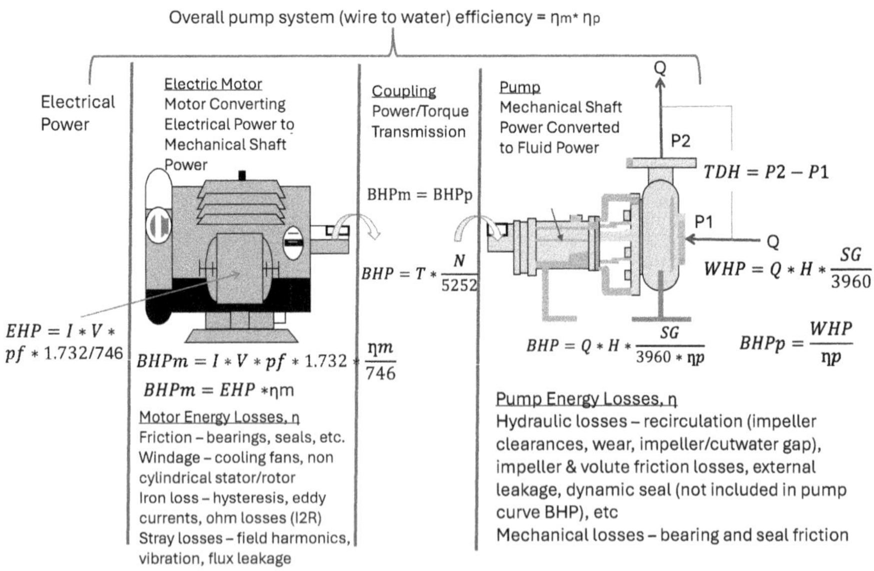

FIGURE 2.5 Pump BHP from motor to pump (wire to water).

$$\eta S = \eta m \times \eta p$$

where:

ηS = system efficiency (wire to water)
ηm = motor efficiency
ηp = pump efficiency

Electrical power is the source power for an electric motor. For three phase power, the electrical horsepower, EHP, can be expressed as a function of current, voltage, and power factor as shown in expression below. The electric motor is a machine that converts the electrical power to mechanical shaft power and includes motor efficiency. Motor efficiency accounts for motor energy conversion losses.

Energy efficiency losses include mechanical friction such as bearings and seals. Windage losses may include cooling fans, non-cylindrical rotor, or stator. Iron losses may include hysteresis, eddy currents, ohmic losses (I2R). Stray losses may include magnetic field harmonic losses, vibration, noise, and flux leakage.

$$\text{EHP} = I \times V \times pf \times \frac{1.732}{746}$$

$$\text{BHP}m = \text{EHP} \times \eta m$$

where:

I = current, amps
V = voltage, volts
pf = power factor
1.732 = 3 phase constant
746 = watts to HP conversion

While there are some very small coupling friction losses, the BHP at the shaft couplings is basically the same. Once power goes into the pump shaft, the pump proceeds to convert this mechanical shaft power to fluid energy. The WHP is simply the energy change in the fluid, which involves the total dynamic head that the pump has added to the fluid, fluid flow, and specific gravity. The TDH is the difference between the pressure at the pump suction and the pressure at the pump discharge, both referenced to the pump centerline as shown in Figure 2.5. The BHP of the pump can be quantified by the expression below.

$$\text{BHP}p = Q \times H \times \frac{SG}{3{,}960 \times \eta p}$$

In this energy conversion, there are also many losses – mechanical and hydraulic. The mechanical losses are mostly friction from bearings and seals. Hydraulic losses include fluid friction losses from impeller and casing, internal leakage (recirculation) from impeller clearances, wear, and cutwater gap. If the pump is equipped with a dynamic seal, this also requires energy and is not counted in the BHP on the pump curve.

In evaluating the BHP from the motor end or the pump end, the BHP is essentially equal in the connected pump machine set (motor and pump). Evaluating the BHP of the pump system is a good way to find the pump operating point and verify field data as they should both be close to the same amplitude.

$$\text{BHP}m = \text{BHP}p$$

There will be many examples of pump efficiency application within the various case study examples throughout the book. Pump efficiency not only affects pump HP but also the flow and head the pump can develop in a given system. Other than higher energy cost, pump efficiency may or may not affect process operation but can also have other implications. For example, pump performance data is used to predict/ estimate the flowrate delivered by the pump.

A specific example might be where a pump system uses a pump timer in a batch pump operation to log how many gallons are being processed to an external waste-water facility. This average pump flow might be used to bill for X gallons being processed at a wastewater plant. As the pump efficiency decreases, the pump will run longer. If the pump efficiency isn't taken into account, it will appear that the total gallons pumped is increasing but it is not. The efficiency is lower so the pump will have to run longer to pump the same gallons. The user will then be paying for more gallons of treatment since the pump rate was not changed.

PUMP SIZING AND SELECTION

Selection and sizing of a pump are two similar but different tasks for a pump system. Sizing has to do with doing hydraulic analysis of the system at the process flowrate to determine the system head and pump total dynamic head. Selection has more to do with all the specifics for the pump that has been selected for the application starting with the type of pump.

Pump sizing is the first task to evaluate on a pump system. The pump sizing may eliminate many styles of pumps before the selection process begins. A step-by-step process for sizing a new pump for a system follows five key steps as outlined below.

1. First, the end user provides the system design requirements for the process system. How much flow is required? Pumping from where to where? Any special system requirements such as special process equipment in the system (screens, filters, control valves, etc.). What process control strategy is desired for the system – pump on/off control (batch), recirculation control, adjustable speed drive (ASD), inline control valve, etc.? These inputs are supplied by the end user for the system being designed and the pump being selected. The process drives the selection criteria and what is required to satisfy the process design conditions.

Determining the flow required can be as easy as knowing the process flow into and out of a tank or other process. Maybe a 10,000-gallon tank needs to turnover in 5 minutes, so 2,000 gpm would work. In pulp and paper, the driving factor might be the tons per day (T/D) needed through part of the process. The paper stock gpm is a function of stock consistency and T/D by the expression below.

$$Q = \frac{\text{TPD}(16.67)}{C}$$

where:
Q = flow in gpm
TPD = oven dried tons per day
C = consistency in % (not decimal)

2. Layout a process flow sheet for general system understanding. This will include much of the detail from step 1. The process diagram will give a visual graphic for the system and the foundation for the system designer and engineer as other calculations determine pump sizing.
3. Size the piping in the system which includes suction piping and discharge piping. Pipe sizing is based on target fluid velocities for the design flow required. Also get a rough pipe run isometric, if possible, for a way to know the elevation head, pipe lengths, pipe elbows used, etc. This is needed to do detail friction head calculations.
4. Calculate the pump TDH required based on the system layout from step 3 and based on the design flowrate. Calculate the TDH by knowing all the components of Bernoulli equation, such as the total static head, pressure head, velocity head, and friction head. Also size the control valve for the system if it has one. This is just a good time to do that so a more accurate head results from the specific control valve pressure drop in the system. Output of this step is the pump total dynamic head, TDH.
5. With the flow and total dynamic head of the system, select the pump and motor from the pump design point (flow and head). In general, select a pump nearest to the best efficiency point, BEP.

Much of the calculation work is in the friction losses. There are two main types of friction losses – major losses and minor losses. The name does not mean one is necessarily larger than the other. Major losses are pipe friction losses, while minor losses are fittings, valves, etc. System friction losses can depend on several factors including surface roughness of the pipe, length of the pipe, and the type of flow (laminar or turbulent) inside the pipe. The type of flow will be determined on the flow and size of piping which will be the average fluid velocity. Friction loss in piping systems is directly proportional to fluid velocity as head friction loss is a square of the fluid velocity.

There are several different methods to determine friction loss in piping systems. Many use several different equations, tables, or curves to determine the friction loss. The scope of this book will not allow a detailed look at each of these methods but will summarize each below. I suggest acquiring a good hydraulic reference book, such as *Cameron Hydraulic Data* by Ingersol Rand, for friction loss calculations. This book has been a credible reference since its first publication in 1926 and is republished every five to ten years. Also, today there many software products available on the market for these calculations and system modeling. Three common manual methods for determining friction losses are summarized below.

1. Darcy-Weisbach formula: This can be applied to any fluid calculating major head losses on new piping. Variables include fluid velocity (flow), pipe diameter, length, friction factor, Reynolds number, and pipe surface roughness as shown in the expression below.

$$H = fLV^2(D2g)$$

where:
H = head loss, ft
D = pipe ID, ft
f = friction factor
L = pipe length, ft
V = fluid velocity, ft/s
g = acceleration, 32.174 ft/s^2

2. Hazen and Williams formula: This method only applies to fluid with water properties at 60° F in ten-year-old pipes. Any pipe system different from this reference will result in some error in the calculations. Variables for this method include flow, pipe diameter, length, and friction factor. Friction factor is typically obtained from a table depending on the material and type of pipe.

$$Hf = 0.002083L\left(\frac{100}{C}\right)^{1.85} Q^{1,85}/d^{4.8655}$$

where:
C = Hazen Williams friction factor
Q = flow in gpm
L = length of pipe, ft
d = pipe inside diameter, inches

Friction losses in valves and fittings can be calculated using the resistance coefficient, K. This coefficient can be found in various tables from manufacturers. Many manufacturers also may supply their own pressure drop curves.

$$H2 = K\left(\frac{V^2}{2g}\right)$$

where:
K = head loss factor for valves, etc.
V = fluid velocity, ft/s
g = acceleration gravity, 32.174 ft/s^2

3. Equivalent Length Method: This method may be the easiest for combining major and minor friction losses. This method converts every friction loss into a length of pipe estimate. When the system total length is compiled then a simple head loss per ft of pipe constant can be used to get total friction loss of the system. Tables are referenced for piping, valves, and fittings to equal an equivalent length of pipe. This equivalent length of pipe can be used in other friction loss formulas. If you are using friction loss tables, many will show the friction loss per 100 ft of equivalent length of pipe.

   If using equivalent length method, pay attention to the friction loss tables as to what kind of pipe is being modeled. The friction factor or surface roughness can significantly alter results. When using water in steel or cast-iron pipe, for example, there could be significant difference in stainless steel pipe friction losses. The *Cameron Hydraulic Book* does not list tables for stainless steel pipe but does for steel pipe, cast iron pipe, and copper pipe. I found that using copper friction loss tables provides a more accurate calculation when compared to stainless steel piping. The friction factor and surfaced roughness is similar (both had a shiny, mirror-like finish on the pipe ID).

Other fluids may require special friction data to calculate friction losses in the pump system. One of those is pulp and paper stock. Up to 2% paper stock consistency can be treated like water for friction loss calculations, but above 2% consistency, specialized friction loss curves or calculation methods must be used.

Once all the head components are known, the total dynamic head, TDH, that the pump must produce can be calculated using the modified Bernoulli equation. The flow and head will represent the pump design point where pump selection can be made. Therefore, when a possible list of pumps is available, which pump should you select? Which criteria is the best to make the pump selection?

There are a number of potential areas to compare such as pump efficiency, design point percentage-rated BEP, full curve runout, relative position to left or right of BEP, NPSH, NPSHmargin, pump-specific speed, suction-specific speed, or rise to shutoff head. One thing to note is moving to the right of BEP, NPSH, and cavitation issues become more of a concern so avoid that area if there is a pump just to left of BEP. How much NPSH margin is covered in Chapter 5 on pump suction issues.

Pump hydraulic efficiency is achieved when the flow enters and exits at vector angles that match the impeller vanes. Alter pump flows and these inlet and outlet angles don't match the impeller vane angles, resulting in lower pump efficiency and other issues. Pump efficiency is becoming an increasingly important factor in pump design and selection. There are now more stringent DOE standards for clean water pump efficiency.

Even if an efficient pump is selected, the pump and system design and fit are the major factors in achieving actual efficiency and power savings. An efficient pump does not matter if the pump is not operating at an appropriate point on the performance curve. Efficiency must also be kept in perspective with each application. A 1% efficiency gain on 10 HP is 0.1 HP, therefore, it is not a big energy savings; however,

Pump Type	Maximum Efficiency, %
Double Suction/Split Case	87
Large End Suction	85
Axial Flow	83
End Suction ANSI	81
Multistage Boiler Feed	80
End Suction Stock	75
End Suction Sewage	72
End Suction Slurry	66
Self Priming Suction Lift	58
Recessed Impeller	51

FIGURE 2.6 Pump efficiencies by pump type.

a 1% efficiency gain on 1,000 HP is 10 HP. Always reference the pump-specific curve information for not only efficiency data but all other performance data. When specifying pump efficiency, note that each type of pump has its own maximum potential efficiency. Some pump types and their maximum efficiencies are shown in Figure 2.6.

It is generally not good to pick just one area to make pump selection, as there are usually compromises on good fit for overall pump performance, efficiency, and reliability for each specific application. For pumps with low duty cycle (not running very often), efficiency may be a less significant factor. More reliable pumps are preferred over more efficient ones due to the risk of production interruptions and the cost of rebuild or replacement. However, there is some good correlation between pump efficiency and overall reliability. Efficient pumps also tend to be reliable pumps.

The total pump system efficiency involves pump efficiency, motor efficiency, and any power transmission efficiency. Motor efficiency will depend on motor size and also how loaded the motor is operating. Below 50% motor load, the motor efficiency tends to drop in most electric motors. Power transmission efficiency is the other mechanical components that connect the motor to the pump, such as a coupling, belt drive, or gearbox. Refer to motor manufacturer tables, curves, or nameplates for specific motor efficiency, but most NEMA motors will have a motor efficiency in the 93–96%. For power transmission, a well-maintained V-belt will have 95% efficiency, gearing at 98% efficiency, and a coupling will be at 99%, just to name a few.

$$\text{Total Pump System Efficiency} = \text{Pe} \times \text{Me} \times \text{PTe}$$

where:

Pe = pump efficiency
Me = motor efficiency
Pte = power transmission efficiency (coupling, belt drive, gearbox, etc.)

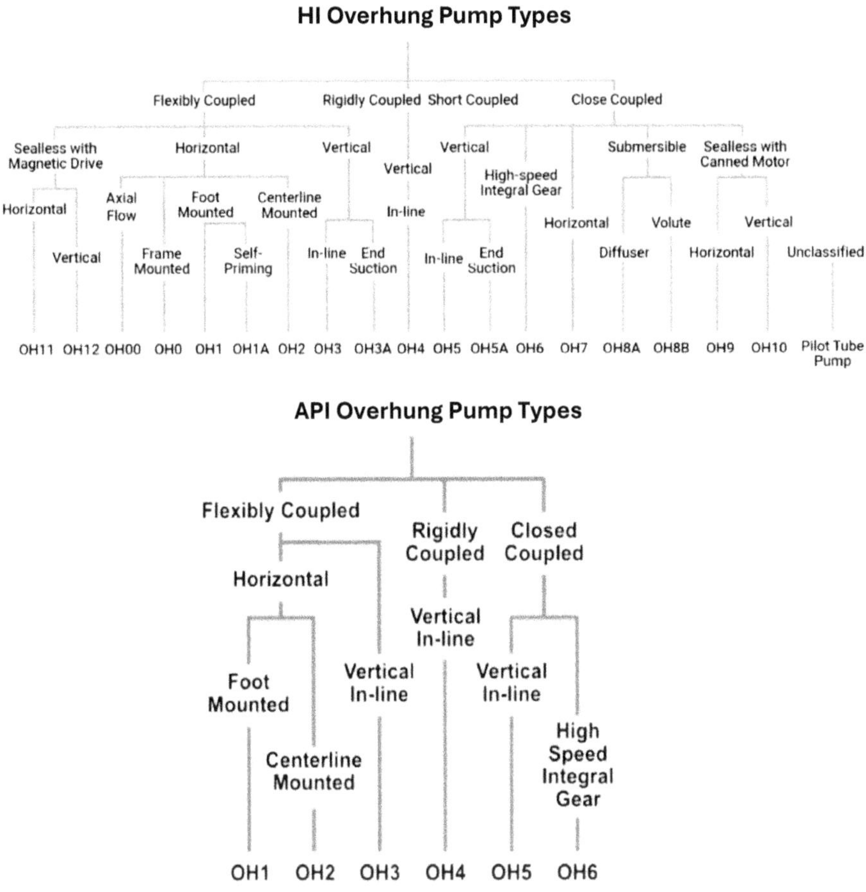

FIGURE 2.7 HI and API overhung pump classifications.

Pump selection will also choose between many different styles or classifications of pumps. The Hydraulic Institute (HI) and American Petroleum Institute (API) each have their own pump classifications. The HI classification is similar but has more pump classification. The HI and API overhung pump classifications are shown in Figure 2.7.

For many process industries, the OH1 is the workhorse pump and OH2 is also very popular in petrochemical manufacturing. Overhung pumps are pumps in which the impeller is mounted in an overhung position on the shaft assembly. They will have a pump flow capacity up to 10,000 gpm and heads up to 1000 ft. Drive HP may also go up to 700 HP. OH1 pumps are generally used for services $< 300°F$ but may be used at higher temperatures with cooling and offsets for thermal growth on shaft alignment. OH2 pumps are preferred at higher temperatures since the thermal growth is nullified with center-supported casings. Most of the remainder of the book focuses on discussions and case studies on OH1-type centrifugal pumps.

REPLACEMENT PUMP SIZING

For replacing an existing pump in service, there are several things to verify. Perhaps the most important thing to satisfy is that the hydraulic performance meets the process demands of the application. The worst thing that could happen when replacing another pump is for the new pump not to perform at the level as the previous pump (see Case Study 2).

There are a couple of approaches for the replacement situation. First would be to just match the existing pump design flow and head design operating point. This is a safe approach which will typically deliver at least the same or similar performance and reliability as the previous pump.

The second and probably preferred option is to study and determine the required pump for the operating process system. This may be different from the original design and may provide a great opportunity to improve the future of the new pump for improved reliability, energy usage, and process performance. Opportunities to

CASE STUDY 2 PUMP SELECTION LESSONS TAUGHT BY THE SCHOOL OF HARD KNOCKS

About a year out of engineering school, I was a young engineer assigned one of my first major pump projects. I had worked on some small pumps and piping (positive displacement pumps), but I had not yet sized and installed a large centrifugal pump. The project was straightforward, and the basic scope was to install a larger pulper pump on a batch pulper system that feeds a paper machine. I was instructed to make it the same size as another similar pulper, assuming that will be sufficient. Ok, sounds simple enough. Remove the older, smaller pump and install a newer, larger pump. Copy the pump that was on the other pulper. Got it. A good first project for a new engineer. Hit the easy button. What could go wrong?

I went straight to the engineering files looking at the specifications on the other pulper pump that the production managers had asked me to replicate. I got all the specifications and proceeded to get the pump quoted for the other pulper. It was basically a similar piping system as the other pulper that was sitting in the same location. I don't recall doing a lot of calculations, after all the systems were basically the same and I was putting the same pump from one system to the other. As a new learning pump engineer, I should have gone overboard on calculations just for the learning part if nothing else. I was also learning how to manage capital projects too, so there was a lot of new information to digest at once.

The project progressed and a short outage allowed installation of the new pump. I had doctored up some basic flowsheets and P&IDs. The tie-ins were easy, so I just let the piping contractor field route the short inlet and discharge sections. The first surprise came when the discharge piping would not fit the existing pipe. This mill was around 100 years old at the time and some systems, as I learned, had fractional-sized pipes. I don't recall the exact size, but it was around 15.5″. The piping contractor had to make some adjustments to his plans but was able to make it work. These pipe fitters can make any engineer look good.

The rest of the project installation went in as planned. As a young engineer just coordinating all the installation was a pretty good accomplishment with only one minor issue on the piping. The second surprise occurred after startup. Startup of the pump went smooth, and I walked away feeling pretty good. The pump started up and it pumped stock from pulper to the machine.

Then I received a call that the pump was not performing as expected and was taking too long to empty the pulper. It was not pumping like the other pulpers as expected after the pump upgrade. After some investigation, I found out that I had not installed the correct size pump. There was a pump upgrade on the other pulper that did not have the equipment history files properly updated, so I copied a smaller pump than what was currently on that pulper.

So, one lesson right there is about how important equipment records are, but also never blindly trust equipment or computer records alone without verification. Always verify in the field what is actually installed. I was a young engineer and thought I had done enough investigating, but I had not. As an inexperienced pump engineer, I made many mistakes and there were a lot of things I could have and should have done. Looking back now as an engineer with many years of pump experience, following are four of the things I could/should have done.

1. Get serial number of the pulper pump that I was to copy. Then I would know the target flow and head of the other pulper. I could also have measured the pumping time for the pulper I was to copy, to determine the flowrate it was pumping out. This would have been an easy exercise in a batch process.
2. Get the pulper volume and ask production manager how fast he wanted the pulper to empty and this would set my flowrate. They likely knew how fast they needed it to pump to keep up with the process. Would have been good if the process manager had communicated that but then again, as the project engineer, I should have asked too. Again, another easy datapoint to compare as I sized the new pump.
3. Related to above, the ton per day, consistency, pulping time, and pumping time could all be calculated to determine the target pumping rate. After all, the production manager didn't really know a pump flowrate either, he just liked how the other pulper emptied out faster.
4. Model the pump system so I would know the exact head for the correct flowrate to size and select the new pump. This modeling could have been just calculating the total head of the system or even using the piping affinity laws at least to determine what new head I might be looking at with a larger pump.

As a young engineer, I'm sure I was assigned this project due to its simplicity and not much could go wrong, but those are often famous last words. Thank God management didn't give up on me and continued to give me pump system projects.

This pump system failure had pump system design as the root cause. Due to poor engineering records and insufficient field work in the scope, the pump was underdesigned for what was needed on the project. In the end, the pump underdelivered to achieve its purpose. A redesign would be required to attain an increase in pump performance.

Fast forward almost 30 years and another replacement project came up to be scoped out. A paper stock pump (#2 pump) in a floatation cell system needed to be replaced as there was no spare pump and the original pump was now obsolete, and parts were not available. When drawings were pulled up, the pump specifications showed the pump to have a design point of 8,500 gpm at 70 ft. There had been no major changes on this system since it was newly installed 20 years earlier.

After 30+ years of experience and remembering how easy an error can happen, the latest process drawing information was not solely used to scope and size the replacement pump. Taking a field trip to verify the pump specifications before moving forward. Instantly an issue was discovered. The existing pump was sized for 7,925 gpm at 37 ft per the pump manufacturer nameplate design data. The other floatation pumps (3,4,5) were also the same size as #2. There was no technical reason for #2 pump to differ from pumps 3, 4, and 5, as the systems were essentially identical.

Looking deeper at the specification, the process drawing on #2 ended up matching the specifications for #1 feed pump, which was in the basement. This explained having a higher head on the pump specification. The process flowsheet verified that there was a typo in the plant process drawing as someone copied and pasted #1 pump specification to #2. While the larger pump would have been more expensive on the front end, it would have run at a lower flow and head condition; however, the system inefficiency and wasted energy that would have resulted would have had a significant long-term impact. The lessons in my youth had taught me some things about verifying what is actually in the field. A lesson that stuck with me for decades.

All this reminded me of a story about learning from experience. A young learner asked the experienced veteran, what is the key to success? His answer, two words – good decisions. The learner asked, how do I do that? The mentor answered, one word – experience. Well, asked the learner, how do I gain experience? Mentor answered, two words – bad decisions. I made a bad decision on my first pump system project, and it taught me a valuable lesson.

Getting experience can come from mentoring/shadowing a seasoned veteran in the business or learning it yourself. I have had some of both in my career and been blessed by those who shared their learnings with me. Whichever way learning happens, learning from past mistakes and applying those to future pump system projects and improvements is the result and benefit of real-life experience. It is the intent for sharing these learnings from the case studies shared in the pump reliability examples in this book.

Case Study 2 Learning – Verify pump specifications and performance from more than one source to increase the probability that the pump will deliver expected results. Always verify in the field and do some basic sizing evaluation.

make life-long pump and system changes don't come along very often, so make the most of the opportunity when replacement becomes available.

With either option there are some things to find out on the pump to be replaced. Below is a short list to gather if the information is available.

- Existing pump curve and pump performance (operating points).
- Impeller type, size (diameter).
- Pump system operating pressure.
- Pump motor load – amps, voltage, and speed.
- Current pump reliability and maintenance history.

If the pump is expected to fit on the existing pump baseplate, then all the physical dimensions will need to be checked. If it is a small process American National Standards Institute (ANSI) pump, then the ANSI dimensions should already be a fit if the replacement pump is an ANSI-built pump. The ANSI pump standard was established in 1977 to apply to end suction pumps (OH1 class). The more common dimensions are ANSI standards, such as pump height, nozzle locations, pump shaft size, overall length and height, hold down bolt locations among a few others as shown in Figure 2.8. As seen in Figure 2.8, not all dimensions on the pump conform to ANSI standards, so additional dimensions should be verified before blindly replacing.

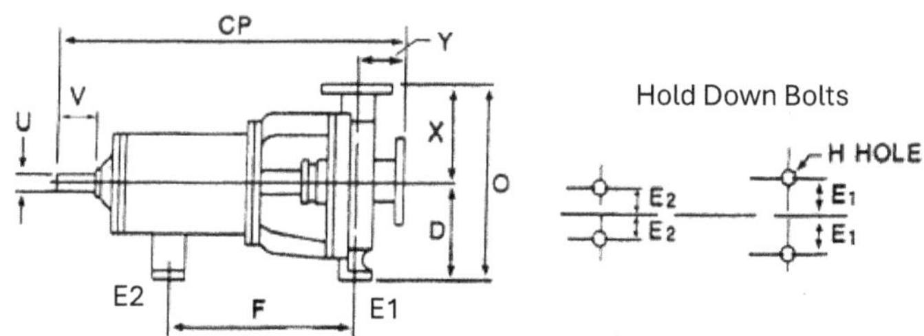

Dimension designation	Suction × discharge × nominal impeller diameter	CP	D	2E₁	2E₂	F	H	O	U Diam.	U Keyway	V, min	X	Y
AA	1½ × 1 × 6	17½	5¼	6	0	7¼	⅝	11¾	⅞	3/16 × 3/32	2	6½	4
AB	3 × 1½ × 6	17½	5¼	6	0	7¼	⅝	11¾	⅞	3/16 × 3/32	2	6½	4
A10	3 × 2 × 6	23½	8¼	9¾	7¼	12½	⅝	16¾	1⅛	¼ × ⅛	2⅝	8¼	4
AA	1½ × 1 × 8	17½	5¼	6	0	7¼	⅝	11¾	⅞	3/16 × 3/32	2	6½	4
A50	3 × 1½ × 8	23½	8¼	9¾	7¼	12½	⅝	16¾	1⅛	¼ × ⅛	2⅝	8½	4
A60	3 × 2 × 8	23½	8¼	9¾	7¼	12½	⅝	17¾	1⅛	¼ × ⅛	2⅝	9½	4
A70	4 × 3 × 8	23½	8¼	9¾	7¼	12½	⅝	19¼	1⅛	¼ × ⅛	2⅝	11	4
A05	2 × 1 × 10	23½	8¼	9¾	7¼	12½	⅝	16¾	1⅛	¼ × ⅛	2⅝	8½	4
A50	3 × 1½ × 10	23½	8¼	9¾	7¼	12½	⅝	16¾	1⅛	¼ × ⅛	2⅝	8½	4
A60	3 × 2 × 10	23½	8¼	9¾	7¼	12½	⅝	17¾	1⅛	¼ × ⅛	2⅝	9½	4
A70	4 × 3 × 10	23½	8¼	9¾	7¼	12½	⅝	19¼	1⅛	¼ × ⅛	2⅝	11	4
A20	3 × 1½ × 13	23½	10	9¾	7¼	12½	⅝	20¼	1⅛	¼ × ⅛	2⅝	10½	4
A30	3 × 2 × 13	23½	10	9¾	7¼	12½	⅝	21¼	1⅛	¼ × ⅛	2⅝	11½	4
A40	4 × 3 × 13	23½	10	9¾	7¼	12½	⅝	22½	1⅛	¼ × ⅛	2⅝	12½	4
A80"	6 × 4 × 13	23½	10	9¾	7¼	12½	⅝	23½	1⅛	¼ × ⅛	2⅝	13½	4
A90"	8 × 6 × 13	33⅞	14½	16	9	18¾	⅞	30½	2⅜	⅝ × 5/16	4	16	6
A100"	10 × 8 × 13	33⅞	14½	16	9	18¾	⅞	32½	2⅜	⅝ × 5/16	4	18	6
A110"	8 × 6 × 15	33⅞	14½	16	9	18¾	⅞	32½	2⅜	⅝ × 5/16	4	18	6
A120"	10 × 8 × 15	33⅞	14½	16	9	18¾	⅞	32½	2⅜	⅝ × 5/16	4	19	6

FIGURE 2.8 ANSI B73.1 pump standard dimensions.

MECHANICAL COMPONENTS AND MATERIALS

A pump is a mechanical machine with some basic components. Most pump sizing and selection has component design choices decided by the pump manufacturer, but there are many mechanical component options to consider after the pump has been sized for the system from a hydraulic standpoint.

Once the hydraulic parameters (flow and head) are determined, the pump specification is just the beginning. The type of pump, positive displacement versus centrifugal, must be decided. Head, flow, shear-sensitive fluids are just a few factors to consider, but for the scope of this discussion, only centrifugal pumps will be addressed. For centrifugal pump, is it end suction or double suction? Is self-priming pump needed? Is suction lift required? API versus ANSI design?

Next, pump speed and pump size must be considered as part of the pump selection process. Is pump a fixed speed or VFD? If fixed speed, what speed will satisfy flow and head – 900 rpm, 1,200 rpm, 1,800 rpm, 3,600 rpm, etc. If a VFD is chosen, it should have $>$ 50% of head loss due to friction loss for feasible use. In a high static head system, the pump will have to run a high speed just to overcome the static head, so no real energy savings will be realized.

Looking at suction-specific speed (NSS), it is desired to have NSS $<$ 8000 (NSS $>$ 12,000 high suction energy and may be problematic). Choose a lower speed pump if all other items are equal as there are so many benefits from bearing life, seal life, and hydraulic. One drawback at lower speed pumps is larger size pumps so the installation location must support that larger size pump.

Pump-specific speed is a pump characteristic that we may not use directly in selecting a pump, but the impact of this parameter reflects the pump hydraulics, impeller shape/design, and pump curve shape. Thus, all these things will play a major factor in which a pump is selected. Maximum pump efficiency across the spectrum will be with the specific speed (Ns) ~2500 while head and flow are at BEP. Pump-specific speed is calculated by the equation below.

$$Ns = \frac{N \times Q^{0.5}}{H^{0.75}}$$

where:

Ns = pump-specific speed
H = head at BEP
Q = flow in gpm at BEP
N = speed in rpm

The Ns mostly has to do with the impeller design. Low Ns pumps are high head and low flow with mostly radial flow. High Ns pumps are typically those above 6,500. High Ns pumps are high flow, low head, and axial flow pumps. Medium Ns pumps are mixed flow design, operating between low- and high-type flow conditions.

Again, most of the other hydraulic conditions will dictate the pump selection but sometimes the pump-specific speed can also affect how the pump interacts with the

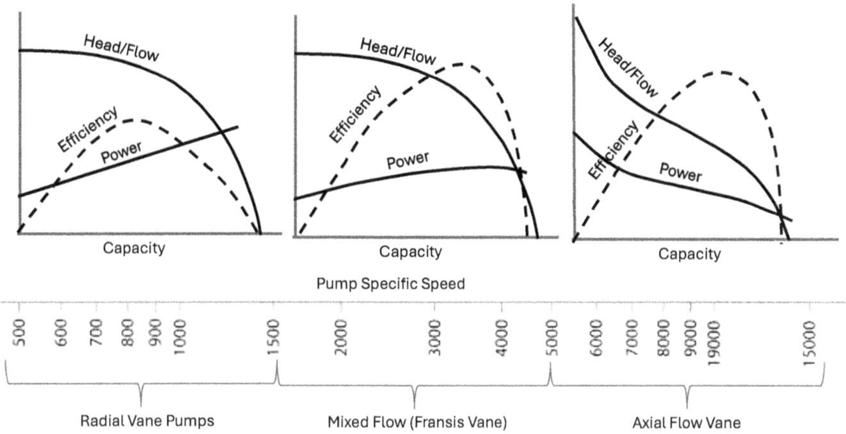

FIGURE 2.9 Pump impeller types and pump performance.

system. The pump curve shapes are also different with different pump-specific speed pumps as shown in Figure 2.9. The curve shapes will also affect process controls in a pump system. A steep pump curve (rise to shutoff) will take a significant head change to change flow; however, a flat pump curve shape will be very flow-sensitive to head changes. That may mean very tight control tuning to manage the process. A few examples have been provided in Chapter 6 on pump and system curves.

The pump driver choice is typically a simple one and mostly with an electric motor. Although, some plants may have excess steam and choose to use a small turbine which also generates lower pressure steam for other applications, such as process drying. There are many choices for motors, such as an induction motor, which may be an open frame or closed frame, with the most popular being totally enclosed fan cooled (TEFC) option, or maybe a rare synchronous motor. If it is a motor used on a VFD, then make sure the motor is sized appropriately for proper turndown. The motor has sufficient insulation rating for low speed and potential higher temperature operation.

The integral HP rule from DOE for motor designs requires that motors meet premium efficiency standards per NEMA. A best practice is to typically select a pump driver (motor) with full curve runout to eliminate operational issues and ensure the best turndown capacity for the selected pump. This ensures that no matter how the pump system changes, the pump driver will not be overloaded. Figure 2.10 shows where a 7″ impeller would require a 7.5 HP motor for full runout operation at any system head. Like Figure 2.10 shows, many times this will even allow for a slightly larger impeller with no other equipment modifications.

If the pump driver (motor) is 200 HP or less, it is likely a motor manufactured to NEMA standards, where many of the key dimensions are standardized in manufacturing. Not every dimension is standardized but the main ones are shaft height, motor feet spacing, shaft diameter, and many others. This allows for any motor manufacturer to supply a motor to NEMA standards that will basically be a drop-in replacement based on NEMA dimensional standards.

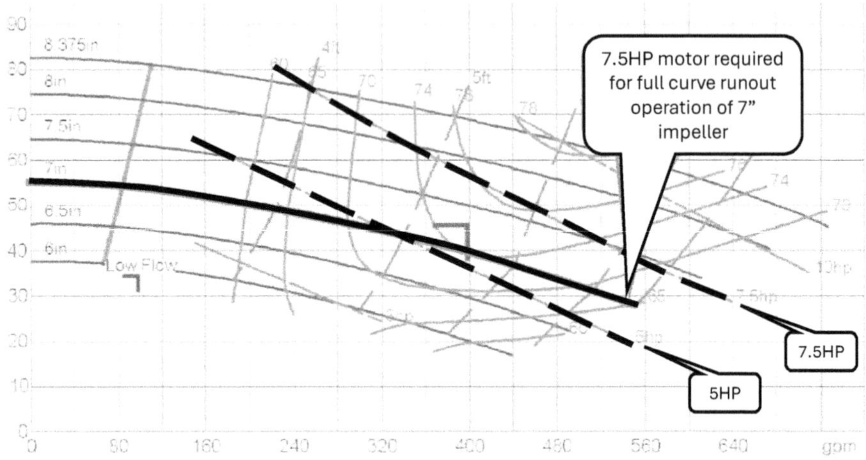

FIGURE 2.10 Pump curve with motor providing full curve runout.

The other basic motor selection part which related to the pump installation is starter size and motor cable size. Figure 2.11 shows a quick reference for NEMA motor HP and starter size.

Selection of the baseplate may include several choices from cast iron, steel fabricated, or a polycast material. This is typically the order of cost as well. A polycast can have great chemical resistance and can decrease installation time as there is no cure time for the polycast material. As with selecting a pump that has full

NEMA SIZE	Maximum Horsepower Three Phase Motors											
	Full Voltage Starting			Auto Transformer Starting			Part Winding Starting			Wye Delta Starting		
	200V	230V	460V 575V	200V	230V	460V 575V	200V	230V	460V 575V	200V	230V	460V 575V
00	1 ½	1 ½	2	--	--	--	--	--	--	--	--	--
0	3	3	5	--	--	--	--	--	--	--	--	--
1	7 ½	7 ½	10	7 ½	7 ½	10	10	10	15	10	10	15
2	10	15	25	10	15	25	20	25	40	20	25	40
3	25	30	50	25	30	50	40	50	75	40	50	75
4	40	50	100	40	50	100	75	75	150	60	75	150
5	75	100	200	75	100	200	150	150	350	150	150	300
6	150	200	400	150	200	400	--	300	600	300	350	700
7	--	300	600	--	300	600	--	450	900	500	500	1000
8	--	450	900	--	450	900	--	700	1400	750	800	1500
9	--	800	1600	--	800	1600	--	1300	2600	1500	1500	3000

FIGURE 2.11 NEMA motor HP and starter size.

curve runout, it is also a good practice to size the baseplate to accept the next size motor in case a process change is made later, a new baseplate will not be required. It also gives the baseplate some extra mass for vibration dampening. If the pump is operating with high temperature fluids, then baseplate springs may be needed to minimize pump nozzle loads that result from thermal growth of the piping system.

The motor portion of the baseplate should be designed so that at least 1/8″ shims are able to be placed under the motor for completing alignment. The baseplate should also be equipped with motor jackbolts to aid in alignment activities. There should be horizontal adjustments on each foot on both sides as well as axial adjustments.

In general, pump-specific design components are mostly set by the pump manufacturer, but there are many areas where choices exists and other aspects need to be understood even if selection options are limited. But if you are comparing in selecting pumps, comparing some different pump design elements may provide some different choices such as shaft stiffness in two different designs.

Shaft stiffness is a simple ratio of shaft dimensions that indicates the shaft's resistance to deflection under load. Other factors (such as shaft material strength) can affect shaft deflection, but it is a good measure assuming that all shafting is similar, stiffness provides a good measure of the shaft's rigidity.

The lower the shaft stiffness number, the stiffer the shaft design is. What this means is that the shaft will hold its shape better under higher shaft radial loads. Low shaft stiffness and shaft deflection may result as shown in Figure 2.12, affecting shaft sealing performance and ultimately pump reliability.

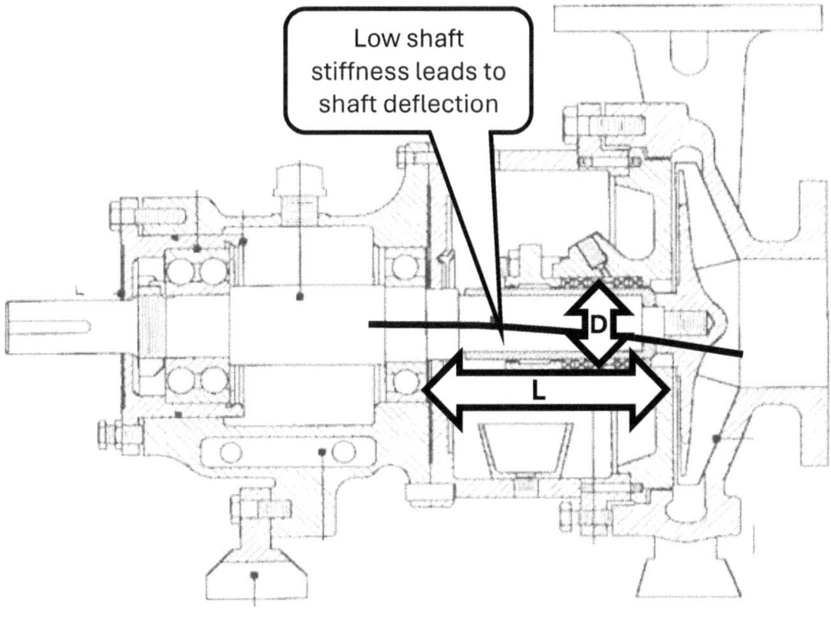

FIGURE 2.12 Shaft stiffness effect.

If a shaft has a sleeve, it does not add to shaft stiffness (only the core shaft diameter does that). A shaft sleeve is a must when using a packing sealing arrangement, but if shaft stiffness is an issue and a mechanical seal is used, some pump designs will also make a solid shaft to the diameter of the sleeve and thus increase its stiffness. This can all be decided upfront on pump design and specification, so evaluate shaft stiffness for your design.

$$\mathrm{SSR} = \frac{L^3}{D^4}$$

where:

SSR = shaft stiffness ratio
L = unsupported shaft length, in
D = shaft diameter, in

The industry standard shaft stiffness design target is < 55. The American Petroleum Institute (API) had a shaft stiffness of 147 in the fifth edition but increases this to 40 in the ninth edition. Is this enough or will a higher shaft stiffness work? The answer is: it depends on the pump and system design and maybe budget. The other side of shaft stiffness is the radial force on the shaft. If the pump and system design have the pump operating in the POR, then low radial loads may not make shaft stiffness as critical, but if the pump and system design has the pump operating over a large range of conditions then a better shaft stiffness may be desired.

If you wanted to determine if there is enough shaft stiffness, then knowing the resultant radial load on the pump and calculating the shaft deflection is a good start. Compare the actual shaft deflection to what is allowed. ANSI B73.1 max shaft deflection at the stuffing box is 0.002″ over the allowable operating range (AOR) of the pump. Also, the maximum TIR = 0.001″ for a solid shaft and 0.002″ for a sleeved shaft.

BEARINGS

Pump bearing selection and sizing is determined by the pump manufacturer, so typically not much for the end user to specify here, but the end user must understand pump maintenance and service requirements once the pump is in operation. Also, if pump selection involves comparing different pump designs, the bearing design may also influence the selection for the application. Bearings may be a plain journal/sleeve bearing or an anti-friction roller bearings (ball, spherical, and cylindrical). Bearing will be determined based on type and size of loads and operating speeds that the pump manufacturer has determined for the specific pump design.

Bearings are one of the limiting components for overall pump reliability. As many bearing experts have witnessed, many bearings don't die but they are murdered. Killed by poor lubrication, contamination, or poor installation. Bearings in pump (or any other equipment) design and selection does not factor in poor lubrication or contamination or poor installation. They assume those things will occur correctly. So, if a bearing has a short service life, something has fallen short of expectations.

Most pumps have single or double row deep groove and angular contact ball bearings. Single row angular contact bearings have had faces ground so they can be used in tandem arrangements. Angular contact bearings are typically 40° contact angle bearings. API 610 recommends combination of two back-to-back mounted 40° angular contact bearings. Special bearing sets can also be used if bearing skidding is an issue. Matched sets of 40° and 15° angular contact bearings (SKF PumpPac) are designed to avoid or reduce skidding. This is because with back-to-back bearings, only one bearing is loaded at a time during operation (one during startup and the other in normal operating conditions) as the thrust load changes. The unloaded bearing may have a tendency to skid or smear in an unloaded condition.

So how long do pump bearings last? L10 bearing life for centrifugal pump is typically in the 17,000–25,000 hrs at rated conditions. A lower rating would result in maximum radial and axial loads. API standards generally have L10 bearing life at 25,000 hrs, while ANSI B73.1 has L10 bearing life at 17,000 hrs. Good to remember that L10 bearing life is only bearing life based on bearing fatigue. Bearing failure from water contamination or poor lubrication or installation have nothing to do with L10 life. L10 bearing life assumes all those areas are executed properly. It is also important to remember that most bearings don't fail from fatigue, so bearing life and reliability is about so many other factors – mostly related to lubrication and bearing installation. I've seen reliable pumps operate for 10–15 years with the same rotating assembly and no maintenance other than maybe changing oil. If pump bearings are not lasting at least three years, then likely there is a considerable underlying issue.

Some pump bearing power end frames may come equipped with bearing housing cooling options. In general, don't use pedestal or bearing housing cooling water with rolling element bearings unless the application really needs it. Most pumps don't need it. This can cause the bearing clearances to be drastically reduced, which elevates operating temperature. Moisture condensation can also be a problem when a power end cools on shutdown.

Bearing housing shaft seals can be several design choices such as contact seals or non-contact seals. Bearing housing shaft seals function is to keep the lubricant in and keep the contaminants out. Lip seals were common most of the early years on pump designs but in later years the non-contact labyrinth seals are more common. Lip seal materials may consist of Buna, Nitrile, Silicon, Polyethylene, or a Thermoplastic. In extreme conditions of heat or temperature, Flouroelastomers may be used up to 600°F.

Lip seals work very well when they are new but have a pretty short life typically of three to six months, depending on shaft surface speed, before they become worn and begin losing contact with the shaft. They still offer some sealing but it is not as effective. Another drawback to lip seals is that they will grove the shaft. Some lip seal designs may offer a grease port to provide lip seal lubrication to decrease shaft scoring and lip wear.

Labyrinth seals are non-contact and thus have infinite life. Original designs had large gaps, but more modern designs have intricate passages and very tight clearances. These modern seals provide excellent sealing with infinite life, but they do not typically withstand any misalignment. Modern labyrinth seals are more expensive than lip seals but can provide a good life cycle cost return as they are reusable. Magnetic seals are also an option in today's pump market. These are a contact seal but don't have the excessive wear that a lip seal does. One drawback of magnetic seals is they will attract metal particles that can accumulate on seal faces and cause sealing issues.

LUBRICATION

Upfront on pump specification, there are probably only a few things to decide and know about pump lubrication. Most of the lubrication items will be addressed in the pump maintenance chapter. Bearing lubrication will either be grease or oil. Some pump designs will only give you one or the other and some designs will let you use either. There are pros and cons to each.

In general, I prefer oil, if possible, for these reasons. Oil can easily be tested and even visually inspected for condition. Oil can easily be changed. While grease is 80% oil, oil is 100% oil. It is the oil that provides the lube film and does the bulk of the lubrication. Grease has the thickener to hold the oil. Oil can separate off some water. Grease tends to trap water. Grease tends to channel and requires manual replenishment, which can be missed. Channeling will not allow fresh grease to always get to the critical bearing areas. Grease does have some advantages in that it can act as a barrier to contamination ingress. Grease will also stick in bearing housing, making it less prone to leakage.

Oil lubrication can be several different types such as oil ring/flinger, flooded sump, or oil mist. Oil mist is considered the best lubrication method, but it is the most expensive to install and justify. Due to that they are also not very common in most industrial applications. Flooded sump is probably the most common lubrication method used. Oil ring designs are also used when the conditions warrant the need for it. Some pumps may utilize an oil flinger. These decisions again are typically designed and decided by the pump manufacturer, but it may be a factor if you are choosing between many different pumps in the selection process. One negative for an oil flinger or oil ring is during static conditions the bearing is sitting outside of the oil so more vulnerable to static corrosion. Also keep in mind that oil flingers will have a minimum speed so if using on an ASD, do not go below the minimum speed for the pump oil flinger design.

The speed factor rating (Dn) shown in the equation below can give some guidelines to which lubrication method might be best. If the speed factor rating dn $< 160{,}000$ then a flooded sump lubrication method is suitable. If the speed rating factor dn $> 160{,}000$ then the bearing temperature will increase, and slinger rings are typically required. For grease bearings there can be what is called lubed for life where the bearings come pre-greased and sealed and do not require re-lubrication. Generally, I would only consider this if the pump was very small and a throwaway pump. Lube for life bearings typically may last only three to five years. A pump bearing with a speed rating factor, $Dn > 80{,}000$ should not have bearings grease lubricated seal for life designs.

$$Dn = D \times n$$

where:

Dn = speed factor
D = bearing bore, inches
n = shaft speed, rpm

For pumps with an oil ring, the bearings sit outside the oil. The oil ring picks up the oil from the sump and flings the oil into the environment to lubricate the bearings. With oil level there may also be some choice with the type of oil level setting or verification. Most pump bearing housings will come with a bull's eye sightglass. Some may come or choose to use a tube sightglass or a constant level oiler. These items will be covered further in the pump maintenance chapter.

SHAFT SEALING

For the seal chamber (or stuffing box), there are several choices for each pump: the standard straight bore stuffing box or a newer design taper bore stuffing box. The straight bore stuffing box may house a mechanical seal or packing, while the taper bore will only accept a mechanical seal. For a packing arrangement there are several designs for the latern ring (ring that allows seal water to packing and shaft). There is the standard latern ring H design or there are special bushing/bearing-type latern rings which take up the inner two rows of packing and lantern ring area.

Packing arrangements are typically a 2L3 – meaning 2 rows of packing, latern ring, then 3 rows of packing and packing follower (see figure in Chapter 8 on pump maintenance). Packing offers a cheap sealing method but does have significant seal water leakage rates compared to mechanical seals. Packing can have various hookup plans to either quench and drain or product flush into the product stream with many kinds of variations with hookup. Packing has a wide range of speed in application with typical being 200–3,600 fpm surface speed application range.

Packing typically has a square cross-section with ½″ being a common size for a lot of pumps. There is some packing called anti-keystone packing, where the cross-section is wider at the top compared to the bottom. This is to offset some of the keystoning effect on packing as it is wrapped around a shaft. Because the OD circumference is longer than the ID circumference, square packing will keystone (top of packing narrows compared to bottom) in response, leading to increased pressure points and uneven wear on the shaft sleeve. Packing must be adjusted routinely and periodically replaced.

Packing materials are another item which may need specification, depending on the fluid type and application. For general application, most nomex yarn packing is an accepted selection. High speed and high temperature may want to consider a graphite-type packing material to improve heat dissipation. Applications for chemical attack may need to consider a packing with a Teflon-type material. For abrasive applications, consider aramid or phenoic materials. Latern rings are typically a polymer material, Teflon, or a soft metal such as brass. Always wise to consult the packing supplier/manufacturer to better choose the right packing material for the application as there are many variations and special-made materials.

Mechanical seals are the other main sealing option, which are also becoming popular with more emphasis on environmental leakage control and energy efficiency. Mechanical seals are environmentally friendly with respect to using seal water and leakage to the environment. They are also more energy efficient and do not wear out a shaft or shaft sleeve. Mechanical seals are available in a wide range of designs such as single seals, double seals, one piece cartridge seals, or more complicated component seals. Most pumps can be configured to accept packing or seals. However, an exception applies to fire water pumps, where NFPA standards do not allow mechanical seals and require the use of packing for shaft sealing.

Mechanical seals have primary and secondary sealing elements. The primary sealing elements are a rotating seal ring and a stationary seal ring. These rings may be made from a ceramic-metal composite or carbon-graphite materials. Secondary sealing elements seal off all other leak paths on the seal. They prevent any leakage and don't require lubrication. Seals can be a pusher seal design (ability to move) or a non-pusher seal (seal can deform). Orings typically used on pusher-type seals and bellows are typically used on non-pusher seals.

Typical mechanical seal element materials are carbon, tungsten carbide, silicon carbide, and aluminum oxide (ceramic). Carbon is the most common seal face material with a wide range of applications including with some corrosive fluids. Tungsten carbide has higher chemical and thermal resistance than carbon. Silicon carbide has many different grades and has low thermal expansion and excellent resistance to corrosive fluids. Silicon has good strength properties at elevated temperatures.

Temperature can play a major factor in which type of seal to consider. Non-pusher seals may be used in very cold ($< 0°F$) or very hot temperature ($> 400°F$) applications. Seals can come in single seal or dual seal arrangements. Dual seals have two pair of seal faces and have unpressurized or pressurized seal environments. Dual unpressurized seals have a fluid between the pairs of seal faces at a pressure lower than the process pressure in the stuffing box. The fluid is called the buffer fluid. Dual pressurized seals have the fluid between the seal faces at a higher pressure than the seal chamber pressure, so leakage will be directed into the process fluid. This seal fluid is called the barrier fluid, as its higher pressure prevents the process fluid from reaching the seal faces.

Dual pressurized seals have a few advantages over single seals. They avoid leakage of the process fluid into the atmosphere (important for hazardous fluids), and they can improve the fluid film on the seal faces since that is coming from the barrier fluid and not the process fluid. Dual unpressurized seals are less complicated than pressurized seals and they do avoid contamination of the process fluid from the barrier fluid.

For simple fluids that are waterlike or can absorb seal water leakage into the pumped fluid, single seals are a very good choice for good reliability. Choosing the proper seal support system plan can be key on these. Seal face and oring material selection is also important.

One other consideration is choosing a cartridge seal or component seals. Cartridge seals do require a larger seal chamber and are more expensive than component seals, but there are several advantages. Component seals are more complicated requiring detail and skill in installation and maintenance. Infant mortality is much higher for component seals typically so the extra cost for a cartridge seal is usually offset from these early life failures. Cartridge seals can also be pressure tested before installation to verify assembly integrity, whereas component seals can't be tested until the pump is started. Again, consult your mechanical seal supplier for application assistance.

Split seals are usually specified on a split case pump so they can be installed without disassembling the pump. Split seals, however, are very complicated and can be challenging to install with all the intricate mating parts. I would recommend having the seal field service folks assist or lead any split seal installation. There are also other challenges with seals on double suction split case pumps (more on that in Chapter 10), so do not assume that using a split seal on a double suction pump is a slam dunk success because they are not. I've seen probably half of them not provide good reliability on split case pumps.

Seal water flush plans can also vary for mechanical seals so research this with the seal supplier for the application to ensure that the correct plan is in place. There are many plans, and this is not a detail discussion on any of the flush plans but just a summary of some of the common ones used. Figure 2.13 shows a few of the more common flush plans. Plan 11 provides flush water from the discharge of the pump to the seal chamber. It typically needs an orifice or valve to throttle the flow to the seal. Flow through the seal removes heat and also adds flush water to the pumped fluid. Plan 21 is similar to Plan 11 but a heat exchanger is added to cool the product fluid before it goes to the seal chamber. Plan 23 recirculates fluid from the seal chamber in a closed loop but through a heat exchanger.

Plan 52 is similar to Plan 23 but the circulated buffer fluid from the seal chamber goes through a pressure chamber in a closed loop system where the pressure is set lower than the seal chamber pressure. Plan 53 is the same piped-up plan as Plan 52, but the pressure is set higher than the seal chamber pressure, so it is a barrier fluid. Plan 54 is also a barrier fluid plan, but the pressure comes from an outside source and not in a closed loop fashion as Plan 53.

Plan 32 utilizes an external seal water supply source to the seal chamber. Typically, a pressure gauge and throttle valve are used to ensure that sufficient seal water pressure is supplied to the seal chamber. Flush water is typically flushed into the pumped fluid. Many times, a throttle bushing to is used create back pressure to the seal chamber.

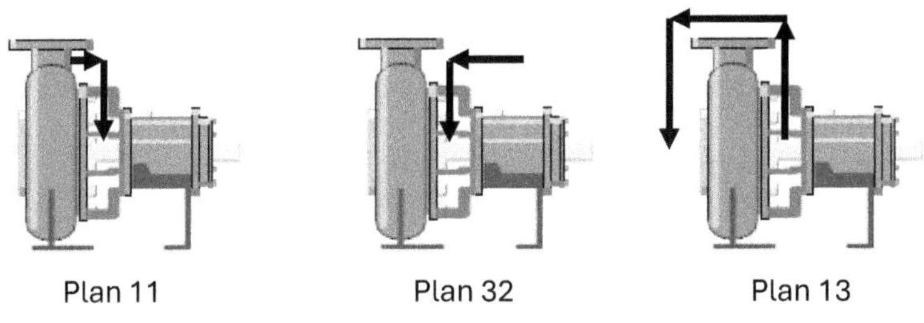

FIGURE 2.13 Common seal water flush plans.

Plan 13 has recirculation from the seal chamber back to the pump suction. An orifice may be needed to throttle the seal water to the appropriate level. This creates a flow over the seal to remove generated heat. It may also reduce the seal chamber pressure. A few more details on seal water setups are included in the pump maintenance chapter.

Mechanical seal face materials are something to discuss with your mechanical seal manufacturer as this is selected for each pump application. There are several different materials from which to select such as carbon, silicon carbide, and tungsten carbide. For clean fluids, start with carbon seal faces. These will run cooler and less brittle than the hard seal faces. For abrasive fluids, go with silicone carbide or tungsten carbide. I generally start up with a more standard seal material and design before advancing to higher-end materials or designs due to cost and complexity. If it isn't needed, then no justification for the upgrade.

If choosing a mechanical seal, keep in mind that for many pump manufacturers there can be a small difference in shaft sleeve dimensions for using packing versus using a mechanical seal. Also consider shaft sleeve material differences for a seal versus packing. For packing, the shaft sleeve should be a harder material or have a hardened tungsten carbide coating for wear resistance. For a mechanical seal, a softer material is desired so that the setscrews can bite into the sleeve for good holding capacity and wear resistance is not needed for a seal application.

Another type of shaft sealing for a centrifugal pump offered by some manufacturers on some designs is a dynamic seal. These are not as common in the field, but I have had very good success in a handful of applications where they have been used. A dynamic seal uses an expeller or repeller on the shaft to dynamically seal the shaft seal area once the pump is running at full speed. Dynamic seals are essentially infinite life-type seals as there are no real wear parts. Dynamic seals typically last the life of the pump and requires no seal water during normal operation.

Dynamic seals do use 2–2.5X more shaft power as mechanical seals. BHP for a dynamic seal is not accounted for on pump curves so factor this into motor sizing and selection if a dynamic seal is chosen for your pump. Usually, there is a separate BHP curve or table for the dynamic seal. In general, I would only consider dynamic seals on fairly clean fluids. I have used dynamic seals on fluids with up to 2% solids or consistency but might not go much more than that. A flush line might be recommended for the application. Discuss with pump manufacturer for the application. Dynamic seals are typically not allowed in ATEX explosive environments so choose another sealing option.

There are several limitations for dynamic seals. One is a maximum suction head that a dynamic seal can function. On high suction head systems, the repeller may not generate enough head to seal off from suction pressure to the chamber. The head that the dynamic seal will produce is also a function of shaft speed. Also, the dynamic seal does not produce any backpressure for seal chamber when pump is down, yet suction pressure is likely still present. This should be closely checked on an ASD application that the minimum operating speed does not go below the minimum speed to produce more head than the suction head on the pump.

IMPELLER, WEAR PLATE, AND CASING

The pump impeller for the most part comes with the hydraulic portion of the pump selection, but there are some additional items to take into consideration around details of the impeller. In pump selection, it may give you pump options for open, semi-open, and even enclosed impellers. Enclosed impellers are for clean fluids but may handle consistencies up to 2% on some applications. For higher consistency fluids or fluids with some solids, open or semi-open impellers are required. The number of vanes on the impeller also relate to the hydraulic performance and pump selection. However, some special consideration may be given to the number of vanes to reduce pressure pulses to the process fluid, vibration effects, or size of solids that can be passed. Less vanes generally indicate the impeller can handle larger solids.

Impeller materials generally fall into the same type of discussion as pump casings, therefore, refer to the discussion on pump casings that also apply to pump impeller. One special material consideration is around bronze impellers where a failure mechanism called dezincification may occur as shown in Figure 2.14.

For many years, SAE40 bronze was used for pump impellers which had low zinc content. It was discontinued in the 1990s due to 5% lead content. Silicon bronze and other alloys were used which had 16% or more of zinc. If chlorine residuals are present as low as 2 ppm, dezincification can occur where the impeller erodes. If bronze impeller is used, ask for nickel or aluminum bronze to eliminate dezincification.

The impeller size is mostly dictated by the hydraulics on pump selection; however, there are times where there are choices in impeller diameter which may give a choice between different pumps for a fixed speed application or when choosing impeller diameter for an ASD. A smaller impeller diameter is more hydraulically inefficient as there is more internal recirculation between the casing cutwater and the smaller impeller (larger impeller cutwater gap). Smaller impellers may also have

FIGURE 2.14 Bronze impeller dezincification.

slightly higher NPSHR. To minimize impeller vane pass, choose an impeller gap with 6% of max impeller diameter.

Choosing an impeller diameter for a VFD is about making sure there is enough pump turndown for the full range of operating conditions and to ensure there is enough motor torque available at the lower speed. This is discussed further in Chapter 6 on VFD pump systems but note that impeller diameter has extra weight on a VFD application.

The suction wear plate on an end suction pump typically only has a few choices mostly around material specification. The suction wear plate many times is the same material as the impeller. So a 316SS impeller typically has a 316SS suction wear plate. There are some other special-made suction wear plates that incorporate cutting edges or slots intended to cut strings of plastic or other material as they pass through the pump, helping to prevent clogging or roping as shown in Figure 2.15. These do not help with wear issues from abrasives but only with breaking up strings or lumps in the pump fluid.

The function of the casing is to convert fluid velocity that is discharged from the impeller to pressure energy as it exits the volute. Most pump casings are made for the specific pump model and do not offer much in the way of options other than materials and a few finishing details. Most end suction pumps may have a single volute, single cutwater but some larger end suction pumps may come with a double volute. A double volute pump eliminates or reduces radial loads from the impeller if the pump is operating away from BEP. Double volute casings are also more inefficient as hydraulic losses are increased with two cutwaters. If a cutwater wears unevenly or the cutwaters are not the same, then some radial load will result, and the force profile will not be balanced in that condition.

The selection of pump materials may vary throughout the pump as a mixture depending on the issue. A pump may have a 316SS impeller, wear plate, and stuffing

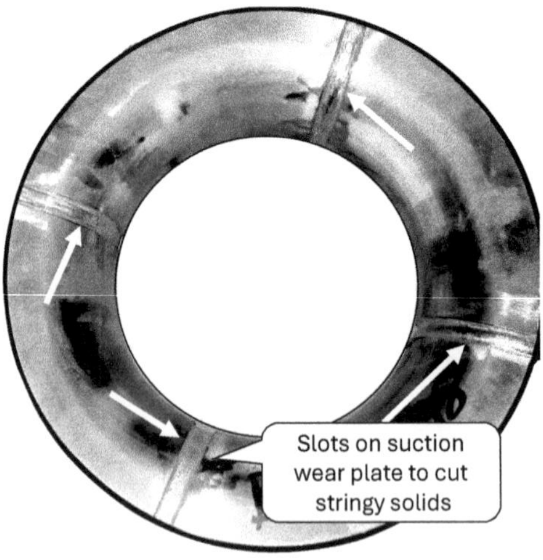

FIGURE 2.15 Suction wear plate with slots.

box but have a cast iron casing and be sufficient in that application for reliable operation. For every component on the pump, the material may have to combat several reliability life factors. Outside of cost, which is always a factor in pump specification or selection, some of those factors are as follows.

1. Material strength particularly for casings for rated pressures
2. Casting properties for cast components
3. Corrosion resistance
4. Abrasive wear resistance
5. Cavitation resistance

Cavitation can affect several parts of the pump such as the impeller, suction wear plate, stuffing box behind impeller and casing. Of course, for cavitation the goal would be to design and select the pump and system so that cavitation is small where pump damage is not the result. In cases where cavitation may be an issue, proper material selection for the pump components can pay great dividends, but you don't want to overdesign unnecessary upgrades unless they are required. Material life factors shown in Figure 2.16 can be referenced to improve cavitation resistance.

Abrasive wear can be another factor in pump selection. While some factors that make a fluid more abrasive may be obvious; others are not. Factors that make a fluid abrasive might include the following.

- Larger particles are more abrasive.
- More particles are more abrasive.
- Harder particles are more abrasive.
- Higher pump head (TDH) systems are more abrasive.
- Lower fluid viscosity is more abrasive as particles move easily through the fluid and to wear surfaces.
- When particle density is more than fluid density, the particles move to the outside (casing) to the wear surface easily due to centrifugal forces.

Material	Life Factor
Nickel Aluminum Bronze	8.0
Titanium	6.0
Bronze	4.0
Stainless Steel 300 series	4.0
Stainless Steel 400 series	3.0
Monel	2.0
Cast Iron	1.2
Mild Steel	1.0

FIGURE 2.16 Material cavitation life factors.

- Higher speed pumps are more abrasive which is why most abrasive-type pumps are larger and slower speed. In general, pump wear is proportional to the cube of the speed. Higher speed means higher particle velocity and impacts to the pump surfaces as well as centrifugal forces of the particles on pump casing walls.
- Low impact angle of particles on the pump surface is more abrasive. These areas would be casing walls, impeller surfaces, and suction and discharge nozzles. For high impact angles like cutwater, surface coatings will not provide a long-term solution but reshaping the cutwater will lower the impact and abrasive wear as shown in Figure 2.17.

Abrasive pump casings may be metal or rubber-lined. Rubber-lined are typically used in lower head applications (< 150 ft), smaller solids applications (< 0.250″), and applications with rounded particles. Metal casings are typically used for heads > 150 ft, larger solid sizes, and sharp or jagged-edged particles.

To counter hard-wear particles in the fluid, increasing pump component material hardness can be a way to reduce the wear and increase the life. Figure 2.18 is a list of

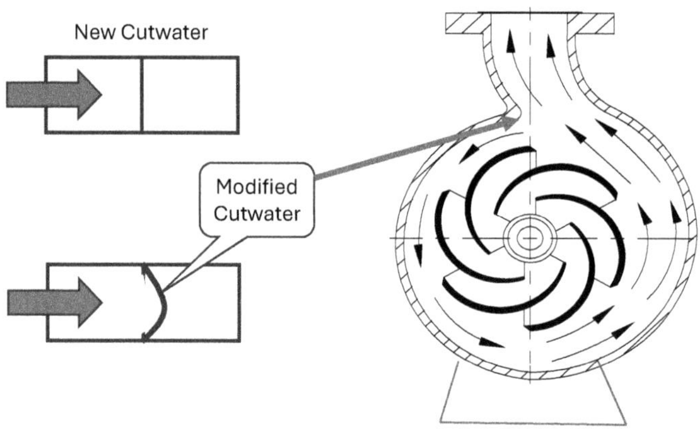

FIGURE 2.17 Cutwater mods for abrasive wear conditions.

Material	Brinell Hardness
Cast Iron	190-220
316SS	150-180
Alloy20	150-180
CD4MCu (Duplex SS)	250-300
28% Chr	400-420
416SS HT	350-400
HC600	600-650
17-4PH SS	450-650

FIGURE 2.18 Common pump material hardness list.

possible pump component materials that might be available in the pump selection process and their corresponding Brinell hardness. This is just a general reference, therefore, always verify with the pump manufacturer their material specifications as they may have some special heat treating or finishing that may affect material properties. Also, not all pump components will be offered in all materials, so verify with the pump manufacturer their available material options.

Surface hardening or coatings can provide some wear resistance to pump components. Coatings such as various types of hard ceramic coatings or ion nitriding and even tungsten carbide coatings may be added to pump casings, nozzles, or piping. In one abrasive pump application, I've used a hard ceramic coating as the wear sacrificial coating. The ceramic coating lasted for about 18 months and then we would recoat for another 18 months. While this seemed excessive, a $3,000 wear coating at the time versus purchasing a $30,000 casing was a pretty good return.

If a pump is being replaced, inspect the previous pump if there is erosion from abrasive fluids and the areas in the pump that are experiencing the abrasive erosion. These are areas where a different approach should be considered before replacing the pump casing or other components. If wear from abrasive materials is not acceptable with cast iron, then upgrade to a duplex stainless like CD4. If 316SS is not giving an acceptable life, then maybe go all the way to 17–4 PH stainless steel.

An example might also be if a pump casing is experiencing erosion around the drain port in the bottom of the volute due to abrasives collecting or impacting that area, then request a new casing without the drain ports as shown in Figure 2.19. Abrasive suspensions or particles in the fluid stream can easily catch and impact the casing drains much faster than anywhere else on the pump casing. While removing casing drains may solve a localized casing erosion issue, there must be an alternative location to drain the pump system (typically in suction piping). Small casing drains will typically plug on fluids with foreign material or abrasive material.

Addressing corrosion might require a much deeper look than any discussion in the scope of this book but just a glance as some obvious things. Chemical material

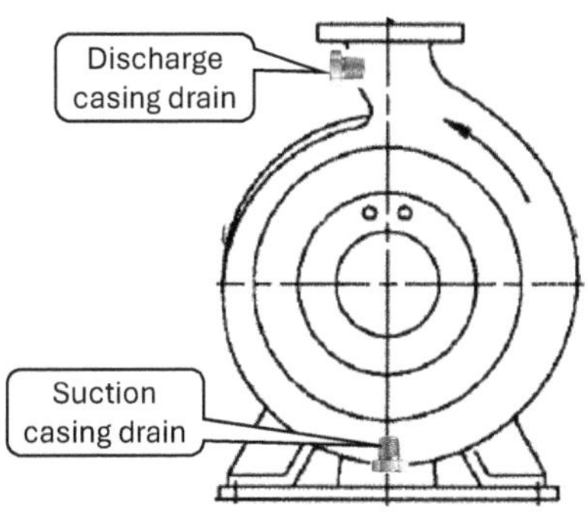

FIGURE 2.19 Casing erosion around drain plug area.

Fluid	Casing Material	Impeller/Wear Plate Material	Impeller Type	Shaft Sleeve
Alum	316SS	316SS	Open	316SS
Ash Sluice	HC600	HC600	Recessed	316SS
Caustic (3-50%)	316SS	316SS	Open	316SS
Chlorine Dioxide 6-8 g/l	Titanium	Titanium	Open	Titanium
Clay Slurry	Carbon Steel	316SS	Open	316SS
Color Dyes	Ductile Iron	Ductile Iron	Open	316SS
Condensate	Ductile Iron	Ductile Iron	Open/Closed	316SS
Filtrate, Extraction	316SS	316SS	Open	316SS
Filtrate, Chlorination	317SS	317SS	Open	317SS
Hydrogen Peroxide	316SS	316SS	Closed	316SS
Lime Mud Slurry	HC600	HC600	Recessed	316SS
Black Liquor, 0-25%	316SS	316SS	Open	316SS
Black Liquor, 25-65%	CD4	CD4	Open	316SS
Green Liquor, 5-15%	316SS	316SS	Open	316SS
Whilte Liquor, 10%	CD4	CD4	Open	316SS
Weak Wash Liquor	316SS	316SS	Open	316SS
Methanol	316SS	316SS	Open	316SS
Pulp, Brown & Bleached	316SS	316SS	Open	316SS
Pulp, Chorlinated	317SS	317SS	Open	317SS
Pulp, Contaminated Recycle	17-4PH SS	17-4PH SS	Open	316SS
Pulp, Clean Recycle	Ductile Iron	316SS	Open	316SS
Sodium Chlorate 46%	316SS	316SS	Open	316SS
Sodium Hydrosulphide	316SS	316SS	Open	316SS
Sodium Hypochlorite	317SS	317SS	Open	317SS
Spent Acid	Titanium	Titanium	Open	Ceramic
Cooked Starch	316SS	316SS	Open	316SS
Starch Slurry	316SS	316SS	Open	316SS
Sulphur Dioxide, 0.5%	316SS	316SS	Open	316SS
Sulphuric Acid, 93-98%	Carpenter SS	Carpenter SS	Open	Carpenter SS
Tall Oil	316SS	316SS	Open	316SS
Titanium Dioxide	CD4	CD4	Open	316SS
Turpentine	Ductile Iron	316SS	Open	316SS
Water, Chilled	Cast Iron	Bronze	Open/Closed	316SS
Water, White	Cast Iron	316SS	Open/Closed	316SS

FIGURE 2.20 Pump material sample chart for some pulp and paper fluids.

compatibility charts are easily available from many pump manufacturers so I would reference that for starters. One may look like the one in Figure 2.20 for common pulp and paper fluids.

There are many types of corrosion such as general corrosion, pitting corrosion, galvanic corrosion, crevice corrosion, or erosion corrosion. In general, a fluid's pH will give some indication on how corrosive the fluid may be as either acidic or alkaline. The pH scale is from 0 to 14 with 0–6 being an acid and 8–14 is alkaline. Seven is considered neutral. Therefore, if the fluid pH is in the 10–14 range or 0–4, then a corrosion resistant alloy will be needed. If the pH is in the 4–10 range, then cast iron, carbon steels, bronze, and stainless steels are all valid materials.

CASE STUDY 3 PUMP CASING AND SUCTION PIPING WEAR/EROSION

A high-density cleaner feed pump was experiencing very high wear rates due to abrasives in stock from the deink process stream. The abrasives were mostly small pieces of metal and glass contained in the process fluid (recycle paper stock). The pump was seeing high wear rates on the casing, suction nozzle, and even suction piping near pump suction nozzle. Wear was also occurring on the impeller, stuffing box, and wear plate. Figure 2.21 shows some of the wear areas on the pump components. The pump casing suction nozzle would leak stock once the wear advanced through the pipe wall.

The pump only operates at 1,200 rpm and the impeller is in the midrange for the pump size. The pump was a 10 × 12–22 with a 19.8″ impeller. The pump hydraulic design was 3,700 gpm at 160 ft. The pump was properly sized for the cleaner system and operated within process design specifications. There was not any room to modify anything with the pump or system to reduce the erosion on the pump components.

The initial evaluation of the pump material selection revealed it was deficient for the abrasive application. The original system designers must have assumed that the abrasive contamination load at that point in the process would not be that high. The casing was supplied as cast iron material which would be fine for a clean paper stock pump, but this application was on the front end of the deink system and had a very high contamination load full of abrasives. The impeller, wear plate, and stuffing box were all 316 stainless steel. The stainless steel held up on those components fair but still had a considerable wear rate with frequent adjustment and replacement. The suction piping was also 316 stainless steel schedule 10 pipe.

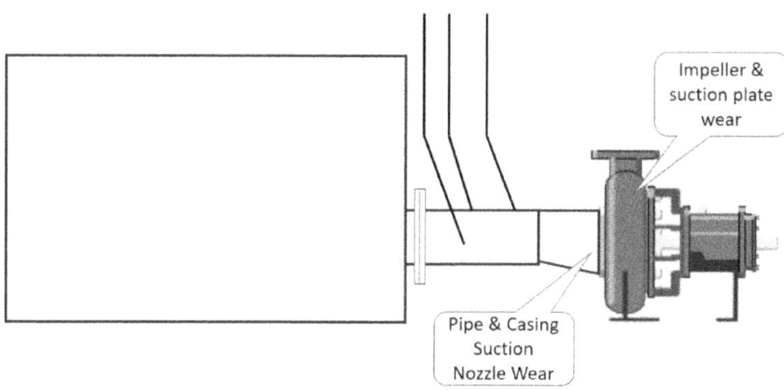

FIGURE 2.21 Pump component wear areas.

The suction piping and suction nozzle on the pump would only last for about six to nine months until it would wear through, and a leak occurred. A repair epoxy metal would be applied to stop the leak until a replacement casing and pipe could be repaired or replaced. The first upgrade on material was a 316 stainless steel casing. This improved the life of the casing to about 18 months. The new casing also was ordered without the piping taps, which was another location where erosion was worse with contaminants catching in the port location.

The pump had a mechanical seal originally. Due to pump impeller and suction wear plate wear, the impeller clearance had to be adjusted regularly throughout the year usually about every six to nine months. While there is a procedure to secure the mechanical seal for impeller adjustment, this often resulted in the mechanical seal eventually leaking. This led to increased seal usage and routine maintenance of the pump.

The second round of improvements to the pump components was to replace the casing, impeller, wear plate, and stuffing box with 17–4 PH stainless steel. The 17–4 PH material is about three times harder than the 316 stainless material. During this time, a shaft sealing change was also made away from a mechanical seal to packing. This eliminated mechanical seal failures from impeller adjustments after component wear. These changes finally resulted in a very good pump equipment life and reliability.

The final pump system wear issue was the suction piping entering the pump suction nozzle. This section was the thinnest and typically wore through the fastest. Some brush on ceramic coatings were tried but that did not improve reliability. The end solution ended up re-piping the suction nozzle with a flanged section where that section was coated with a tungsten carbide coating as shown in Figure 2.22. The tungsten carbide was very hard (1,800 Hv30 hardness). As of this writing, this has run for around seven years without any

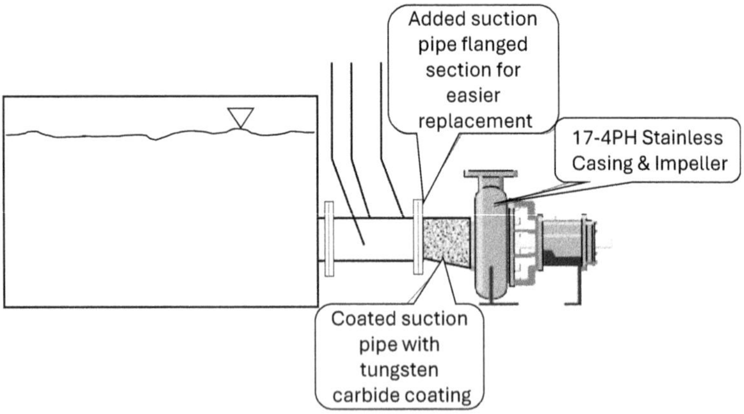

FIGURE 2.22 Pump and piping modification to improve wear.

issues where previously the pipe needed repairing about one to two times a year with holes from abrasive wear. The suction piping would have to be replaced every 12–15 months. During periods of pipe erosion, significant paper stock losses would occur.

The pump suction was also sectioned so the high wear zone had a flanged section for ease of changing out and installing a new section with wear coating as shown in Figure 2.22. The original design was welded pipe from the flange to the tank with multiple other pipe nozzles welded into the pipe, which made repair and replacement difficult and expensive.

This pump system failure has design as its root cause as the pump material was not sufficient to handle the abrasive erosion from the process fluid. The suction piping was also not a good design to handle abrasive wear. Redesigning the pump system materials and wear coatings was the solution to achieve an acceptable pump system reliability.

Case Study 3 Learning – Evaluate material compatibility and upgrades when trying to solve pump abrasive wear problems.

DRIVE COMPONENTS

One last item to select and specify for pump selection is the components connecting the pump and driver shafts. This may be a coupling or a belt drive. A wide variety of couplings are available that will work well in any centrifugal pump. Most choices are a preference of the pump specifier or preference at a particular operating plant, but there are differences and some specifics to note about coupling choices. Couplings may be a flexible coupling, rigid coupling, or a resilient coupling. Couplings may be lubricated or non-lubricated in which there are obvious advantages and disadvantages.

Some selection factors for choosing shaft couplings are torque capacity, torsional stiffness, misalignment capability, variable loading, temperature, unbalance, and lubrication. If the pump supplier is providing the coupling with a full pump, motor, and baseplate assembly, then all they need to know is what type of coupling is desired. If the end user is selecting the coupling, then it is always a good practice to engage the coupling manufacturer in the selection for your pump system.

In the event that the end user is left to select or specify a coupling, then below are a few steps to complete proper coupling sizing and selection.

1. Choose the coupling style. Some common pump coupling types are gear couplings, steel flex (grid) couplings, disc couplings, shear jaw elastomer couplings (Falk Wrapflex, Atraflex), or other elastomer-type couplings (paraflex). Most couplings will work up to 3,600 rpm but some work better at higher speeds. Disc couplings are very good at high-speed applications, while an elastomer coupling is typically limited to speeds around 3,600 rpm.

Drawbacks to disc couplings are they are very sensitive to misalignment and can easily fail from bending fatigue if not aligned well.

Torsional stiffness may be a factor in coupling choice and the gear coupling has the highest torsional stiffness and torque density. Due to their low torque density, elastomer coupling can become large in high torque applications, requiring excessive space. Elastomers also have lower temperature limits. Elastomers also can generate high thrust loads on equipment, so may want to avoid in high torque applications. Elastomers can have a dampening effect which can be beneficial. Elastomer and disc couplings are lubricant-free so that can be an advantage.

Also consider the coupling ability to enable maintenance. For example, spacer couplings are typically used for end suction pumps (back pull-out assemblies) so the pump rotating assembly can be changed without moving the motor. The spacer is removed, and the pump can be changed with the motor remaining in place. Alignment must still take place but less work to realign is generally the result.

For adjusting impeller clearance also consider the ability of the coupling to handle small amounts of axial adjustment and not sacrifice coupling reliability. Disc couplings in particular can be used but must have a way to axial adjust without leaving axial stress on coupling elements, resulting in fatigue failure. Shrink disc can be used on disc couplings for impeller adjustment or a slight clearance fit on one coupling hub. Steelflex (grid couplings) and gear couplings are also popular for pumps as they can handle small axial movements (up to 10% gap target) from impeller clearance setting without greatly affecting coupling reliability.

Shear jaw couplings are similar to disc couplings in that the coupling gap is very important and must be adjusted or checked after impeller clearance is set. Other couplings can be used but extra care must be taken to maintain coupling gap as impeller clearances are set.

2. Choose the coupling service factor. A service factor is just a way to factor in the type of service conditions (loading, environment, etc.) to ensure good reliability. Most of the time a coupling is good with a 1.5–2.0 service factor but always good to consult the coupling of choice manufacturer service factor charts as they can vary depending on the coupling style. Shock loads and variable loading will increase service factors, so a proper understanding of the application loads is also important.
3. Calculate the application torque requirements with the service factor from the equation below. Choose the coupling that has a torque capacity greater than calculated for the application. The service factor is already figured in so no additional capacity is needed in selection. The coupling chosen will have some amount of torque capacity > application torque requirement so the actual SF will be slightly greater than the design SF used.

$$T = \text{HP} \times \text{SF} \times \frac{5252}{N}$$

where:
T = torque, ft-lbs
HP = horsepower
SF = service factor
N = speed, rpm

4. Verify that the pump and motor shaft diameters are less than the maximum bore listed by the coupling size. The maximum bore for a coupling hub can be found on any of the coupling specification sheets for the style coupling being used. Many times, the bore size may drive coupling sizing and not the torque capacity. Another word, the coupling may have to be increased in size to not exceed the maximum bore for that coupling size. The motor shaft is typically larger than pump shafts and may drive this decision but always check both shafts to make sure.
5. Verify other coupling dimensions such as the shaft stickout will fit the hub depth and fit inside coupling guarding. Torque density will drive the element size. While a coupling hub can be overhung, it is generally not desired or industry best practice to do so. Typically, the coupling hub is flush with the end of the shaft.

If the coupling selected is a greased coupling, be sure to use a coupling grade grease and not a general purpose grease. Coupling greases are specially formulated greases to work under very high centrifugal forces seen in couplings and must have controlled oil bleed to work properly. Do not use a general purpose or bearing grade grease on greased couplings.

For ease of maintenance, ensure that larger couplings have puller holes for hub removal. Pullers that grip the outside of a coupling hub flange may permanently bend the outer hub upon removal.

If a belt drive is part of the drive design, there are several considerations. Either V-belts or synchronous belts can be used. V-belts transmit power by friction, therefore, if the pump is not operating in a relatively clean area free from contamination, a synchronous (timing) belt may be a better choice. Synchronous belts are more efficient than V-belts so if overall pump system efficiency is important then synchronous belts are also a good choice.

A few watchouts for belt drives. Synchronous belts exhibits minimal stretch compared to V-belts, therefore, care must be taken to avoid over-tensioning. V-belts, on the other hand, work on friction and stretch much more so under-tensioning can sometimes be a problem on larger V-belt drives. The solution to both is precision belt tensioning with proper belt driving modeling and use of precision belt tension methods, such as deflection force method or ultrasonic belt tensioning.

While alignment is not as critical as direct coupled drives, it is still important. Synchronous belts are more sensitive to alignment with a limit of ¼ degrees. V-belts have an alignment limit twice that of synchronous belts at ½ degrees.

Ensure that the resultant belt radial load is within pump design limits for bearing loads and pump shaft limits. You'll need some belt drive calculations and obtain pump shaft loads limits. High radial shaft loads may require roller bearings on the drive end of the pump or motor instead of a ball bearing. Engage the manufacturers for belt drives and pump to ensure these limits are not exceeded. During assembly, slide the belt sheaves as far back on pump and motor shafts as possible to reduce the resultant overhung shaft and bearing loads.

Other factors to consider in pump and component selections involve spare parts. Select pumps for a particular plant where spares are common and available. You don't want to install a one off-type pump with no spares and no skill or knowledge available to maintain or operate properly. Your chances at good reliability have just decreased with the odd ball choice of pumping equipment. Sometimes the application will force you to do this and if it does, put more time on having proper spares setup in plant storeroom and necessary maintenance training to prepare the team for success.

LIFE CYCLE COST

When comparing different pumping options or systems, a good tool to use in the process for helping to make the best decision is a life cycle cost analysis, LCC. Life cycle cost compares the total cost of ownership of each option over some time period. Typically, I would use at least five- to ten-year time period to get a good sample rate to show differentiation in options. The LCC is taken at net present value dollars, NPV, so the cost impact at year ten are much smaller; therefore, going much further than that may not change the LCC that much. You do need to go far enough to capture all the cost. For example, if normal maintenance is not done until year three or five, one needs to go far enough to capture two to three maintenance cycles. If you did a LCC at three years but maintenance didn't occur thereafter, then the maintenance cost will not be captured in the LCC. Also understand that each pump system LCC analysis may have different components that may be larger factors, but the big buckets are typically energy cost and maintenance cost when it comes to pump systems.

$$\text{LCC} = \sum(Cp + Ci + Ce + Cm + Cdt + Cd + Co + Ces)$$

where:

Cp = purchase cost
Ci = installation cost
Ce = energy cost
Cm = maintenance cost
Cdt = downtime cost
Cd = disposal cost
Ces = environmental and safety cost

CASE STUDY 4 TOTAL PUMP COST (LIFE CYCLE COST)

There were three main options for an upgrade modification to a stock cleaning pump system which consisted of a simple level control on a tank with a pump transferring to another tank. The target flow range for the redesign was 100–400 gpm with a typical flow ~240 gpm. There was an existing system which was being modified and there were three pump system options considered: (1) reusing existing pump system, (2) relocating and repurpose an abandoned pump, and (3) new pump and VFD sized for process needs.

Option 1. Reuse the existing pump on the system as shown in Figure 2.23. The existing pump was a 6 × 10 × 16 end suction stock pump with a 50 HP, 1200 rpm motor. The existing pump was originally sized for 1,800 gpm at 72 ft. The minimum continuous flow for this pump to stay in the allowable operating region (AOR) was ~650 gpm as shown on the pump curve in Figure 2.24. Since the pump is oversized for the new process design, a recirculation control scheme would be necessary to keep the pump flow operating within the pump AOR.

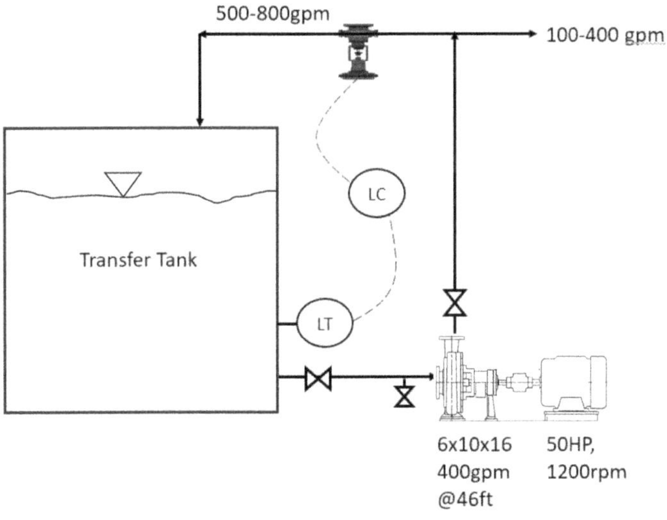

FIGURE 2.23 Process diagram for option 1 existing pump with recirculation control.

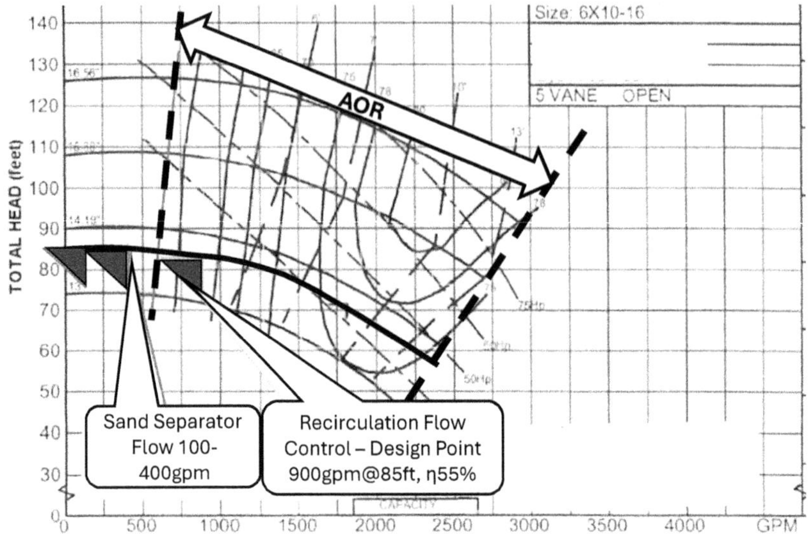

FIGURE 2.24 Existing pump curve.

With the largest pump of the three options, the operating point would contribute to lower reliability of the bearings and seal. Also, the energy cost at the higher flowrate would be inefficient. The pump hydraulic efficiency was the lowest of all three options; however, the low initial cost was viewed favorable by the management team as they would not have to invest as much capital to execute this option. This option would include utilizing the level control recirculation valve.

Option 2. There was another smaller pump with similar hydraulic envelop of what was needed on the new process that had recently been shut down. This pump was a 4 × 6 × 14 end suction stock pump with a 15 HP, 1200 rpm motor. The original design point for this pump was 600 gpm at 60 ft. This pump still operates left of BEP inside the AOR for about half the flow range as shown in Figure 2.25. Therefore, the recirculation-level control would also be used to keep the pump within the AOR for the entire flow range. The pump does have a relocation cost from its previous installation location, but much lower operating cost and some lower maintenance cost due to size of the pump. This option would be more efficient since the recirculation flow would be much lower.

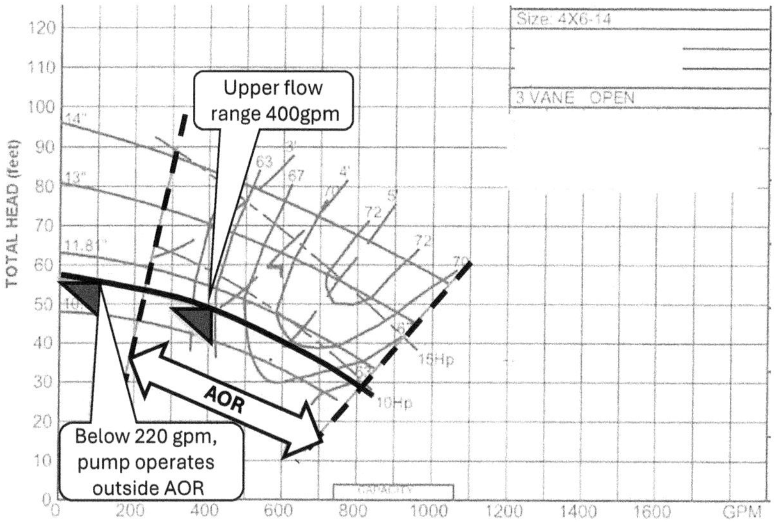

FIGURE 2.25 Pump curve option 2 repurposed pump.

Option 3. Includes a new pump and motor with a VFD correctly sized for the application. The VFD would operate off level control so a control valve would not be needed as shown in the process diagram in Figure 2.26. The pump size would be optimum for the application and the VFD would allow the pump to operate within the AOR the entire operating flow range. Since the VFD is on level control, a control valve is not needed. The new pump would be a 3 × 4 × 8 end suction pump with a 7.3125″ impeller with a 10 HP, 1800 rpm motor. The pump would operate 1300 rpm to 1800 rpm to cover the 100–400 gpm flow range for the system as shown in pump curve in Figure 2.27.

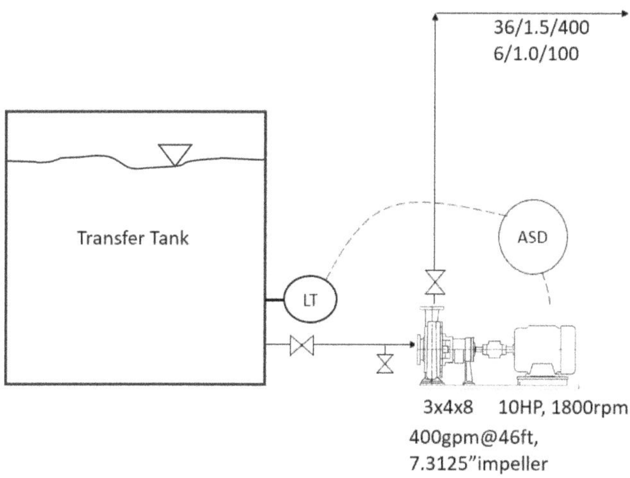

FIGURE 2.26 Option 3 new VFD pump process diagram.

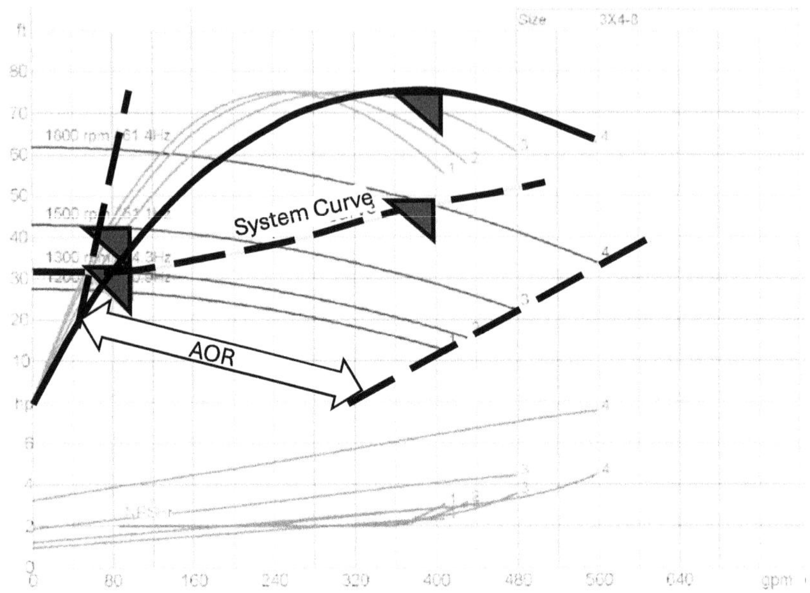

FIGURE 2.27 Option 3 VFD pump curve.

A hydraulic summary of the three pump options is shown in Figure 2.28. The pump efficiency improvement and overall BHP can be seen in the summary. Option 1 requires the most drive HP and wastes the most energy with the oversized pump and large recirculation flow but has a very low installation project cost initially. Option 2 is a happy medium from a hydraulic standpoint and project cost standpoint. Option 3 is the best option from a hydraulic fit and energy standpoint but employs the most capital investment to install. Energy cost is similar on options 2 and 3.

Pump Option	Control Strategy	Flow@Head @ Eff	Total BHP	Annual Energy Cost	Comments
1. Existing Pump	Recirculation Level Control	900gpm@85ft @55%	35.3	$12,600	Decreased reliability with pump operating point
2. Repurposed Pump	Recirculation Level Control	400gpm@46ft @60%	7.8	$5,600	Improved operating point
3. New Pump	VFD Level Control	400gpm@46ft @75%	6.2	$4,400	Best pump fit, operating point and reliability

FIGURE 2.28 Hydraulic summary of three options.

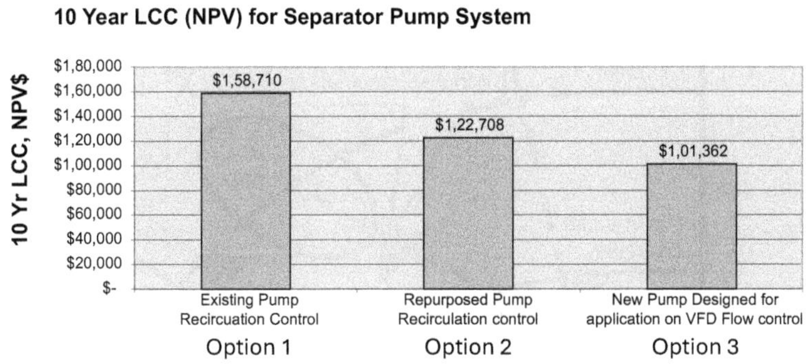

FIGURE 2.29 Ten-year LCC of three pump options.

With the total ten-year life cycle cost shown in Figure 2.29, it is clear that option 3 is the best low-cost option for total cost of ownership even though the initial investment is more. The first two options have lower upfront cost but use more energy to keep the pumps operating within the AOR. Option 3 is more efficient and operates more in the AOR and within the POR for much of the flow range; therefore, it delivers the best LCC. Energy and reliability go the full ten-year life cycle analysis while the higher upfront project cost is only in year 1. Since it is NPV, the initial cost is still a significant part of the LCC comparison, but it becomes less important as more years of energy and reliability costs are included in the analysis.

Case Study 4 Learning – Utilize LCC to evaluate financial impact of pump system options which reveal the total cost over pump life. This can remove the emotion of making hasty pump project decisions.

PUMP CURVE DESIGN POINT SELECTION

Along with all the other selection factors for choosing a pump for an application, there are many things to balance in making the selection of best pump. Sometimes there may be one main item that drives the selection but there are trade-offs on which pump is best. Assuming there is not some over-riding element that decides pump selection, one more final comparison to look at on the pump curves and where the design operating point is for each.

The best point on the pump curve is obviously at best efficiency point (BEP) but that is not an option as a design operating point most of the time. Pump manufacturers do not make pumps with BEP for every possible operating design condition.

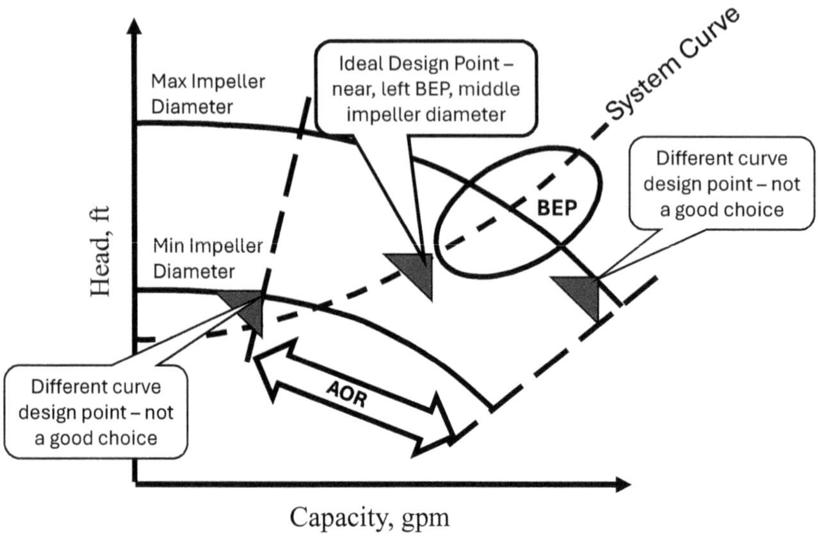

FIGURE 2.30 Pump curve optimizing pump selection.

They make pumps to fit in a range. This range will fall somewhere in the allowable operating region (AOR). If the choice is available, choosing a pump with a design point just to the left of BEP with an impeller in the middle range of its diameter would be the best fit as shown in Figure 2.30.

Here are some reasons to choose this optimum point. There have been some studies showing lower pump vibration with an operating point just to the left of BEP. If there is any need for additional flow or more head from the pump, being left of BEP will allow for that sizing error or redesign on the pump. Left of BEP is also lower NPSHR, which means lower cavitation potential. A middle impeller diameter means there is room to modify pump in either direction in case pump system hydraulics change over time or if sizing has some error. A very small impeller is less efficient, but a large impeller has a small cutwater gap, and vibration may be slightly higher. The pump base can be made to accept a hydraulic upgrade (larger impeller and pump motor) where if the largest impeller pump is chosen, this is not an option.

3 Pump System Piping

The pump system most of the time can determine what kind of reliability and performance a pump will produce. The foundation of the pump system is the piping. Much of the focus of pump piping centers around the suction and discharge piping around the pump itself. If each part of the system piping is not done correctly then it will make the pump system more vulnerable to process issues and increase the probability of being unreliable.

One of the key tasks in sizing pump system piping is pipe size which is determined by flow and fluid velocity targets. Some targets will be discussed in the suction and discharge piping sections. The smaller the pipe diameter or larger the flow, the higher the fluid velocity inside the pipe. There is obviously a balance between the pipe size, friction head, and energy consumption as shown in Figure 3.1. If the pipe is too small, then energy cost is high. If the pipe is too large, then the pipe installation cost is high.

Larger pipe has a larger upfront cost not only with the pipe itself but also its installation. Smaller pipe has a higher energy demand with higher friction losses. This also affects the pump sizing and required motor power, making long-term life cycle cost (LCC) a major factor. The pipe size element comes more into play when the major head loss is due to piping friction. Small piping and higher fluid velocities can also affect pipe vibration and suction issues, particularly on the suction side of the pump. I have found that properly sizing the piping adhering to fluid velocity guidelines typically results in a reasonable ROI, placing the system near the optimal balance between the pipe size and energy consumption.

The basic equation for fluid velocity in a pipe is shown below. Keep in mind that the pipe diameter is the inside of the pipe diameter. Depending on the type of pipe being used and pipe thickness (schedule piping), the inside of the pipe diameter will vary. Refer to piping manufacturing tables for specific pipe dimensions, but one general note is that standard piping of up to 12″, the inside diameter (ID) is close to the nominal pipe size. Thicker walled-pipe reduces the inside diameter of the pipe so fluid velocities will increase as pipe schedule increases. Do not use the nominal pipe diameter if accurate results are desired. From Figure 3.2, let's look at the different elements of a good pump system piping on both the inlet and discharge sides of a pump.

$$V = \frac{Q}{2.45 \times d^2}$$

where:

V = fluid velocity in ft/s
Q = fluid flow in gpm
d = pipe inside diameter in inches

DOI: 10.1201/9781003729525-3

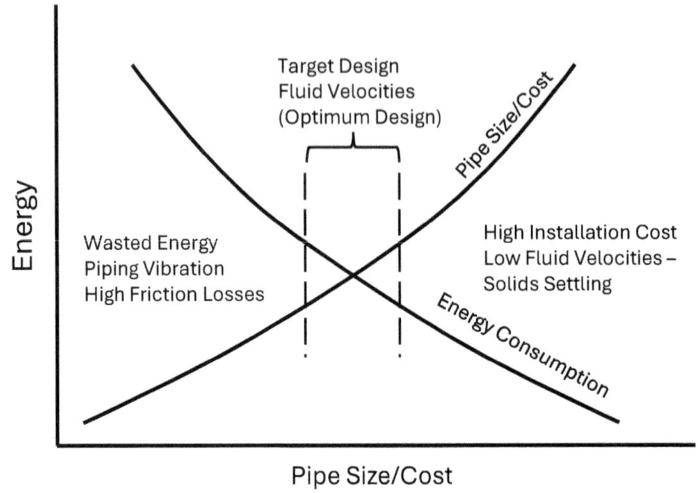

FIGURE 3.1 Cost optimization of pipe sizing.

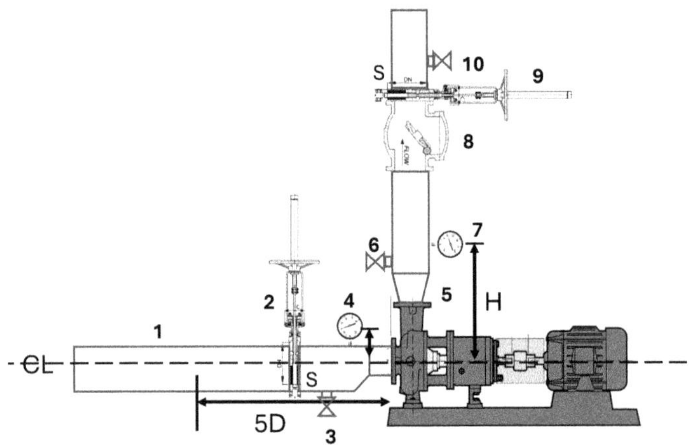

FIGURE 3.2 Pump system piping elements.

SUCTION PIPING

Suction piping is typically at least one pipe size larger than the pump suction nozzle and many times it is several sizes larger. An exception for suction piping is for self-priming pumps where the suction piping is the same size as the suction nozzle on the pump. This is to keep the priming time smaller with the smaller volume suction pipe. However, pipe sizing is done based on target fluid velocities for the application.

For a waterlike fluid, a typical target fluid velocity for suction piping is 3–5 ft/s. For pulp and paper stock applications, < 2% pulp consistency is typically treated like water. For 2–5% pulp consistencies, the design suction fluid velocity should be 1.5–4 ft/s. For 10–15% pulp consistencies, the target suction fluid velocity should be 0.25–0.5 ft/s. A small percentage of centrifugal pump systems are slurry applications, but one should refer to specific fluid recommendations for a minimum fluid velocity to ensure that the solids to not drop out of suspension.

Suction piping should be as straight as possible. If any elbows are used on suction piping, long radius elbows are preferred as long as minimum straight pipe length into the pump suction is maintained. If an elbow had to be placed on the suction, one in vertical plane is better than horizontal plane. Figure 3.3a shows an end suction pump where the straight run pipe was not satisfied. The fluid will not stay streamlined into the pump impeller eye with an elbow too close to the pump suction. Suction vortices and air pocket in the elbow can cause pumping issues in severe situations.

A double suction pump can have an unbalanced rotor due to suction elbow too close to pump suction as shown in Figure 3.3b. Suction piping runs should be as short as possible. Suction head at the pump suction nozzle is very important and every added amount of friction system loss creates a potential suction issue at the pump. Many pump issues originate in the suction piping before affecting the pump, or the rest of the system.

To ensure one-dimensional flow in the pipe, uniform flow of pathlines feeding impeller eye, and balanced velocity profile, a minimum straight length of pipe is required into a pump suction. A typical, minimum straight length of a pipe is 5–10 pipe diameters (D) from the pump suction nozzle shown as 5D in Figure 3.2. High-suction energy pumps may require 10D straight pipe and low-suction energy pumps may be reliable at 5D straight pipe. This can be especially harmful to a double suction pump if there is an elbow too close to the pump suction; the double suction pump will not be evenly loaded, and pump performance and reliability can be altered when operating with an unbalanced rotor.

Since the piping is typically larger than the suction nozzle on the pump, a reducer must be installed to transition to the pump suction nozzle. This reducer should be an eccentric reducer (flat side on top) as shown in Figure 3.4. This is to keep an air pocket from forming on the top and entrained air from the pump. The suction piping

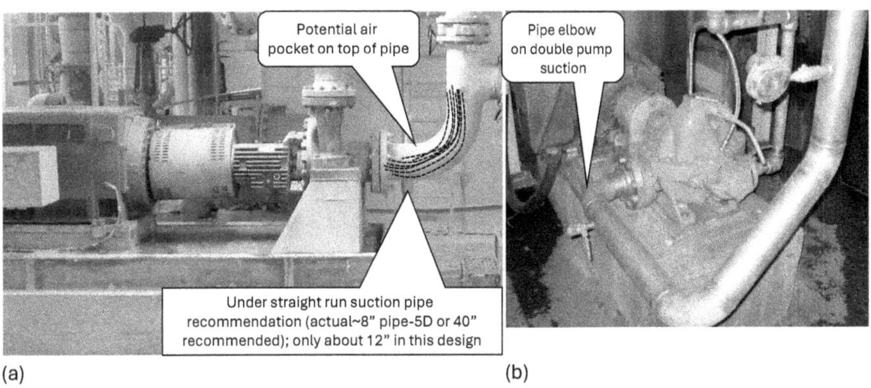

FIGURE 3.3 Insufficient straight run piping into pump suction: (a) end suction pump and (b) double suction pump.

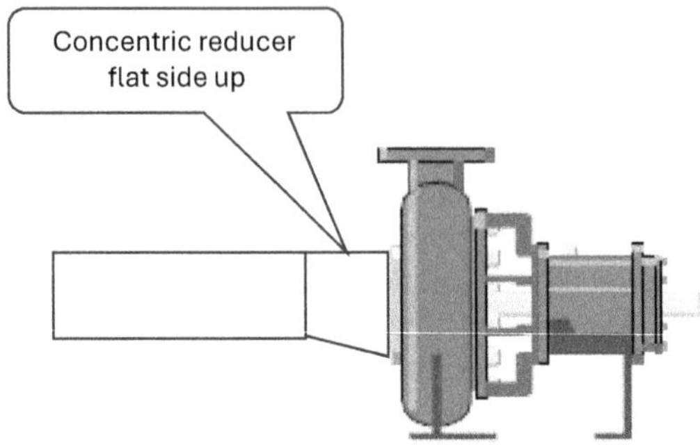

FIGURE 3.4 Correct suction eccentric reducer connection to pump suction nozzle.

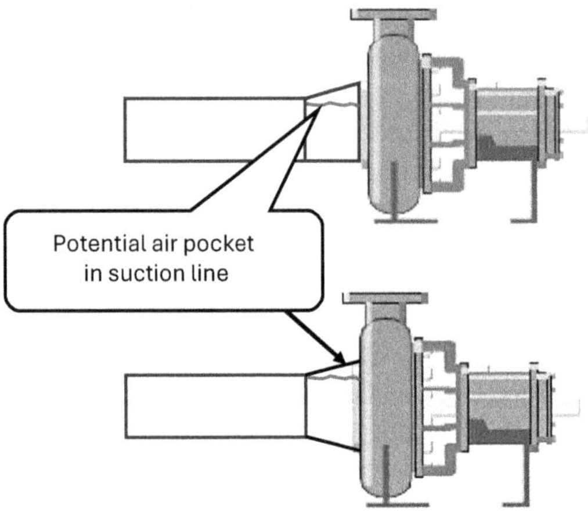

FIGURE 3.5 Possible pump suction reducer errors.

should also be slightly sloped up to the pump. A downward sloped suction piping can also create an air pocket in the suction. Figure 3.5 shows incorrect configurations for pump reducers with an inverted eccentric reducer and using a concentric reducer.

Another note on suction line sizing, specifically in pulp and paper applications and similar fluids, is it relates to percent consistency or solids. For pulp stock-type pumps, they can generally pump up to 6% consistency paper stock at full performance and up to 8% consistency with decreasing performance. For ANSI pumps, the suction size will limit the paper stock consistency that it can handle. A general rule of thumb is consistency should not exceed pump suction line size of up to 4″. Another words, a 4″ suction line should only be used up to 4% consistency and so on. A 6″ suction line size is slightly lower and can only handle 4% consistency. Another example, a pump with a 2-inch suction nozzle should not be used for paper stock with a consistency higher than 2%.

CASE STUDY 5 PUMP SUCTION PIPING TAKING THE HARD ROAD

Pipe tees can also impact the fluid flow similar to an elbow and they should also be included in the straight run pipe to pump suction. Any interruption in the flow of the fluid straight into the pump impeller eye will disrupt proper hydraulics. Small trim dilution control may be acceptable closer to the pump suction, but larger lines should be avoided inside the straight run pipe section. Figure 3.6 is an example of what not to do. The system was installed with all the tank inlet lines going into a small pump suction line instead of entering the backside of the tank from the top. It appears to have been done to save installation cost (additional piping) but this cannot be confirmed.

The flow from all the inlet lines was supposed to flow back away from the pump suction into the tank. The small pump was used as a sweetener pump for another part of the process. This was a blend tank, and another issue in the original design was that the different process flows were not mixing properly for this pump. However, the issue driving the need for a solution was that the pump would lose flow or have inconsistent flow output.

Flow performance issues happened when flows were increased on the lines closer to pump suction (which were also the largest). The flow vectors and velocity profile were not streamlined entering the pump leading to inconsistent

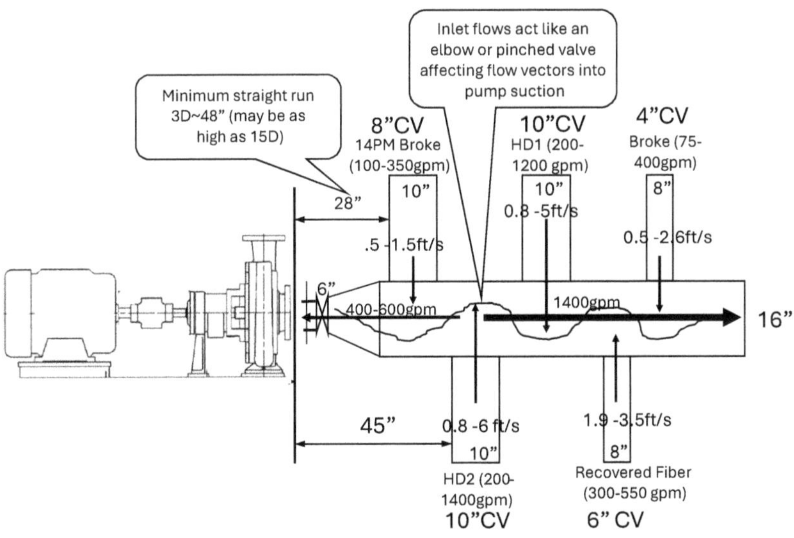

FIGURE 3.6 Poor suction piping into pump suction nozzle.

pump performance. Even though the straight pipe run was somewhat satisfied, the flow vector pathline into the pump suction was disrupted due to the large tee inlets. That is really what the minimum straight run pipe is there for, but this is erased when large inlet flows perpendicular to the straight run flow vectors into the pump suction were interrupted. The resultant flow was also back into the tank. The large flow perpendicular to suction flow was basically a hydraulic gate blocking good flow into the impeller eye.

The solution to the problem had several options: (1) re-pipe suction inlet lines and (2) relocate small pump to different location. Both of these were expensive options and not something the company wanted to execute. So, another option was to evaluate if there was another pump option to eliminate the small pump. Another larger pump on the same tank had just enough capacity to move the demand to that pump and shut down the smaller sweetener pump. This required a small piping job and nothing else. The solution also eliminated another piece of equipment to maintain and resolved the original problem of erratic flow from the small pump.

The root cause of the pump system issues was suction piping design in this case study. The original design was likely chosen for this system as the better alternative was more expensive. The system designers were not aware of the pump system operational issues that might result from this design. This is another good example of the importance of proper pump suction design for pump performance.

Case Study 5 Learning – Design pump suction with proper straight run piping (laminar flow) to prevent flow inconsistency.

If the pump is a suction lift situation (where the liquid level is below the pump centerline), then a suction check valve or foot valve may be necessary to keep the suction line full of fluid and primed for easier startups. Without this check valve in the suction line, pump priming may not occur, or it will be extremely long.

If the process fluid has a tendency to plug the pump suction, then a flanged cleanout port may be a good idea to avoid pump disassembly to remove a plug. Usually, this is located on top of the line with a blind flange large enough to allow access for reaching in near the pump suction nozzle to remove the plugged material.

For some pump suction designs, a suction bell is designed at the suction inlet into a tank or pit as shown in Figure 3.7. This is to reduce the entrance fluid velocities into the suction line. It is also important to have proper spacing between suction lines and from sump or tank walls also shown in Figure 3.7. Many horizontal suction lines in a tank may also be equipped with a long radius elbow at a certain height above the floor. When an elbow is used, the pipe cut across the elbow creates an elliptical shape on suction inlet acting similar to a suction bell. This reduces air entrainment resulting from fluid vortex at surface of fluid in the tank. The suction bell diameter can be estimated from the following equation.

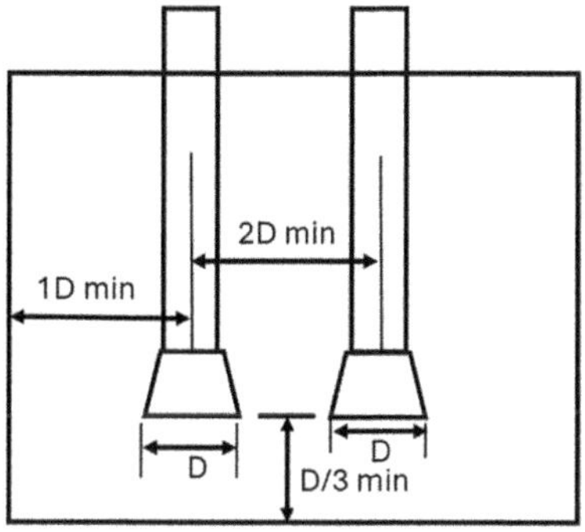

FIGURE 3.7 Suction bell dimensions in tank.

$$D = (0.0744 \times Q)^2$$

where:

D = suction bell diameter in inches
Q = flow in gpm

SUCTION ISOLATION VALVE

The suction line should have a suction isolation valve. Typically, a gate valve or a ball valve may be used on smaller pumps. The suction valve should be a full port valve to lower the pressure drop and restrictions to pump suction. The suction valve should also be placed before the eccentric reducer and away from the pump suction. Some system designers or installers may be tempted to save money by putting the valve after reducer close to the pump or even mounted on suction nozzle of pump to save money on the valve (smaller valve size).

This can cause several problems such as valve reliability issues (fluid starts to rotate at suction nozzle), such as valve erosion. Also, the valve can interfere with pump casing maintenance being in the way. The valve at that location will also cause a small amount of flow turbulence right before it enters the pump suction. If an impeller inducer ever needs to be installed on the pump later, the valve at the suction nozzle could prevent that option.

If a unidirectional flow isolation valve is installed, make sure the valve seat is on the pump side so the feed pressure will seat the valve for pump isolation and maintenance. Valve seat location is also shown in Figure 3.2 with an 'S' on side of the valve seat. A good specification for these isolation valves is to purchase valves with lockout ability built into the valve.

SUCTION DRAIN VALVE

The suction line should have a drain valve located between the suction isolation valve and the pump nozzle. It should be at the low point of the piping to ensure drainage of all fluid in the piping and pump. The valve should also be a full port valve of ample size to handle the fluid being pumped and type solids in the fluid, so it doesn't get plugged easily. Adequate room to the floor should also be provided for this valve to allow rodding in case of plugging. This valve is usually required to be left open as part of lockout.

SUCTION PRESSURE GAUGE

It is a good practice to provide a port to install a pressure gauge on the suction line. This can be used for troubleshooting later and as part of Total Dynamic Head (TDH) verification of the pump performance. This pressure gauge can also help troubleshoot a plugged suction line. It is also a good practice to install a quick isolation valve before pressure gauge so it can be changed while the system is running. The pipe connection should be at least a ½″ pipe size and maybe larger if plugging is an issue. A pressure gauge with a diaphragm may be needed for high consistency stock or other high solid fluids.

If the pump is in a suction lift situation, then a compound gauge (absolute pressure gauge reading vacuum and pressure) or a vacuum gauge will need to be used. For a suction lift pump, a priming port may also be required. The same port as the pressure gauge could be used, but better to have a separate port due to the volume needed for a priming port location.

DISCHARGE PIPING

Discharge piping is typically at least one pipe size larger than the pump discharge nozzle and several sizes larger in some instances. Like suction pipe sizing, discharge pipe sizing is also done based on target fluid velocities. For a waterlike fluid, a typical target fluid velocity for discharge piping is 6–11 ft/s. For pulp and paper stock applications, $< 2\%$ pulp consistency is typically treated like water. For 2–5% pulp consistencies, the design discharge fluid velocity should be 3–8 ft/s. For 10–15% pulp consistencies, the target suction fluid velocity should be 0.5–1 ft/s.

The pipe transition from the pump discharge nozzle to the system piping is typically accomplished with a concentric reducer. For a vertical pipe discharge, there is no concern of air entrainment at this point in the piping. The straight run section is not as critical on the discharge side, but many professionals will still recommend 5 pipe diameters straight run even on discharge piping. I've only witnessed one instance that had a strange elbow on the pump discharge (see Figure 3.8), and it did have some pitting/erosion issues on the pump discharge nozzle. It also had some other issues, but it is always good to have some straight run pipe from pump discharge too.

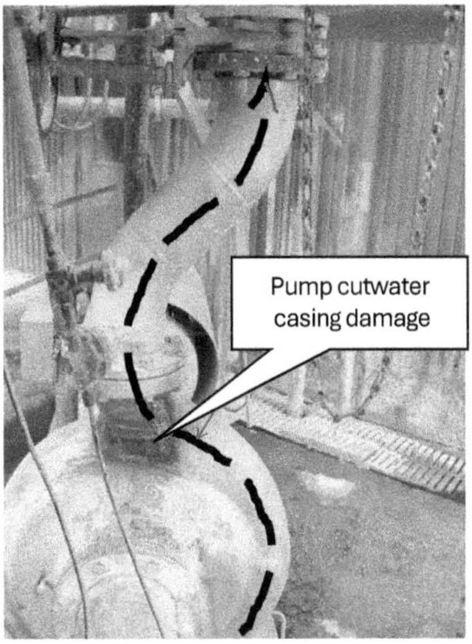

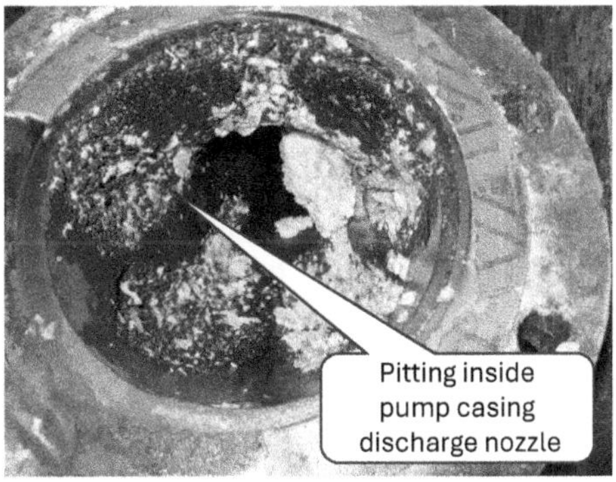

FIGURE 3.8 Discharge piping straight run error and resulting casing damage.

CASE STUDY 6 PIPE SIZING RELIABILITY IMPACT

A process water pump system shown in Figure 3.9 was having several reliability issues including piping failures causing significant process downtime. The original pump system had two pumps with a total system design flow at 2,600 gpm. The piping was sized for a 10″ schedule 10 piping so by using pipe velocity equation, the design fluid velocity was 10 ft/s, which is within the target velocity range for discharge piping design. Several years later,

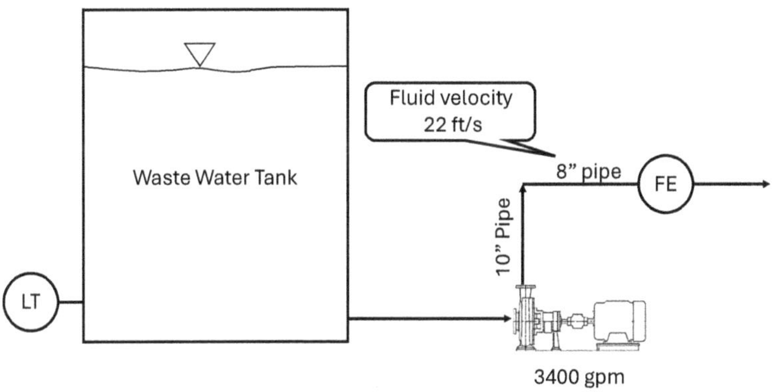

FIGURE 3.9 Process diagram of pump system piping.

the process system had a redesign which increased the flow to 3,400 gpm. Over time, the piping began to have failures. Some were poor-quality welds on the piping and after multiple repairs, it was determined that the line needed to be replaced. Failures recurred even after replacing the entire line, indicating the project did not resolve the main issue.

A failure investigation was initiated based on the failure mode of pipe fatigue particularly at welded joints. Several possible causes were investigated such as weld porosity (weld quality), water hammer pushing air out of the system, pipe resonance, effect of backflush procedure on the pipe system, and poor hanger design. While weld quality could have been an issue on a few occasions, over the years that was not seen as a major root cause since the piping was replaced.

Water hammer and backflush did not occur frequently to be considered the main root cause. While pipe hangers were not ideal and many were missing or damaged, these were replaced or upgraded during the first piping replacement, yet failures still occurred. The piping replacement was carried out as a like-for-like replacement as it was believed that poor weld quality and pipe assembly were the main issues.

The impact of the increased fluid velocities on the piping stress when combined with normal piping elbows and flow path was not recognized. The pump system upgrade of 3,400 gpm increased the fluid velocity to ~13.8 ft/s. At times the flow was even higher than that. The piping also had an 8″ section in which the fluid velocities increased to ~22 ft/s. This combined with the existing pipe elbows and hanger design led to excessive pipe vibration and pipe stress to where the pipe was cracking in multiple locations. The force created by fluid flow is proportional to square of the fluid velocity, so the high velocity caused tremendous force on the piping and constant excessive vibration.

The issue was resolved by installing larger piping, which lowered the fluid velocities down to a maximum of 10 ft/s in all sections and most of the time even lower than that. The redesign also involved breaking up the flow between two lines as an older abandoned line was refurbished which provided a very feasible solution to the problem. After this, the system, which had been experiencing four to five failures a year at a minimum, has now operated for eight years without a single pipe failure. The last and only design change done was getting the fluid velocities down to proper design targets.

The pump and system upgrade are all too common where the process and pump are upgraded but the piping remains the same due to cost pressure to execute the project. Don't forget to execute proper system analysis, evaluating the pipe design on any process or pump system upgrade. The root cause of pump system reliability issues in the piping was a design flaw introduced during the first pump system redesign. The first pump system redesign created another problem that is discussed in a case study later. It would require further pump system redesign to address the other issues.

Case Study 6 Learning – Design piping systems to target fluid velocities.

DISCHARGE PIPING DRAIN VALVE

The discharge line should have a drain valve (6) between the discharge line check valve and the pump nozzle. It should be as low as possible to ensure drainage of all fluid in the piping and pump. The valve should also be a full port design of ample size to handle the fluid being pumped and any entrained solids, so it doesn't get plugged easily. There should also be a drain valve (10) above the check valve so the system can be drained. Some check valves may be equipped with a drain and in some cases can function as the drain valve for the system. This valve should be larger than the other drain valve (6) to accommodate the higher drainage volume.

DISCHARGE PRESSURE GAUGE

It is a good practice to provide a port for installing a pressure gauge on the discharge line. This can be used for troubleshooting later and as part of TDH verification of the pump performance. It is also a good practice to install a quick isolation valve before pressure gauge, so the gauge can be changed during system operation. The pipe connection and valve should be at least a ½″ pipe size. One note about pressure gauges and pump TDH. If the pump TDH is calculated from suction and discharge pressure gauges, a correction to the pump centerline must be applied. Also, if the discharge pressure gauge is located 4 feet above the pump centerline, then 4 feet must be added to the measured head to correct it back to the pump centerline for an accurate TDH calculation.

CHECK VALVE

A check valve will prevent back flow when the pump shuts off. A check valve will prevent the pump from running backward, which is helpful especially if the pump were to come back on while it is turning backward. This can cause shaft torsional fracture under the right circumstances. The check valve will also prevent runaway speed for a pump after it is shut down and draining backward through the pump. For pump specific speed pumps $< 10{,}000$, the pump runaway speed is typically < 2 times normal operating speed. A pump head $>$ BEP head has the potential to cause reverse runaway speed.

The check valve will also keep fluid in the line and will shorten time for system head development upon startup. Startup can cause pump to run outside AOR for a short period of time as well can cause significant pipe vibration as fluid fills the line as air is pushed out of the system. Keep in mind that a check valve is not a good option if handling heavy solids, paper stock, or some slurries. Check valves are directional by function, so pay attention to position during installation. Also note if the check valve can be installed in horizontal and vertical positions. An exception to using a discharge check valve is fluid with high consistency (paper stock) or trash load which will lodge in check valves. Swing check valves are popular in simple systems but note that the life of a swing check may be as little as five years before an internal wear or failure may render the valve unserviceable.

DISCHARGE SHUT OFF VALVE

The discharge line should have a discharge isolation valve. The discharge valve should be a full port valve to lower pressure drop and restrictions. If a unidirectional flow isolation valve is installed, make sure the valve seat is on the pump side so system pressure will push on valve seat as shown in Figure 3.2 or marked with an S. Generally, discharge valves are used either fully open or fully closed. If a valve is used for throttling, the seat should be located on the downstream side to ensure proper gate support. A good specification for these isolation valves is to purchase valves with built-in lockout capability.

CASE STUDY 7 WATER HAMMER AND POOR PIPING SUPPORTS

A disk filter pump system was experiencing several failure issues in the piping and inline flowmeter. The piping was vibrating severely especially on pump startup. Hangers were getting loose or breaking, and flange gasket failures occurred on the pump discharge when the flanges blew out. The pump system was pretty simple with white water from an atmospheric tank with pump delivering the fluid through a flowmeter and control valve to the disk filter as shown in Figure 3.10. The piping system is large consisting mostly of 20″ piping with a 16″ control valve.

Upon shutdown the piping tends to drain back down as there is no check valve in the system. The control valve is after the flowmeter and this control valve also closes after the pump is shut off. Much of the fluid still drains back down but much slower. On startup the piping rapidly fills with fluid. The pump is a 16 × 16 × 19 large end suction pump sized at 8,830 gpm @ 90 ft. The pump has a suction-specific speed of 10,831. The pump is driven by a fixed speed 250 HP, 1200 rpm motor via a starter, causing it to reach 100% speed immediately.

With the line emptying out, a lot of air must be expelled from the system, while fluid quickly rushes through the piping. The piping has many 90° elbows. The control valve operates on flow control, but the valve is throttled during startup. There is still ~5,000 gpm at startup in a 20″ pipe on 90° elbows so there are tremendous forces exerting on the piping and hangers. Startup and shutdown create a significant water hammer situation. Water hammer happens due to rapid change in fluid velocity of the water column from pump startup and shutdown. This kinetic energy is converted to pressure energy in the fluid and travels down the pipe at the speed of sound in the fluid (water in pipe ~4,720 ft/sec). The pressure energy is transformed into force reactions on the piping system where piping and hangers must absorb this energy.

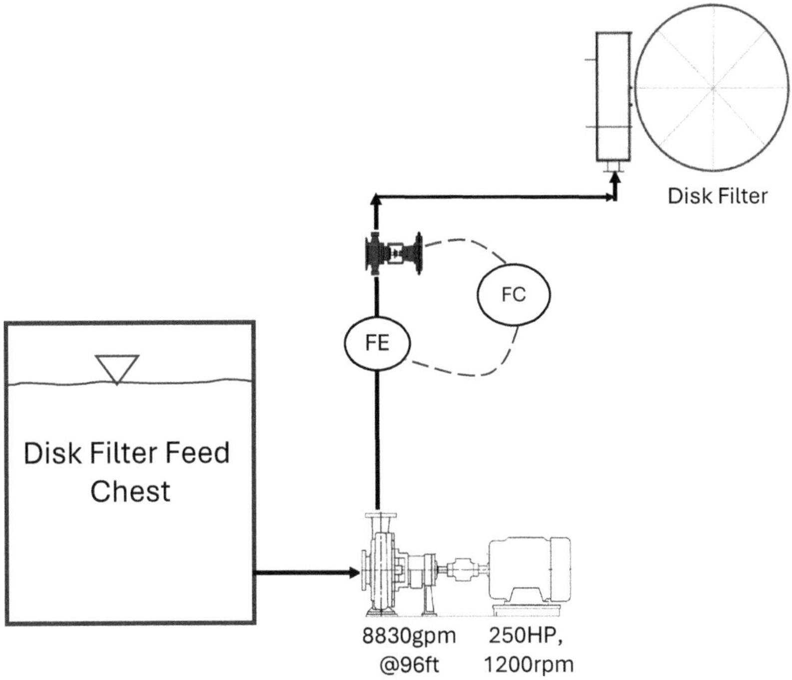

FIGURE 3.10 Process diagram for disk filter feed system.

To assess the water hammer potential and forces on the piping, the magnitude of this system must be considered. The mass of water column in this piping system is ~92 ft of 20″ 316SS SCH10, which is 192 ft^3 or 1,436 gal or 11,981 lbsm of fluid in piping that starts or stops instantly as the pump starts. This may result in a kinetic energy of 30,505 ft-lbf. The momentum force on the piping resulting from an instant start of fluid from rest to full flow is ~37,138 lbf. If the hanger system is not properly designed, significant damage will result which this system experienced.

As a first step to reduce this startup water hammer, the control was set to open slowly about 5% every 10 seconds until target flow was achieved. One issue that prevented this from helping much is the fact that the control valve was toward the end of the system, so a lot of water momentum has already hammered its way through the system and also pushing out the air.

Pipe hanger design was already identified as an important corrective action on this system but the water hammer impacting still needed to be addressed. A check valve was needed to keep the line from draining empty upon shutdown. Also, the pipe hangers needed redesigning as several were insufficient to properly support the pipe with high forces at elbows. A surge buster-type check valve was the corrective action to address the water hammer issues causing pipe damage, flange leaks, and hanger damage. The check valve would also be critical in solving another reliability issue in this piping system. The compressed fiber pipe gasket was also replaced with a spiral wound metal gasket to increase strength to reduce blowout potential.

Another issue occurring on this system was the magnetic flowmeter tube was failing. The Teflon tube was being sucked in toward the inside of the pipe. For many years, the cause of this flowmeter failure was unknown. The cause of this flowmeter failure was vacuum in the line created during shutdown as the line drained.

When the pump shutdown, the control valve would close. The control valve was above the flowmeter. The control valve was not a zero-leakage valve and would leak down over time after it closes. Without a check valve, the piping would drain back into the tank. With the control valve closed, the fluid could not drain without forming a vacuum holding a column of fluid as it drained down. Since the flowmeter was below the control valve, this vacuum would pull in the lining on the magnetic flowmeter causing a flowmeter functional failure. The same check valve identified to solve the water hammer issue would also prevent backflow and vacuum condition at the flowmeter when the control valve is closed.

All these pump system issues had system design as the root cause. The corrective actions included redesigning the pipe hanger system, installing a check valve, and modifying the control valve logic. As of this writing, the pipe hanger and flowmeter have never had any issues since the corrective actions were taken.

Case Study 7 Learning – Never underestimate the impact of water hammer on the piping system. Also understand the importance of check valves and proper pipe hangers.

4 Pump Installation

Pump installation is one of those activities where the reliability can be cemented into the pump life until a major physical change is made to the pump or pump system. Installation can have many elements starting with the foundation and baseplate as well as pump and driver alignment to piping. Pipe stress and thermal growth can produce hidden reliability events to pumps. To address these installation issues, the following fundamentals are essential for ensuring pump system reliability. Just like a house, a good foundation is the starting point as everything is built upon it.

FOUNDATION/BASEPLATE

Any pump installation will start with the foundation and baseplate. While this is mostly a design and installation function, maintenance quality checkoff on new installations can help identify issues before they are set into the machine reliability. The foundation and baseplate provide several functions. First, it supports the static and dynamic loads of the pump and driver. It also dampens the vibration of the pump and driver. It also holds the alignment not only of the pump and driver but also the piping connected to the pump. A poor foundation can be an underlying issue where unreliability begins.

There are three main parts of the rotating equipment support base – concrete foundation, grout layer, and baseplate as shown in Figure 4.1. The mass of the support base should be >3–5X weight of the rotating machine (motor, gearbox, pump, etc.) it is supporting. This mass helps to dampen and absorb the vibration from the pump and motor rotors. Insufficient mass can lead to baseplate and foundation failures as well as pump and motor decreased life cycle.

The concrete foundation must be designed for each application but in general the concrete should have a minimum compressive strength of 4,000 psi. The concrete foundation will typically cure to full strength in ~28 days. Foundations installed over the existing floors should have proper dowels/ties to the floor. Anchor bolts should have sleeves for the concrete foundation. This will allow more stretch in the anchor when the baseplate is preloaded during anchor bolt tightening. Each section of the foundation must have good adhesion between each layer, so no air gaps or looseness is seen.

There are several kinds of baseplates such as fabricated steel or cast iron. Both are grouted baseplates which provides better vibration dampening characteristics and many times better long-term corrosion protection. There are also non-grouted base-plates (polycast) which provides easier installation, but the structural design must be much stiffer than grout style. Structural design must be stiff enough to have any natural frequencies away from equipment operating frequencies. Grout is typically installed in a flowable consistency in small quantities, quickly and continuously. Place grout from one end to ensure driving out all air bubbles and voids. Most of the time there is also

DOI: 10.1201/9781003729525-4

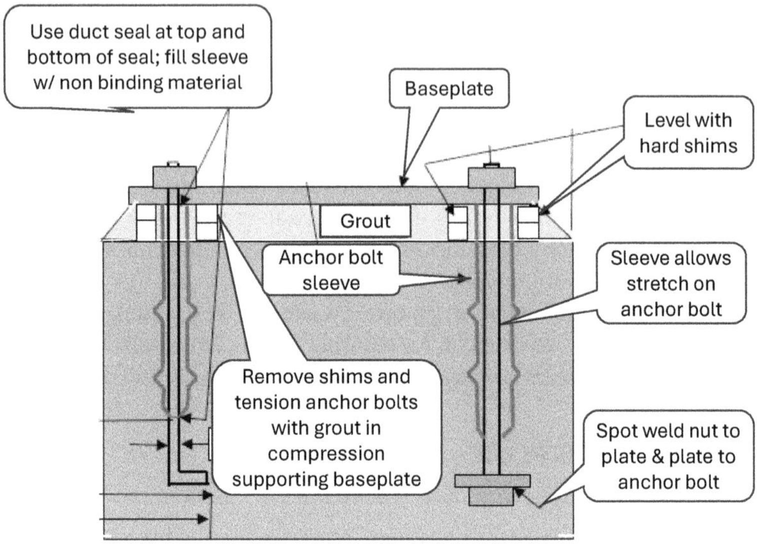

FIGURE 4.1 Typical pump foundation and baseplate.

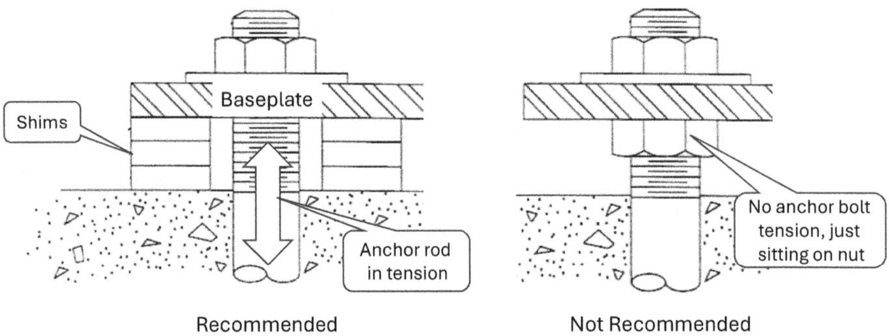

FIGURE 4.2 Pump baseplate anchoring with shims.

openings at the top of baseplate, so baseplate is fully supported by grout letting all air escape. The goal is to not have any voids in the grout under the baseplate.

Baseplates should be shimmed to concrete foundation as shown in Figure 4.2 also using sleeves around anchor bolts. As the grout hardens, the shims are supposed to be removed and anchor bolts tightened down so baseplate is tight against the grout. The anchor bolts should remain in tension, transferring load down into the concrete foundation. This places the grout under the baseplate in compression, which is the condition each element is best designed to withstand. A common error is to install a nut under the baseplate to hold the baseplate up while grouting takes place. The anchor bolt cannot be properly tightened against the grout, and the grout is not placed in compression. The result is there will be looseness between the grout and the baseplate.

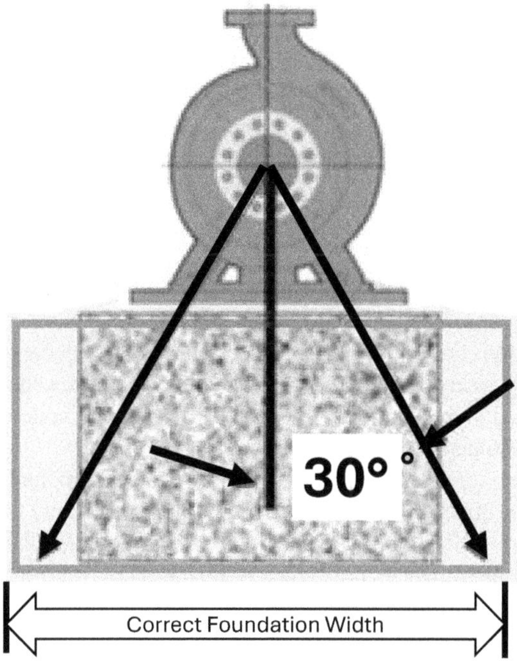

FIGURE 4.3 Foundation size relationship to pump shaft for correct foundation width.

The width of the foundation should be at least 30° from shaft centerline down to the floor to prevent rocking and provide a stable foundation as shown in Figure 4.3. This helps to stabilize the pump and baseplate in operation with a wide base.

There are two types of grouts – cementitious and epoxy. Epoxy grout has a temperature limit of ~200°F (operating pump process). Epoxy grout is better for chemical environments. Epoxy grout tends to soften and creep at higher temperatures. Cement-based grouts can be trimmed before final set, whereas epoxy grouts need to have chamfer strips built into the forms. Use cementitious grout for high temperature applications > 200°F. The grout for the foundation/baseplate should be a non-shrink cement or epoxy-based grout.

Not all pump baseplates may require to be grouted. Small pump assemblies less than 10–15 HP or less than 350–400 lbs typically will work just fine without grout. A stiff steel or cast-iron baseplate can usually offer plenty of mass and stiffness to prevent any measurable reliability issues even on some pump baseplates up to 30–40 HP or 1,000 lbs. Anything 50 HP or larger than a grout system approach should be used. I still prefer to grout all baseplates in installation regardless of HP or size of pump unit.

Anchor bolts shall be tightened to specified torque to have proper preload between baseplate, grout, and foundation. Too little preload will allow stress fluctuations, which potentially leads to fatigue failure in anchor bolts and spalling grout or concrete. Also, the baseplate will be loose to grout, resulting in a soft foot effect, baseplate distortion, and looseness of the machine. This may show up as misalignment, high 1X unbalance, or even looseness on the rotating machines as baseplate flexes in operation. Problems usually get worse the further away from the foundation; therefore, getting the first layer of pump system installation correct is a critical first step toward pump system reliability.

SHAFT ALIGNMENT

Shaft alignment is a key factor in all shaft connected equipment's overall reliability which includes pumps. Poor shaft alignment has many negative consequences such as increased energy consumption, decreased bearing life, decreased seal life, and increased coupling wear just to name a few. Studies have showed a 2% reduction in energy usage by alignment improvements. A 2% energy reduction on a 100 HP motor is about $800 annual energy savings.

There are many types of misalignments which are compounded together for a total misalignment effect on the equipment. Misalignment may show up as offset, angular, and axial. Offset and angular can be in multiple planes (vertical and horizontal). Axial alignment is not typically managed by shaft alignment tools but is important for proper coupling engagement and has a significant impact on the overall equipment reliability.

Vertical plane corrections (offset and angular) are accomplished by shimming. Horizontal plane adjustments are made by horizontal moveable machine (typically driver/motor) movement. Generally, don't shim the stationary machine (pump) but in the case of pipe strain, shimming the pump may be necessary to achieve target alignment. The stationary machine is to remain fixed without any shimming. The pump needs to stay aligned with the piping it was first connected to, while the pump driver (motor) is aligned to the pump.

Before laser or indicator are installed on the shaft, a rough alignment is executed. Rough alignment can be done with a straight edge for offset and tape measure or dial calipers on shaft/coupling for angular alignment. Rough alignments are only to get the shafts closer so a more precision method can be completed. Some may be tempted to think a straight edge alignment is pretty good but rough alignment will never get the shafts as good as required by shaft alignment tolerances.

Shaft alignment tolerances are based on run speeds (higher speeds have tighter alignment targets) as shown in Figure 4.4. The laser equipment has built-in alignment targets that trigger a green or red indication showing if the alignment is within acceptable limits for the specified operating speed. When using the laser, it is important to use the correct operating speed. Shaft alignment affects many things such as bearing loads, coupling wear, energy usage, and operating temperatures.

Shaft alignment refers to aligning the shafts and not the motor feet. The alignment process is for machine shafts where they are connected. It is called shaft alignment and not feet alignment. A long motor compared to a short motor with the same shaft alignment will have drastically different feet alignment shim numbers. There is usually not a significant benefit to achieving alignment tighter than the specified shaft alignment targets given in the table in Figure 4.4.

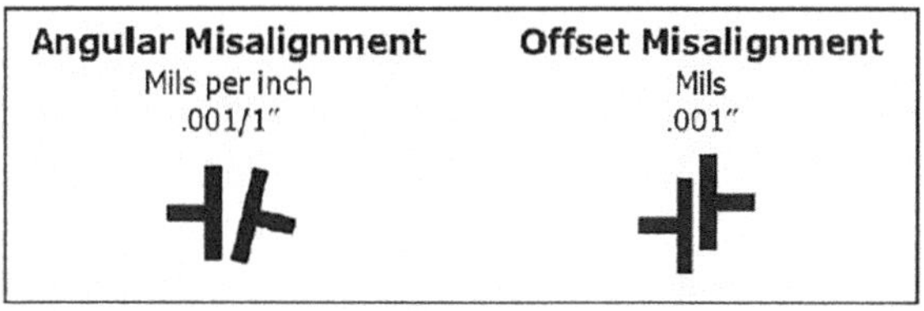

RPM	Excellent	Acceptable	Excellent	Acceptable
3600	0.3/1″	0.5/1″	1.0	2.0
1800	0.5/1″	0.7/1″	2.0	4.0
1200	0.7/1″	1.0/1″	3.0	6.0
900	1.0/1″	1.5/1″	4.0	8.0

FIGURE 4.4 Pump shaft alignment tolerances.

It is also important to note that the static shaft alignment is not the operating alignment. Many factors, such as thermal growth and reaction forces, from the pump operation will change the shaft position such that a rotating assembly with 0.002″ radial internal clearance that operates and the shaft reaction force is up in the 10 o'clock position then the shaft alignment offset will be impacted some amount vertically and not where it was aligned. This is another reason why going tighter on alignment may not yield a return on time investment.

Good reminder that shaft alignment is a human function, and the laser really only assists in the calculations. The laser is just a specialized calculator that calculates based on the users input parameters so use care when entering setup as it will affect output – garbage in = garbage out. Also, too many are quick to report a faulty laser when the alignment isn't going well. The laser is just a calculator so if it powers on, it isn't the laser. Here are a few things to check – laser setup, sights tight and clean, soft foot, or shafts touching. These can all cause erratic behavior of a laser, but the laser is just telling you there is a problem somewhere.

Shaft alignment is really straightforward and just required knowhow and discipline not to get in a hurry. Below are some basic alignment steps.

1. **Clean and prepare baseplate flatness and level:** Some standards specify flatness to be less than 0.005″. Checking pump or motor feet for flatness and level can also be important. Clean up any corrosion on the baseplate and mounting feet. Honing the feet to get out any imperfections can be a quick field task if the problem isn't too bad. Flatness helps to prevent looseness or squishy shims. Level is different than flat. Some specifications will have a level of not more than 0.0005″/ft. A good precision 98 level is a good tool to achieve baseplate level. Unlevel mounting conditions can affect oil level in the pump or if the four motor feet are not level on baseplate, then a soft foot or even worse variable motor frame air gap can occur causing motor performance and vibration issues.

2. **Check shaft runout TIR:** This isn't really a shaft alignment step but if there is a bent shaft, now is the time to correct it. A bent shaft can cause much difficulty during alignment and issues later during operation. Generally, shaft total indicated runout (TIR) should not exceed 0.002″. This is normally executed by installing a mag base setup with dial indicator on shaft to measure radial movement at one rotation. If a coupling is already on the shaft, a good idea to check it radially and on face in case the coupling flange is bent for some reason.
3. **Check and remove soft foot:** Soft foot is where one of the four mounting feet is not properly supported, and an air gap exists. This will cause many issues such as higher vibration (looseness, 1X unbalance), potential higher shaft loads (if soft foot is tightened up and pump or motor is in a bind), and motor air gap issues as frame is distorted. Soft foot can be parallel soft foot, bent foot, spring soft foot (stack of bent or damaged shims), or stress-induced soft foot. Most soft foot conditions are not visible to the eye. To check soft foot, there are several methods such as with laser hooked up, loosen bolts on motor feet one at a time, and see laser (shaft) movement. Also use a 0.002″ feeler gauge and feel under each foot to find a soft foot gap. A dial indicator can also be set up on each foot and bolts loosened to see any movement. A general limit of 0.002″ soft foot is typically allowed but best to get all of soft foot out before getting into alignment as it will cause all kinds of errors. Step shims and other shimming methods can address small amounts of soft foot if the root cause can't be corrected.
4. **Check and correct pipe strain:** Pipe strain is where the suction and discharge piping are either in tension or compression on the pump nozzles. Pipe stress is often assumed to act only as a downward load on the pump nozzle, but flange gap can also induce pipe stress. There are acceptable pump nozzle loads limit but they are difficult to measure, therefore, similar to checking for soft foot, the shafts or laser alignment should not shift more than 0.002″ when the pipe is unbolted from the pump flanges. Piping should be freely supported to handle loads. The pump nozzles are not supposed to be pipe supports. This added stress will affect clearances, alignment, and loads on pump components.
5. **Perform rough alignment:** Using a straight edge or checking shaft/coupling gaps perform rough alignment. This will get the shafts in a good position for precision alignment on next step.
6. **Measure shaft misalignment:** Setup the laser or dial indicators and perform measurements to determine the current shaft misalignment. The stationary and moveable machine will need to have basic operating data entered such as feet spacing, shaft spacing, and operating speed.
7. **Correct vertical misalignment, also adjust for thermal growth:** Vertical alignment (offset and angular) is a good place to start adding and removing shims. Once this comes within tolerance, revise shims to keep the number of shims to less than four shims. Use a pry bar to raise the motor feet, then place the shims. Use full width shims if available or the correct size shims for the motor feet (A, B, C, D). Try to handle the shims by the stick-out

tabs. This is for safety reasons and to eliminate soiling the shims with contamination from your hands or gloves.

8. **Correct horizontal misalignment:** Horizontal alignment also involves offset and angular. Best practice is to have adjustment screws on the baseplate for easy adjustment. Hammers to motor feet are not only hard to control but damage motor feet.
9. **Check axial alignment:** Axial alignment has more to do with coupling engagement. Every coupling will have a target gap. Generally, a +/−10% gap tolerance to target is allowed but consult specific coupling manufacturer recommendations. If out of target, this can not only lead to short coupling life but also additional loads to pump and motor bearings and shafts.
10. **Torque all bolts and recheck alignment:** Before conducting final alignment, ensure that all machine hold-down bolts, pump flange bolts, coupling bolts, and any bolted assemblies are preloaded per specifications. If the alignment changes significantly (out of target), then there may be an issue in the assembly.

PIPE STRAIN

Pipe strain is a unique problem with centrifugal pumps. Tremendous forces can be created from pipe strain either from pump flanges not matching piping, pipe movement during operation (poorly supported), or thermal growth-induced stress. Many pump manufacturers may list flange nozzle load ratings, but we have no way to really measure the actual resultant nozzle load. We typically measure the results of pipe stress in the strain of the pump shaft.

The piping should be self-supporting and not require the pump casing flanges to support the pipe. If a pump casing is being changed and will not come out due to pipe load on casing, then there is too much pipe stress, and it should be corrected. While there are allowable pump nozzle loads, ideal piping support does not load the pump nozzle flanges at all. Caution should be taken if piping is handling hot fluids or has long straight piping runs. Thermal expansion loops or expansion joints may be needed in those cases.

For new pump installation, the pump and driver equipment should be installed, set and aligned and then the piping should be set to ensure stress-free connection. This will relieve most of the pipe stress from the pump as any piping errors and stress can be removed as the piping is installed. When piping is installed to the pump, often times the alignment error is not found until final bolt-up. Suction piping must have tank nozzles aligned with the pump suction nozzle. For large tanks, the tank should be set first and the pump suction nozzle and piping aligned to the tank flange to avoid inducing stress.

Expansion joints are often used to handle some of the pipe strain on pump regardless of the source. The side closest to the pump should be anchored to protect the pump. Tie rods on the joint should also be used to minimize the external forces from the piping.

An easy way to measure pipe strain is to hookup dial indicator on pump shaft and disconnect piping. If the shaft movement is $< 0.002''$ then pipe strain is acceptable. However, it is always best to remove as much pipe strain as possible on every installation. Proper pipe hanger supports are critically important to not only support the piping but to also minimize the pipe movement.

Pipe supports are necessary to hold piping in place during operation. There should be proper pipe support to independently stabilize the piping (without the pump even connected). This support includes the weight of the pipe and contents as well as reaction forces, especially in elbows from the momentum energy of flowing fluids. Significant pipe forces can result from fluid momentum changes in the pump system. Another example of this was on a paper machine water system, significant water hammer occurred during startup as the incoming fluid displaced air from the piping system. When the system was started up, there were still maintenance technicians working in the pipe bridge. The pipe hammered so hard and moved that it broke the ribs of one of the maintenance technicians. The pipe was supported but mostly only in the vertical direction holding up the weight, but there was not any real support in horizontal and axial planes to control pipe movement.

While the pipe can provide some of its own structural stiffness, there is a limit to how much unsupported length of pipe can be allowed. Pipe hanger spacing to support pipe and content weight can be found from many piping supplier manuals. A general rule of thumb for pipe supports and spacing (unsupported span) should not exceed that in the table in Figure 4.5. The larger the pipe, the less support it needs as the pipe itself provides stiffness and support so more space is allowed before the next pipe hanger.

Nominal Pipe Size, inches	Maximum Pipe Span in Feet
1	7
1.5	9
2	10
2.5	11
3	12
3.5	13
4	14
5	16
6	17
8	19
10	22
12	23
14	25
16	27
18	28
20	30
24	32

FIGURE 4.5 Pipe hanger supports spacing guide.

Pipe supports come in a variety of types such as adjustable clevis, pipe and beam clamps, U-bolts, pipe rolls, alignment guides, and spring hangers just to name a few. All these hangers support the pipe in different ways. The clevis hanger only supports the pipe in vertical direction. Pipe and beam clamps also provide vertical support. A U-bolt pipe support usually is a rigid support off a structural beam it is clamped to, therefore, it can provide multiple planes of support. Alignment hangers allow the pipe to move in either vertical, horizontal, or axial directions depending on placement but will limit movement in other planes. Pipe roll hangers provide vertical support of pipe but will allow it to move axially such is in the case of thermal expansion. Vertical pipe movement from thermal expansion may be managed by using spring hangers to absorb the movement.

One common mistake that is found in many process plants in regard to pipe hangers is the overuse of clevis-type hangers. While they are good in supporting piping in the vertical direction, they provide no real support in horizontal or axial directions. With many 90° elbows in piping systems, the fluid momentum change will create high-impact forces on the pipe that will move the piping in an all clevis hanger system. This will not only impact piping reliability but also transfer impact forces to the pump nozzles affecting pump reliability.

An easy clue to see if this is an issue in the piping systems is to notice pipe movement within the clevis hanger. If the pipe has been moving back and forth or the clevis is always in a different location, then additional pipe supports may be needed. There is piping software available today where a more advanced analysis can be done. Contact your project engineering firms to have this type of analysis completed on troublesome piping systems.

CASE STUDY 8 PIPING MISALIGNMENT

Pipe misalignment can cause many issues on a centrifugal pump. Whether misalignment is axial compression or tension on the pump nozzle or offset misalignment, the stress transferred to the pump may cause many issues such as shaft misalignment, extra bearing loads, increased vibration, and ultimately lower pump reliability. There is also a potential hydraulic effect on the pump and piping. An end suction pump in a paper mill was experiencing casing erosion on the discharge nozzle. This was a rare failure mode and location for pump casing pitting and erosion but had been occurring in chronic frequency.

The pipe misalignment will not be obvious to a normal observer, but a closer study of the pump and piping revealed the issue. Closer inspection on the pump casing discharge nozzle clearly showed the pipe offset misalignment with the discharge flange on the pump casing as shown in Figure 4.6. The pipe flange was hanging over in the flow path coming out of the pump discharge nozzle. This was deflecting the exiting fluid velocity opposite the pipe flange protrusion similar to placing your thumb over the end of a water hose nozzle.

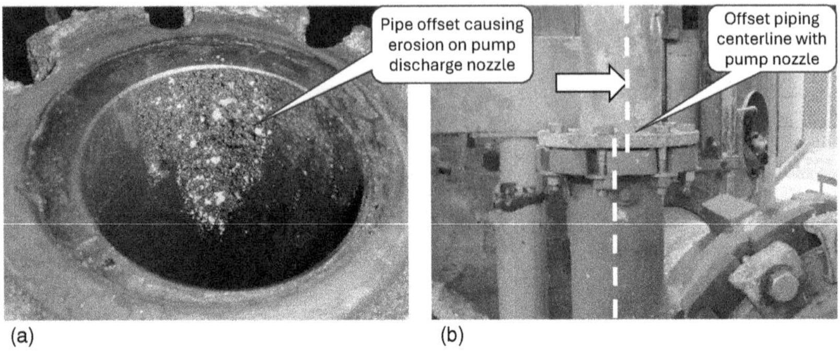

FIGURE 4.6 (a) Pipe misalignment with discharge nozzle. (b) Pump casing erosion on discharge flange.

The root cause of this pump reliability issue was piping installation (part of design). This could have been a case where the pump was installed, and the piping was run to the pump instead of away from the pump. Once they got to the pump, it did not align with the pump discharge nozzle. Maintenance would correct this issue but was not involved in the original design and installation.

Case Study 8 Learning – Check for pipe misalignment on pump suction and discharge nozzles during regular pump maintenance.

THERMAL GROWTH

Thermal growth corrections in alignment are based on the difference between the pump vertical growth of the driven equipment and the vertical thermal growth of the motor. This can be due to the pump handling a hot process fluid or a cold process fluid. The motor size and temperature are also a key factor in thermal growth magnitudes. Some pump manufacturers may supply a thermal growth adjustment table based on process fluid temperature. For example, one pump manufacturer says that for a process fluid temperature of 250°F, put a 0.005″ vertical offset thermal growth allowance. What this means is that the pump assembly will grow 0.005″ due to hot process fluid, therefore, align the motor/driver 0.005″ higher on vertical offset so when the unit is hot it will be within alignment tolerances. This assumes a nominal motor temperature and growth, which is just a guess, so better estimates can be obtained on any specific pump and motor set physical and operating data.

Pump and motor thermal growth calculations can also be completed for a set based on individual actual temperatures and vertical dimensions of the pump and motor. The thermal growth follows the formula below.

$$\Delta H = H\alpha\Delta T$$

where:

H = shaft height on pump or motor, inches
ΔH = vertical thermal growth, inches
ΔT = temperature difference at ambient vs operating temperature, degrees
α = thermal growth coefficient of material, /°F

Perhaps the most accurate way to account for thermal growth is to perform a hot alignment. Hot alignment starts up the pump until it reaches the operating temperature and then shuts down and performs alignment while still hot. This may not be practical or feasible all the time, so the calculations are often relied on for initial alignment. An example of estimating this is shown in Figure 4.7. The thermal growth of a pump and motor in reality is not just one temperature through the entire frame but usually some gradient. The thermal growth can be calculated at several intervals of height on the machine to be more accurate than just using one temperature for the entire machine frame.

Thermal growth can result in increased shaft and bearing loads on the pump and motor drastically reducing bearing life in some cases. Pumps handling hot process fluids like condensate will have the pump casing end grow much more than the pump drive end bearing, which is typically operating much cooler as shown in Figure 4.8. In this example, there is about approximately 145°F temperature difference between the pump casing and power end where the shaft and bearings are located. This can generate a significant thermal growth and can result in stress on the pump and motor if not corrected.

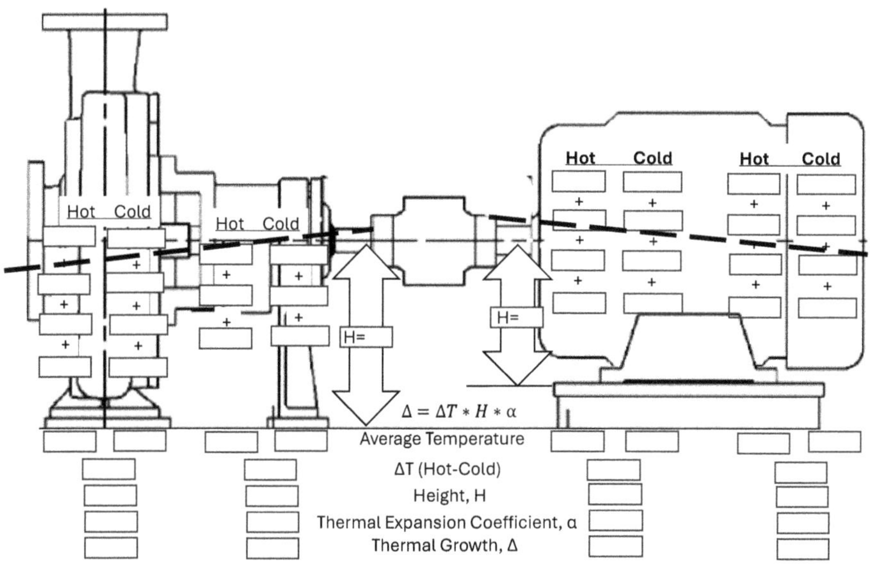

FIGURE 4.7 Thermal growth calculation model for pump and motor (simple machine).

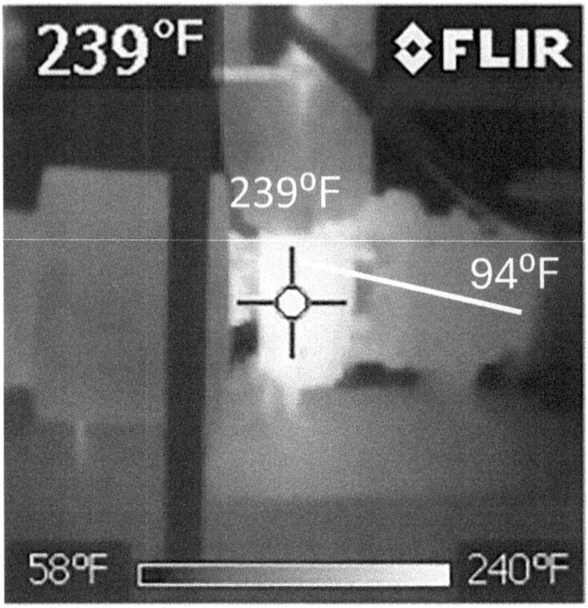

FIGURE 4.8 IR photo showing thermal growth shaft slope of condensate pump assembly.

Most totally enclosed fan cooled motors (TEFC) will be hotter on the drive end due to shaft loads and better cooling on the opposite drive end, ODE, where the cooling fan is located. There is ~35–40°F difference in each end of the motor in Figure 4.9, which creates angular misalignment. A typical motor with a hot condensate pump may at least have the slope of the thermal growth in the same direction even if the magnitudes are different.

After cooling air leaves the fan shroud on back of the motor, it quickly dissipates into the surroundings and air velocities decrease drastically around the cooling fins on the drive end of the motor. This is where motor covers are used to contain and direct the cooling air, ensuring it flows properly over the motor cooling fins. It seems counter-intuitive by encapsulating the motor but keeping the convection heat transfer coefficient high with higher velocity air down the entire length of the cooling fins is critical as opposed to just natural convection cooling. As shown in Figure 4.9, the air will naturally dissipate into the atmosphere without a cover to direct the air velocity down the cooling fins.

With a cold fluid in a pump such as a chilled water pump, the thermal growth effect can be significant as shown in Figure 4.10. In a cold pumped fluid situation, the slope of the rotating axis thermal growth is now in opposite directions between the pump and motor. This condition should certainly be addressed to minimize the impact of misalignment forces on the coupling, shaft, and bearings. The slope angle in this application is in opposite directions so the impact of the misalignment at the shaft interface is much more extreme.

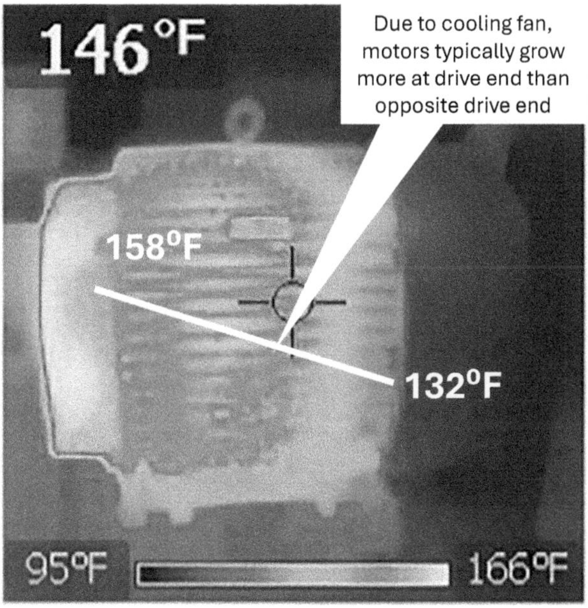

FIGURE 4.9 Typical motor thermal profile (IR photo) showing shaft slope change from thermal growth.

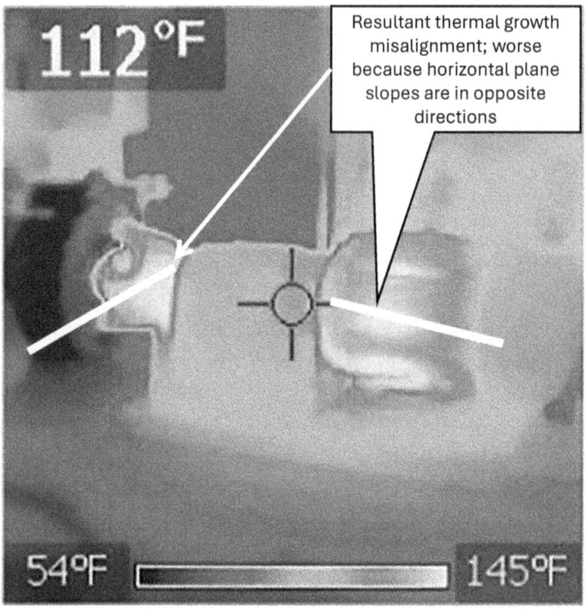

FIGURE 4.10 Typical motor thermal profile (IR photo) of chilled water pump showing shaft slope change from thermal growth.

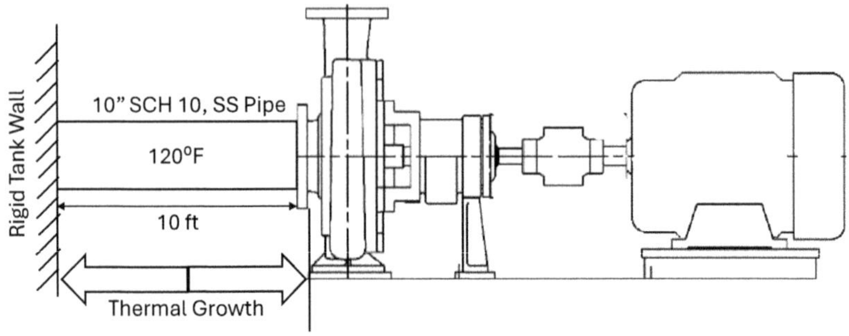

FIGURE 4.11 Thermal growth potential impact on pump suction nozzle.

Thermal growth on piping systems can also have a significant impact on the pump system. Piping is generally fixed connections and if long distances or large ΔT from cold to hot operating conditions occurs, potential large resultant forces can be transferred to pump suction or discharge nozzles. For example, consider a typical pump suction on a process pump operating at 120°F as shown in Figure 4.11. The pump has a 10″ schedule 10, 316SS suction pipe that is 10 ft long with a 30°F ΔT from cold to hot operating condition. Both end connections are fixed by rigid flange connections. The resultant thermal expansion force on a rigid connection can be estimated by the total deformation for an axial-loaded member in the equation below (this is after substituting thermal expansion elongation into basic equation).

$$\Delta = \alpha L \Delta T$$

$$\Delta = \frac{PL}{AE}$$

$$P = \alpha \Delta T A E$$

where:

α = thermal expansion coefficient for stainless steel = 9.6 × 10^{-6}/°F
E = modulus of elasticity for stainless steel = 27,600,000 psi
P = resultant force from thermal expansion, lbf
A = cross sectional stress area of pipe wall, in^2
ΔT = cold to hot (temperature difference), °F

For this example, the resultant force, P from the thermal expansion of the rigid pipe on the pump suction nozzle would be ~45,785 lbf. A pump manufacturer typically has a maximum nozzle load for the suction and discharge nozzles on the pump casing. A sample of one is shown in Figure 4.12. From this example, a 10″ pump suction would have a nozzle max load of 1,870 lbf so the thermal growth force if it was 100% rigid and nowhere to grow to allow for the stress would far exceed the pump design. It might be necessary then to add an expansion joint to allow free expansion and reduce or eliminate the compressive stress.

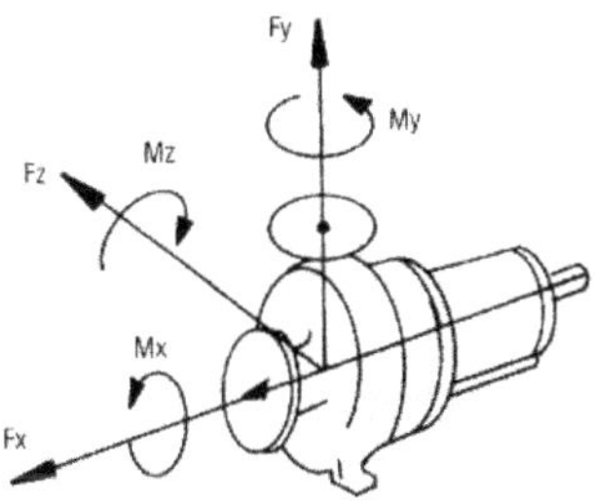

Pump Size	Combined Loading at Center of Pump						Flange Loading Suction	Flange Loading Discharge
	Fx	Fy	Fz	Mx	My	Mz	Fr + (Mr/3)	Fr + (Mr/3)
1½ x 1	120	295	235	590	295	295	325	220
2 x 1	145	365	285	730	365	365	435	220
2 x 1½	165	405	325	815	405	405	435	325
2 x 2	175	435	350	875	435	435	435	435
3 x 1½	220	545	435	1090	545	545	650	325
3 x 2	235	590	470	1170	590	590	650	435
3 x 3	260	650	520	1300	650	650	650	650
4 x 3	325	815	650	1625	815	815	865	650
6 x 4	470	1170	940	2345	1170	1170	1300	865
8 x 6	610	1520	1215	3030	1520	1520	1730	1300
10 x 8	670	1670	1335	3350	1670	1670	1870	1730
12 x 10	730	1820	1460	3640	1820	1820	2080	1820

FIGURE 4.12 Example pump nozzle load limitations.

The reality is that in this common real-world example, the actual thermal growth of the pipe is only ~0.035″. The pipe gasket compression alone will drastically reduce the compressive force transmitted to the pump suction nozzle to a magnitude well below the max nozzle rating for the pump used in the example. There are situations though where this stress can add up quickly and be well above a pump nozzle rating. The impact can lead to pump premature failures on various levels from sealing failures to casing failures.

CASE STUDY 9 CHRONIC PUMP SHAFT FRACTURE

A small ANSI pump started experiencing shaft fracture shortly after maintenance team changed the rotating assembly. The shaft fracture was happening at the drive end on the pump as shown in Figure 4.13. This pump rotating assembly failed three times in less than two weeks with exactly the same failure mode. There was no previous history of pump shaft fractures on this pump or any other application of this particular pump size.

What is happening causing these shaft fractures? It is important to understand the physics of the failure (POF) which will point to the possible causes and eventually to the root cause of failure. The fracture was between the coupling and the drive-end thrust bearing. This was a direct coupled drive so there should not be a high radial load at this location with good alignment. However, the shaft fracture surface exhibited a bending fatigue-type failure – ratchet marks and perpendicular fracture surface.

The investigation revealed some difficulty getting alignment during the pump changeout, especially the first few times. The final alignment numbers looked to be in line with an 1800 rpm machine. It was finally discovered that there was a significant soft foot condition noted on the last pump installation by a different crew. It is possible to still align two shafts with a soft foot condition. Unfortunately, it usually involves forcing the motor into a position it may not naturally assume in order to achieve the alignment, especially when using a laser.

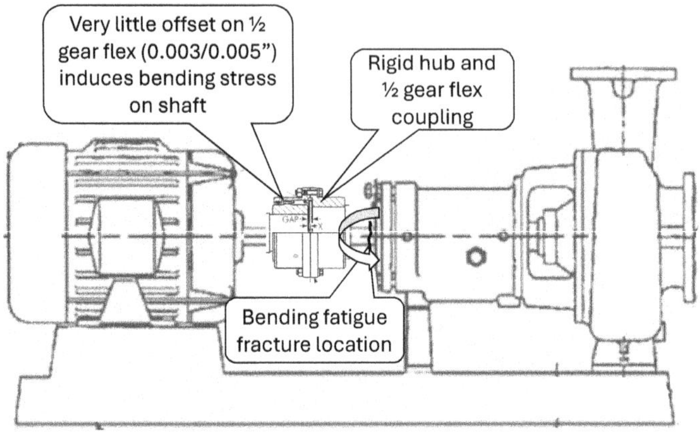

FIGURE 4.13 Pump shaft fracture location.

The soft foot condition was on the motor so why was the pump shaft breaking? The same force on the motor is also transferred to the pump. As a typical motor shaft is usually larger than most pump shafts, so the stresses are much higher on the pump shaft. Since it was a fatigue failure, cyclic bending stresses can be small but at 1800 rpm the stress cycles add up when it comes to fatigue. This was still an unusual type of failure. Most couplings can take some misalignment and not transfer these kinds of forces.

Further investigation revealed that the pump coupling had gotten changed to a different arrangement from the original which had a full flex gear coupling. It now had a rigid hub on the pump and a flex half on the motor. This arrangement and size of the coupling meant that it only had about 0.003/0.004″ tooth clearance in the radial direction to handle any radial misalignment. The soft foot condition exceeded that so when the soft foot was forced to satisfy the shaft position, a considerable bending stress was induced on the shafts. After a short run time at 1800 rpm, the stress cycles approached finite life, and a fatigue fracture occurred.

This failure is an example where two conditions had to be met for a failure to occur. Soft foot by itself will not cause the failure as well as having a very rigid coupling will not cause the failure. Soft foot with the correct coupling (full flex gear) would mean more coupling wear and lower pump reliability. Same with the coupling arrangement: while not ideal, if the shafts had been aligned properly, no failure would occur. It took two failure conditions (the extreme soft foot and rigid coupling arrangement) to create enough cyclic stress on the shaft to cause failure.

This pump reliability issue had multiple root causes. The coupling design was not optimum for proper assembly and reliability. Also, the maintenance execution on the alignment was not executed correctly and allowed a condition in which very high stresses resulted in causing a shaft fracture. Machine stress is hidden to the naked eye but can be a serious threat to pump reliability.

Case Study 9 Learning – Importance of soft foot correction and do not force alignment.

5 Pump Suction Problems

The first part of the pump system is the suction piping. The goal of the suction piping is to provide sufficient liquid (pressure and quantity) to the pump impeller eye. Any breakdown in delivering liquid to the pump suction and pump system issues will result before the pump even has a chance. Pump suction issues are many times the source of our pump problems. These issues can arise from piping problems, which we've already discussed, or from suction-related issues within the pump system. This may include submergence, air entrainment, or cavitation issues. Each one must be evaluated separately to diagnose the issue and resolve.

The most obvious pump suction issue is suction line plugging or buildup in the suction line. This decreases the flow area of the suction pipe, which increases the friction losses on the suction line and leads to lower pressure at the pump inlet. If the blockage is severe, it can lead to starving the pump impeller of fluid and can cause cavitation damage. Figure 5.1 shows an example of this type of suction buildup on the pump suction nozzle. Verifying the pump suction pressure and friction loss compared to a theoretical friction loss for the piping system can help diagnose a plugged suction line. A quick pump suction line inspection is also a good way to verify that the suction line is not plugged.

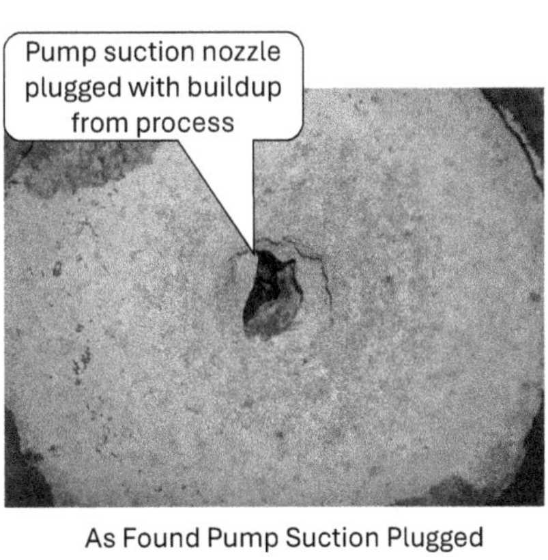

FIGURE 5.1 Pump suction pluggage from process buildup.

DOI: 10.1201/9781003729525-5

SUBMERGENCE

Pump suction submergence is the amount of fluid above the pump suction entrance to prevent free surface air vortices from forming and pulling air into the pump suction. Higher tank level makes it more difficult to get a larger volume (mass) of fluid rotating to form a vortex. Vortex formation can occur in stages as shown in Figure 5.2.

Stage 1. Surface swirl
Stage 2. Surface swirl with center dimple
Stage 3. Vortex formed pulling air into suction intake.
Stage 4. Full developed vortex air core into suction intake

Submergence to prevent vortex can be calculated by the formula below. High flow will increase the necessary submergence to prevent vortex formation. Submergence is not NPSH, nor does it account for NPSH. That is a separate topic and issue to satisfy pump hydraulics. Air entrainment can occur from many causes but one major one is operating without proper submergence. Proper submergence may be calculated and shown by the equation below and Figure 5.3.

$$S = D + 0.574Q/\left(D^{1.5}\right)$$

where:

S = submergence, inches
D = suction bell diameter, inches
Q = flowrate, gpm

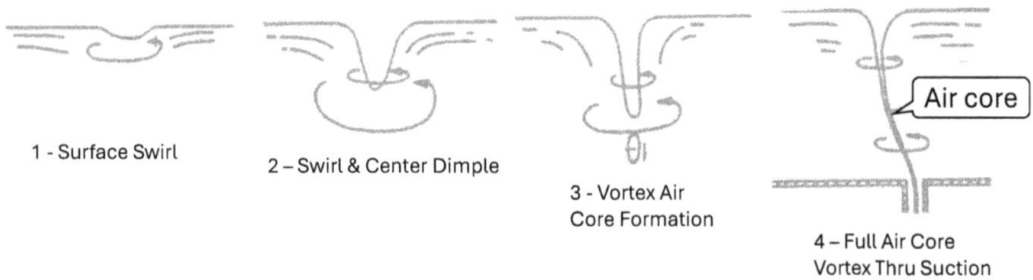

FIGURE 5.2 Vortex formation stages.

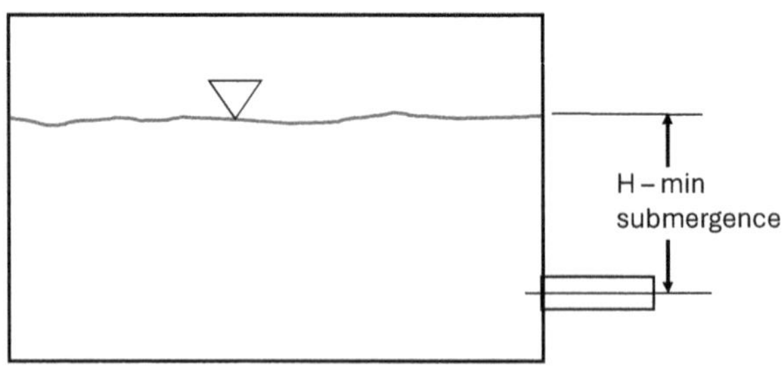

FIGURE 5.3 Tank pump suction minimum submergence.

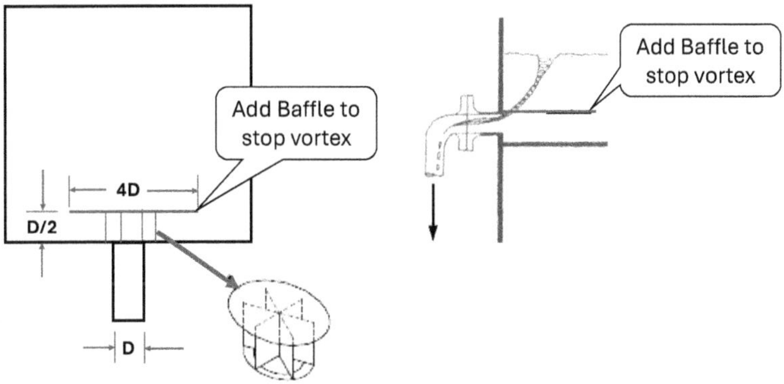

FIGURE 5.4 Types of vortex breakers.

The best solution to air entrainment from vortex formation is to satisfy liquid submergence by raising the tank level. There are situations where the tank level can't be kept above minimum submergence all the time and other methods to control vortex formation may need to be used. There are a number of plate designs for pump suctions from tanks to prevent the vortex pulling air into pump as shown in Figure 5.4. These vortex breakers interrupt the flow path and air vortex from forming in the liquid. One design is a pump inlet pipe baffle design to prevent vortex and air entrainment will follow these dimensions in the equation below.

$$Pd = 4D$$

where:

Pd = diameter of baffle plate
D = suction pipe diameter
Distance of plate from suction pipe = $D/2$

There are some pump designs that incorporate a vortex breaker in the pump suction nozzle. These can be effective in breaking up a vortex entering the pump but there is a caution if the pump has a vortex breaker built into the pump suction. The pump must operate near BEP or at least in the preferred operating region (POR) of the pump curve. If the pump operates outside the POR in a reduced flow condition, this can create suction nozzle erosion as flow recirculation passes over the vortex breaker as shown in Figure 5.5. This particular pump operated outside the POR much of the time and also occasionally operated near or just outside the allowable operating region, AOR. In low flow operation, this vortex breaker may need to be removed if severe pitting of the casing results. This suction vortex breaker was removed and did not create any negative pump system performance issues but did increase the life of the pump casing significantly.

FIGURE 5.5 Vortex breaker in pump suction nozzle pitting damage.

CASE STUDY 10 CONDENSATE TANK VORTEX

A condensate return pump system was having intermittent trouble pumping. The pump was an 8 × 4 × 17 end suction pump with a design point of 600 gpm at 107 ft. The pump had a 40 HP, 1200 rpm motor. The pump had a suction-specific speed of 7,355. Some thought the problem was cavitation as there was some cavitation-type noise from the pump during times when the pump had issues. This also occurred during times of high flow demand which also sounded like cavitation. However, the initial NPSH calculations did not reveal any significant NPSH issues. The NPSHR was ~10 ft, while the NPSHA was ~38 ft.

Further investigation showed that the pumping issues were caused by air entrainment, but the question remained why was it occurring and how to fix it? Although the tank was not very tall, the tank level was run as high as possible and still had issues. With higher tank levels, submergence was not an issue. The pumping issues typically become significant only at higher process flow rates through the tank.

Further study of the piping system on the tank showed that the main inlet pipe header was tangent to the tank wall. The tangent flow into the tank was actually a header that collected condensate from various sources. The tangential flow into the tank created a swirling motion, resulting in a vortex. To make matters worse, the tank was designed with the pump suction from the bottom and in the center of the tank. Therefore, the vortex from the tangential inlet was naturally forming right in the pump suction as shown in Figure 5.6. It was a perfect storm design for vortex formation.

There were only a few options to address this condition and getting a new condensate tank wasn't one of them due to the cost. Redesigning the pump suction and tangential inlet pipe was an expensive option and not very attractive either. The tank was also an ASME pressure vessel making modifications more complicated. The best option was to install a vortex breaker in the tank suction nozzle similar to Figure 5.7. This would break up the vortex from entering the pump suction.

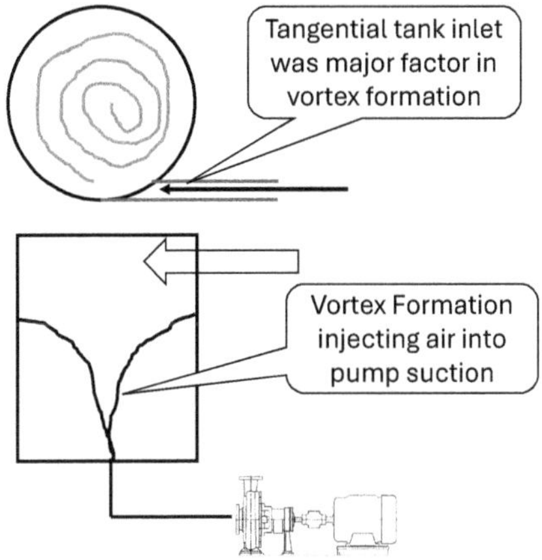

FIGURE 5.6 Tank with vortex condition from tangential inlet design.

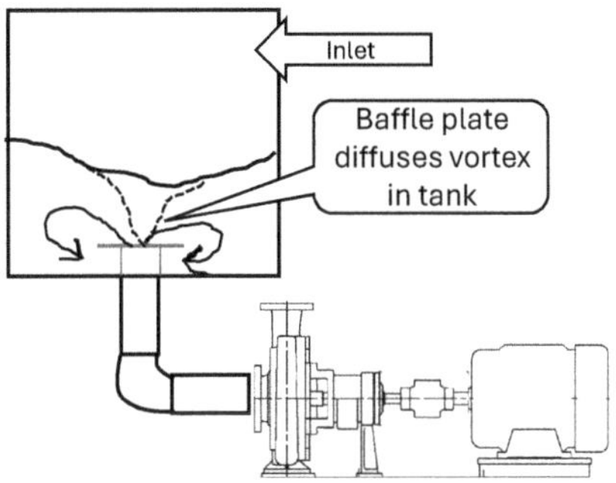

FIGURE 5.7 Vortex breaker solution for suction line from tank.

Once the vortex breaker was installed, the pump performed much closer to target curve performance. One tank and piping design caution is the use of tangential inlet piping. While there may be some reasons for some processes, the inherent result of this can be process-induced vortex inside the tank. In this case, submergence guidelines will likely not be effective to suppress vortex formation. If the tangential inlet is necessary, a vortex breaker may be the only real solution to suppress vortex from affecting pump suction and pump performance.

The root cause of this pumping issue was the design of the tank and suction pipings to the pump. The exact reason for the piping arrangement is not known, but a design change with a vortex breaker on the pump suction resolved the issue and performance was re-established. A redesign of the tank was too invasive and expensive compared to the vortex breaker, which did the job preventing the air core from affecting the pump performance.

Case Study 10 Learning – Even though submergence and NPSHR were both satisfied in this pump system, air entrainment was still an issue so don't stop looking for the cause of air infiltration when air entrainment is occurring.

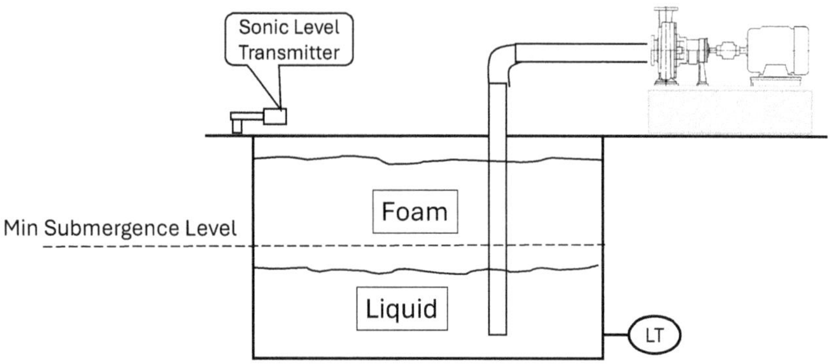

FIGURE 5.8 Sonic level transmitter potential error in sump.

Another issue can occur with pump submergence where the fluid level is not accurate. This could occur due to a level transmitter failure, error, process upset, or application design mistake. One example could be where a fluid is subject to having a foam head on top of the process fluid. A regular-level transmitter will need to be calibrated for a certain foam head amount as the foam density is different than the process liquid. Even then it is an assumption as to how much foam head may reside on top of a process fluid.

Another situation that can occur is if using a top of fluid level transmitter like a sonic level transmitter as shown in Figure 5.8. If the tank level has a foam head above the liquid level, then it will show a high tank level even though the liquid level is much lower than the reading. In automatic, the pump level controls will tell the pump system to run and pump fluid out of the tank; however, the actual pump fluid level is below the minimum submergence, so the pump may not perform as expected.

AIR ENTRAINMENT

Air entrainment in a process fluid can cause significant drop in pump performance. Since a liquid is heavier than air, the liquid goes to the outside of the impeller. The air accumulates near the center of the impeller, at the impeller eye, restricting flow into the impeller and causing a drop in pump capacity. A pump is not a compressor acting on a compressible fluid such as air or any gas. Centrifugal pumps are designed to handle liquids, so any entrained air/gas can choke the fluid flow.

As little as 2% air entrainment by volume can cause a 10% loss of pump capacity and 4% air entrainment may cause as much as 40% loss in pump capacity. A typical end suction centrifugal pump may stall at 10–12% air entrainment. Above 10% air entrainment, a recessed impeller pump will be needed. A recessed impeller pump (sometimes called a vortex pump) can handle air much better than a typical end suction pump as it will not typically stall until 24% air entrainment.

Air entrainment can have significant impact on pump performance but typically air entrainment does not cause damage to pump components like impeller, wear plate, stuffing box, or casing like cavitation can. The mechanical vibration effect of air entrainment may be slightly higher, but it is not present like in cavitation. While cavitation implosions can create very high stresses from vapor implosions on the pump components, air entrainment stresses are much lower closer to normal pump pressures.

While air entrainment is a potential result of insufficient submergence, it is not the only way air entrainment can enter a process fluid. Piping arrangements around tanks can create many conditions in which air enters the liquid as shown in Figure 5.9. Pipe delivering fluid to a tank should not be near a pump suction and should also extend below the tank level.

Another potential source of air entrainment is operating tank level below agitator propeller level as shown in Figure 5.10. The agitator blades need to be fully

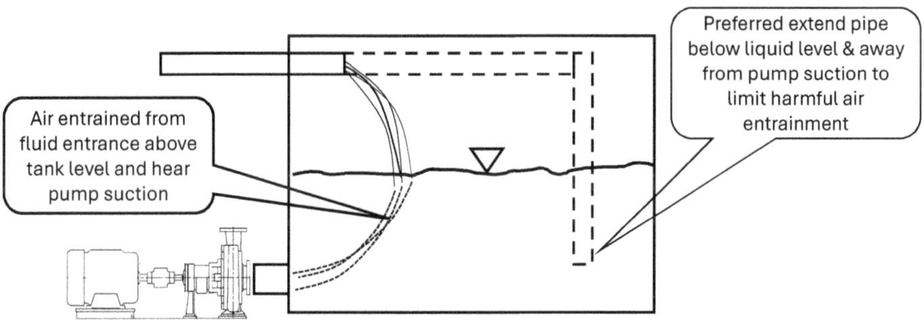

FIGURE 5.9 Air entrainment from tank inlet pipe location.

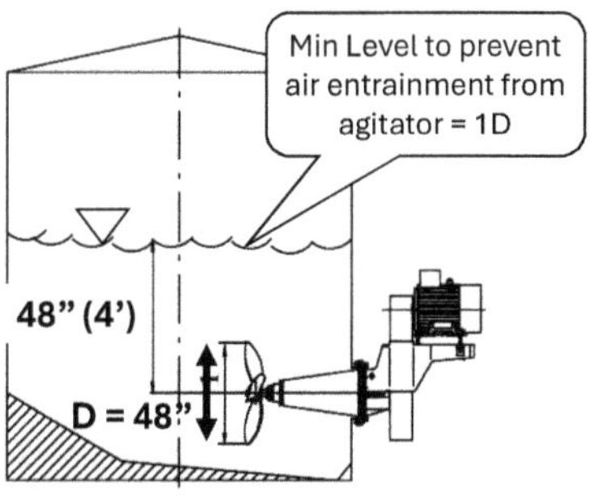

FIGURE 5.10 Minimum tank level with agitators.

submerged in the liquid to where the minimum tank level is one agitator blade diameter above the agitator centerline. I have witnessed some tanks where operators have run normal tank levels to 15–18%, which was near the agitator centerline. Now operators did this partly because the tank was small and issues controlling the process at times overflowed the tank, so they ran it low to give them reaction time. The irony is that running the tank low inducing air made the process harder to control with the feed pump out of the tank. The low tank level not only induced air into the fluid but was hard on the agitator as the blades slapped the top of the fluid. Shock loading agitator assembly is hard on the bearings, seal, and mechanical assembly.

Some process fluids tend to entrain air easier than others. For example, paper stock tends to hold air and the higher the consistency the more air it may hold. Viscous fluids may also tend to not release air easily. In challenging fluids, the best approach is to design the systems so that air will not enter the fluids. This is done with good system, tank designs, and operation.

CASE STUDY 11 LETTING AIR OUT OF PROCESS STALL

An HD cleaner feed pump system had issues with getting feed pressure up to setpoint during startup. The pump was designed to deliver 3,700 gpm with a feed pressure of 61 psig to the cleaner. The cleaners have a recirculation valve and line back to the pump suction to control cleaner differential pressure. The basic process flow diagram is shown in Figure 5.11.

The cleaner feed pressure is ~10 psig low at startup for at least 20 minutes. Operators have also reported that this has persisted since the original startup, nearly 20 years ago. The investigation revealed one likely root cause for the

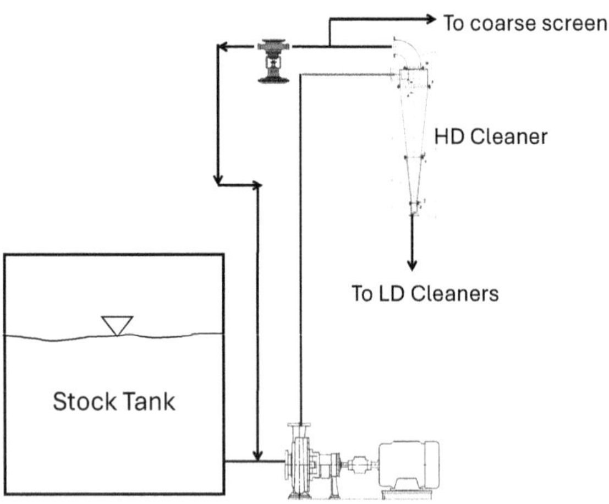

FIGURE 5.11 Process flow diagram HD cleaner feed pump system.

stall during startup – air entrainment. The air entrainment was causing poor pump performance during startup, resulting in longer times to bring the cleaner system to target operating pressures.

There were several potential areas where air may be entrained in the stock. There were several stock inlet lines coming into the feed tank that dumped stock above the tank level. The tank was a relatively small, with a maximum liquid level of only 12 ft. While the inlet lines likely entrained some air, the tank and pump system operated this way continuously, indicating another issue was occurring during startup.

The main issue with air entrainment particularly on startup was the fact that the high-density and low-density recirculation lines returned to the pump suction instead of the feed tank (see Figure 5.11). While this is not an issue once the system is full and in normal operation, during startup it is a big problem. When the system shuts down, all the lines drain completely. On startup the pump must fill all the lines with stock and in turn the air must be vented from the system. The air in the empty lines goes back to the pump suction near the suction nozzle. The supply and return line volume is ~2,000 gallons, therefore, it takes 20–30 minutes to get the volume of air from this piping to vent from the high-density cleaner and vent from the coarse screen.

A short-term improvement was to close the discharge valve on the HD cleaner feed pump, so the discharge line doesn't empty on a short shutdown. The HD recirculation valve is also shut to keep the return lines from emptying. The LD system does not have a valve to shut. The permanent fix was to relocate the recirculation lines away from the pump suction to the tank away from the pump suction as shown in Figure 5.12. The fact that some original operators witnessed this startup air entrainment stall since the process was started up points at a design issue which required a system redesign to correct the pumping problem which was really a system problem and not a pump problem.

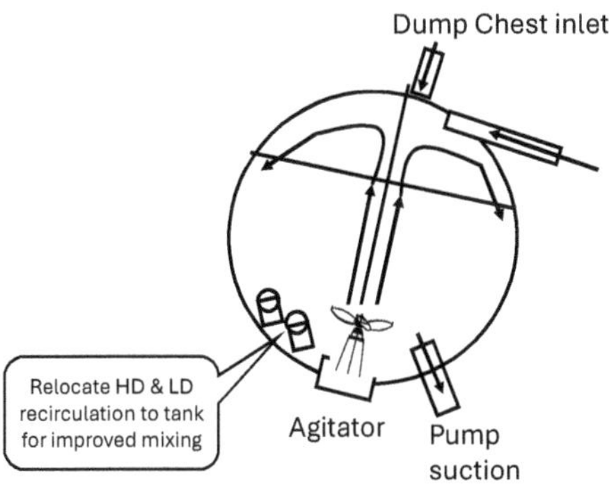

FIGURE 5.12 Feed tank piping modifications.

The system design created the condition which caused the pump performance issues at startup. After the pump system was redesigned, the pump system operated as it should. Maintenance and operations teams discussed how this issue could be fixed, but system design was never considered the cause or solution.

Case Study 11 Learning – Consider where and how fluid enters a pumping system to eliminate air entrainment problems.

Air can enter the process fluid in many different ways. The previous case study is one conventional way in many process systems where air may enter. There are other ways air can find its way into the process fluid other than tank entry positions and tank levels. The next case study will look at a more non-conventional way air may enter a pump system.

CASE STUDY 12 AIRING OUT PUMP PERFORMANCE

A deink de-trashing unit was having issues with shaft sealing using packing. The packing would begin excessive leaking in three to four months much of the time and leak water on bearings and drive belts causing significant equipment reliability issues. An improvement was desired to increase the reliability of the shaft sealing.

The rotating equipment had too much shaft movement to consider a mechanical seal as the rotor clearance had to be readjusted every 3 months. A mechanical air seal design was chosen to trial on the deink equipment shaft seal to improve the seal reliability and life. The process with the de-trashing unit is between the pulper and the pump as shown in Figure 5.13. The pump was a $14 \times 14 \times 16$ end suction pump operated by a VFD on flow control. The design point for the pump was 5,000 gpm at 54 ft. The pump was driven by a 100 HP, 1200 rpm motor.

After the air seal was installed, it appeared to seal the shaft effectively. This type of seal is almost a zero-maintenance labyrinth seal with no wear on the drive shaft. The design itself puts air into the process by its function and there is an air flow control valve to tweak shaft sealing on the application. After several days of operation, it was noticed that the VFD was having to speed up to meet flow setpoint. The pump ran to its maximum speed much of the time so when the air entrainment increased, the flow would not fully support the process. From Figure 5.13, the air from the seal was going directly into the process stream which was feeding the pump suction.

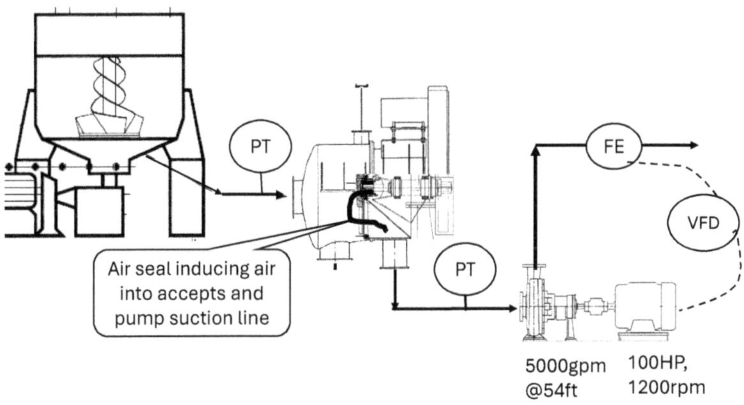

FIGURE 5.13 Process diagram pulper system.

One option was to operate the pump with a larger impeller to offset the negative impact of air entrainment. With a 100 HP motor, the largest impeller that would operate at near maximum speed with full curve motor runout was a 16.75″ × 13.38″ as shown in Figure 5.14. The entrained air will decrease the head and flow from the pump so oversizing the impeller will offset the air impact on performance. However, when correcting for air entrainment, it is always best to consider the BHP the pump would require if no air were present, assuming operation at approximately the same head?

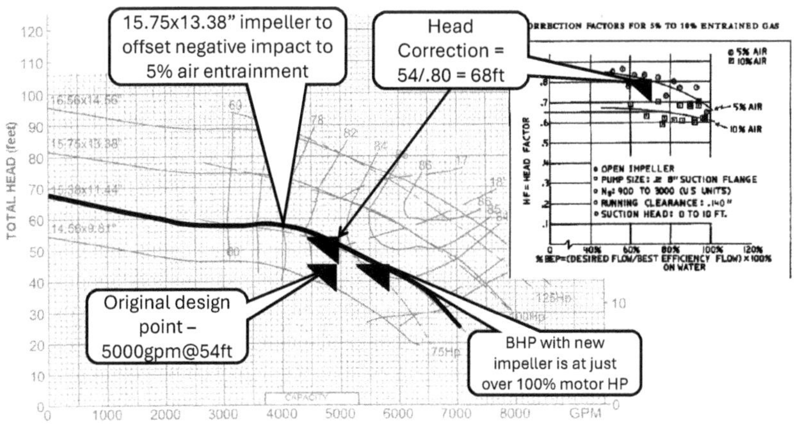

FIGURE 5.14 Pump curve adjustment for air entrainment.

After reducing the air flow, the pump flow responded accordingly. After a few more weeks of experimenting with the air flow and the air seal, it became clear that the previous packing shaft sealing arrangement would need to be re-installed to maintain process flow. The amount of air that properly sealed the shaft was enough to cause significant pump stalling-type events which was not sustainable for this application.

As with most system redesigns, the intent was to solve an issue and not create a new one. In this case, a new pump system problem was created by redesigning what was originally installed. Any air induced into the pump system will cause a drop in pump system performance.

Case Study 12 Learning – Use caution on any induced air into a process system and the negative effects it can have on the pump system.

Many times, there may be no way to rid a fluid of entrained air completely. If that is the case, then there are several ways to solve pumping fluids with air entrainment.

1. Increase impeller clearance from 0.015″ to 0.060–0.090″. The internal recirculation from larger impeller clearance helps to break up the air core at the eye of the impeller. Some of the air then gets passed on with the liquid out of the pump. For this option to be feasible, the pump system needs to have enough design margin (motor HP) to handle lower hydraulic efficiency with higher internal recirculation.
2. Install a larger impeller. This is basically oversizing the pump capacity anticipating the capacity loss from air entrainment. The caution is to avoid overloading the driver if the fluid becomes free of air. If possible, size the impeller assuming the pumping fluid has no air. Basically, the impeller is sized to compensate for the head loss of the pump (flow demanded by process hasn't changed). The correction factors for 2–4% air entrainment can be obtained from the formula and chart in Figure 5.15.[1]

$$\mathrm{HF} = \frac{\mathrm{Hea}}{\mathrm{Hwoea}}$$

$$\mathrm{Hwoea} = \frac{\mathrm{Hea}}{\mathrm{HF}}$$

where:

HF = head factor
Hea = head with entrained air
Hwoea = head without entrained air

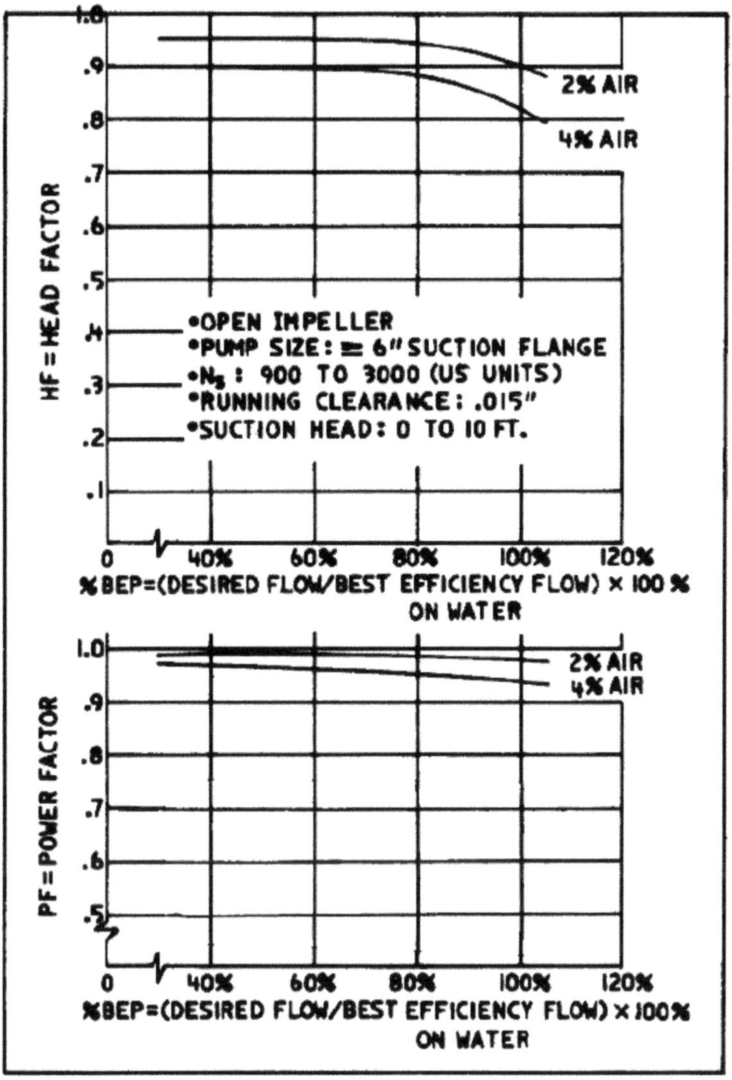

FIGURE 5.15 Entrained air correction.

3. Install an inducer. An inducer improves liquid flow into the eye of the main impeller. Therefore, since the air is collecting around the eye, an inducer helps to break it up and move it through the impeller. An inducer will typically add a small amount of HP to the pump, so consider that while designing the pump. A screw-type inducer will increase the head to the main impeller. A screw type inducers acts similar to two pumps in series or a two stage pump where the head increases at each stage. Typically, this type of inducer is designed for a specific flow rate, which remains mostly

constant since it operates in series. If throttled flow occurred, it would back up at the main impeller; if the flow increases, it would create a pressure drop at the main impeller. So only consider an inducer if the flow does not change much with the pump system.

4. Increase the suction head on pump (tank level). This will inherently increase the total head on the system and help the pump process entrained air fluid (similar to installing a larger impeller). If the pipe can be extended down below the pump fluid level, this will reduce air entrainment. Sometimes this can be accomplished by just raising the tank level. Even if the pipe is not below the fluid, minimizing the drop into the liquid level can reduce the amount of entrained air. Raising the tank level not only may help with submergence issues but also helps other pump suction problems, such as NPSH.
5. Install a lower vane impeller. This increases the fluid passages to pass air more easily. Fewer vanes also have more turbulence at the eye to break up an air pocket. If installing a different vane impeller, ensure to verify the overall pump curve performance and that no other key elements are compromised.

CASE STUDY 13 OVERCOMING AIR ENTRAINMENT

A simple stock transfer system with a flow control valve loop was having some performance issues and not meeting process demands all the time. The basic process design is shown in Figure 5.16. The pump was pumping paper stock with a design point of 1,600 gpm at 88 ft at 4.8% consistency. The pump's original impeller was a 14.8″ diameter and was driven with a 60 HP, 1200 rpm direct drive motor. The consistency would many times be nearly 6–6.5% so the flow demand was near 1,200 gpm.

There were many components of the pump system operation that was affecting the pump performance; this case study focuses mostly on the air entrainment issues. Few of the other issues were discovered during the

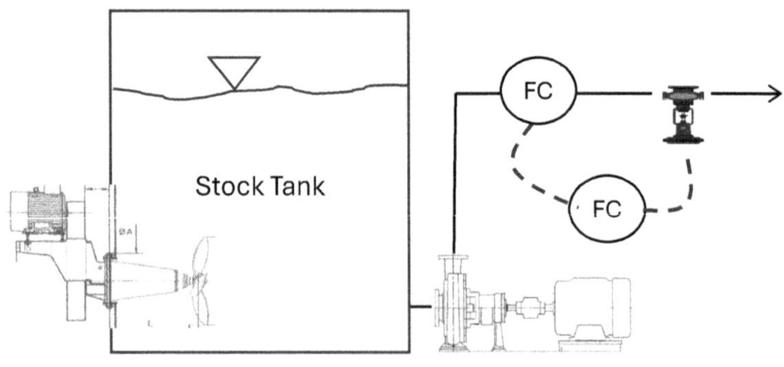

FIGURE 5.16 Stock transfer pump system.

investigation included a discharge gate valve on the pump that was closed 15% for some reason. While it wasn't a large friction loss, it was an unnecessary loss nonetheless. That valve was opened 100%. The pump inspection revealed a worn suction wear plate, so it was changed, and the pump impeller clearance was reset. There were times the stock consistency went high up to 6.5–7%, so pump performance decreased during those events. There were also issues in the system which negatively affected the pump system head on occasion. Some of these type issues will be discussed in other case studies.

Entrained air testing in the process showed ~4% air entrainment in the stock at 4.5% consistency. Higher consistencies entrain and hold more air, so periods of higher consistencies likely contained higher levels of air entrainment. All the sources of air entrainment were probably not understood, but there were a couple of issues that were contributing to the total air entrainment.

The tank level and minimum submergence was checked to be around 4.5 ft. The tank level typically ran around 5–6 ft. The tank wasn't very tall as 100% level was only around 12.5 ft. There were two issues noted on the inlet pipes to the tank: (1) inlet pipe was on pump suction side so the flow could short circuit to pump suction (poor mixing and channeling in the stock) and (2) inlet pipes were not below the tank level. The tank configuration is shown in Figure 5.17. To decrease entrained air from falling stock during stock levelling, it was recommended to increase the tank level to 8–9 ft. Another longer-term recommendation was to relocate the inlet pipe to opposite side of the agitator, so flow doesn't short circuit into pump suction. This will improve tank mixing and allow more time for air to potentially release.

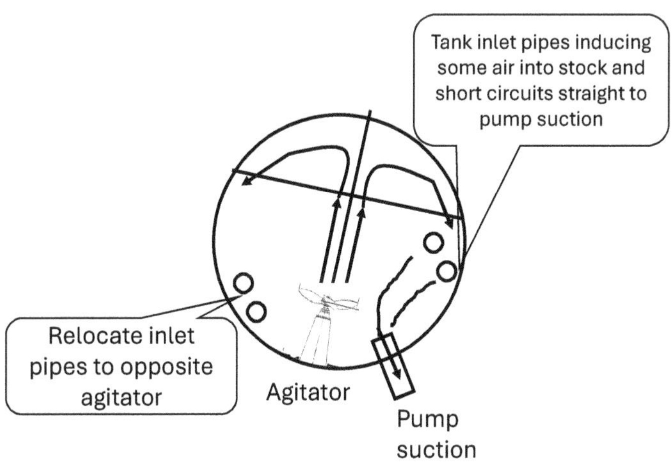

FIGURE 5.17 Tank piping layout with agitator and pump suction.

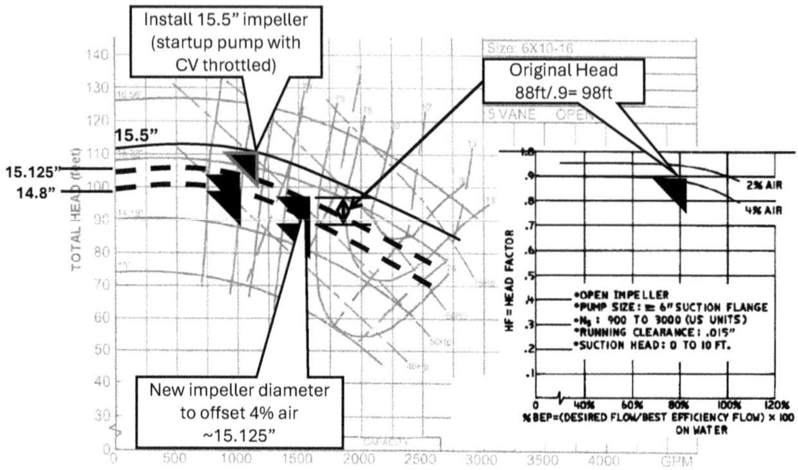

FIGURE 5.18 Pump curve air entrainment impeller correction.

With 4% air entrainment much of the time, a larger impeller can be installed to counter the negative pump performance. Using the head correction factor for 4% air, a head correction of about 10 ft would equate to about a 15.125″ impeller as shown in Figure 5.18. The pump has a 60 HP motor. While the BHP is typically lower with entrained air, an impeller upgrade is preferred to have plenty of motor HP to handle pump curve runout by the curve which is with no air entrainment. Full curve runout with enough motor HP would be ideal but at least need to have curve runout HP to the maximum amount that the process demands. In this case, full curve runout at that larger impeller diameter would be a 75 HP motor. However, the maximum process flow demand was ~1,600 gpm requiring only 55 BHP; therefore, a 60 HP motor should be sufficient to handle a larger impeller under air-free conditions.

The pump and system needed design changes to address the air entrainment issues. The easier modification was the installation of the larger impeller, but modifying the tank inlet piping would resolve the air entrainment. In this case, the root cause category for the issues and the correction was pump and pump system design.

Case Study 13 Learning – When air entrainment can't be avoided, adjust impeller size to overcome negative impact to pump performance.

NPSH AND CAVITATION

Net positive suction head (NPSH) and cavitation is a complicated subject that has volumes written on it and may take years to understand the many aspects of it. The good news is, to achieve pump reliability only a basic understanding is required. The other good thing is that in my 36+ years in the industry, I'd assume only about 5% of the pumps I've worked around had significant damaging cavitation. While many will argue that all pumps have some level of cavitation, it is the damaging cavitation that we are really concerned about with respect to pump reliability. Of this small percentage, I'd also guess about 75% of the cavitation issues were due to a pump system issue and not a pump issue. The system is sometimes easier to modify, making it more feasible to solving the issue. While the percentages of pumps experiencing damaging cavitation represent a small number, the impact from a reliability and cost standpoint can still be significant.

At the end of the day, have enough net positive suction head available, NPSHA, greater than net positive suction head required, NPSHR, with some margin and the pump will perform in a reliable fashion from a cavitation perspective. NPSHR is all about the pump while NPSHA is all about the system. The proper marriage of the pump with the system is what makes a happy pump and system marriage. When things get out of balance, then neither the pump nor the system is happy. As with any complicated subject, it is all about the details and none of this happens by accident.

Early in my career whenever pump systems were evaluated, mostly every engineer calculated the NPSHA and then went to the pump curve and found NPSHR. If NPSHA > NPSHR, then all was considered good in the pump world. This seemed to satisfy most engineers and managers; however, some years later while trying to solve a major pump reliability problem that related to pump cavitation, I learned a few more things about NPSH and cavitation.

The first thing to understand is that curve NPSH is not actually NPSHR. Curve NPSH is the NPSH where the pump is cavitating to the point where it has already lost 3% head. Many times, curve NPSH may be referred to as NPSH3. The Hydraulic Institute established the 3% standard in 1932; it was an acceptable level of pump reliability. Larger and slower pumps as well as the development of reliability advances have moved acceptable NPSH much higher. While many times when NPSH3 is satisfied, there will be reliable pump operation but sometimes even when it is satisfied, pump reliability can be severely compromised.

Many factors can also affect NPSH for a specific pump. Smaller impeller diameters have slightly higher NPSHR. Higher flow increases NPSHR, which follows NPSH3 curves on pump manufacturer curves. Hence, impellers with increased wear ring clearance have a higher NPSHR as well. Air entrainment above 1% increases NPSHR.

Cavitation occurs many times as a result of the NPSHA < NPSHR. Cavitation begins when the fluid pressure drops below the vapor pressure of the fluid and flashing occurs. As the pressure increases, the flashed vapor implodes back into a liquid to complete the cavitation cycle. The implosion is what is typically called cavitation. The pump may sound like it is pumping gravel or making a growling sound. Cavitation energy is a very localized intense event and can produce localized pressures as high as 10,000 psi. This implosion energy can cause significant damage to the pump impeller, casing, and any other pump-wetted components. Cavitation damage is typically shown as pitting in Figure 5.19. Cold water cavitation is typically worse than hot water due to water mass/density although it may be easier for hot water to flash due to lower vapor pressure.

Cavitation can also occur in different areas of the impeller. The most common is on the suction side of the impeller vane as shown in Figure 5.19a. This cavitation is due to NPSHA < NPSHR. Another cavitation can occur with discharge recirculation as shown in Figure 5.19b. Recirculation cavitation can occur on high pressure side of the impeller as a pump runs to the far left of BEP. The pressure drops as fluid recirculates through the cutwater, and increase in pressure implodes the cavitation vapor back into the discharge impeller vane.

One pump parameter that is used to gauge a pump's reliability is suction-specific speed (NSS). This is a characteristic of the pump from the pump manufacturer and the end user can do nothing about it, but it does give a first input into a pump reliability factor. In general, it is desired to have a pump with a suction-specific speed <11,000. Above 11,000, pumps are considered high-suction energy pumps and can be prone to poor reliability. Although, this is not an absolute parameter. I've seen pumps well over 11,000 that perform at high reliability.

Additional information is available from the Hydraulic Institute, including curves showing NSS as a function of NS. For example, at NS = 1000, the allowable NSS = 14,000. At NS = 3000, the allowable NSS = 10,000. Mostly the relationship is linear on this graph. My intent is not to repeat existing standards, so refer to those standards if more detailed information is needed.

For double suction pumps, use only half of the gpm at maximum diameter since a double suction impeller is basically two pumps in one. It is a unitless number defined by the following equation.

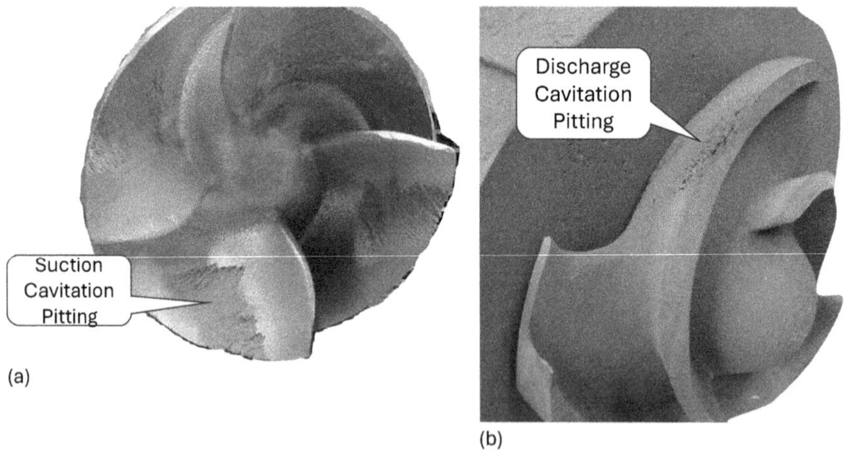

FIGURE 5.19 Pump impeller cavitation: (a) suction side insufficient NPSHA-type cavitation and (b) discharge recirculation type cavitation.

$$\text{NSS} = \left(N(Q)^{0.5}\right)/(\text{NPSH3})^{0.75}$$

where:

NSS = suction-specific speed
N = speed, rpm
Q = flow, gpm at BEP
NPSH3 = curve NPSH at BEP

Impeller tip speed is another way to evaluate potential for damaging cavitation in pump designs. Recommended maximum impeller tip speeds for ANSI-type pumps may follow the guideline below.

Clean water pumps $<$ 145 ft/s
Dirty water pumps $<$ 130 ft/s
Slurry pumps $<$ 100 ft/s
Non-metallic pumps $<$ 85 ft/s

$$V = ND/229$$

where:

V = impeller tip velocity, ft/s
N = speed, rpm
D = impeller diameter, inches

Since satisfying NPSHA > NPSH3 is not always enough, an NPSHmargin must be established. Some just use an NPSHmargin of 1.5 but there are other ways to approach NPSHmargin. In order to determine NPSHmargin, another parameter can be very helpful – suction energy (SE). NPSHmargin can be used in the following equation to establish a more realistic NPSHR.

$$\text{NPSHR} = \text{NPSHmargin} \times \text{NPSH3}$$

where:

NPSHR = net positive suction head required
NPSHmargin = ratio from SE table
NPSH3 = curve NPSH at operating point

Suction energy can be effectively used to analyze NPSH and cavitation potential. Suction energy is another factor that helps determine a safe NPSHmargin to apply to a pump NPSH3 to prevent damaging cavitation. The equation for SE is shown below.

$$\text{SE} = \text{De} \times N \times \text{NSS} \times \text{SG}$$

where:

SE = suction energy
N = pump speed, rpm
NSS = suction specific speed
De = impeller eye diameter, inches
De = ~0.9*suction diameter
SG = fluid specific gravity

Looking at the equation it is obvious to see the major impact to SE is pump speed. SE is a function of a square of the speed (by way of suction-specific speed). Therefore, if a pump can operate slower, the impact to SE and cavitation potential is drastically reduced and thus speed becomes the largest factor affecting cavitation potential. The suction (impeller eye) diameter can be determined from the impeller eye suction area, usually given on a pump curve or estimated by 90% of suction diameter for end suction pumps or 75% for a double suction pump. Once the SE values are calculated, determine if the pump is a low, high, or very high suction energy pump based on values in Figure 5.20. Low suction energy pumps are typically not prone to significant damage from cavitation except if handling abrasive or corrosive fluids.

While changing pump NPSH3 is difficult, there are some things to consider modifying your NPSH3 or NPSHR condition from the pump standpoint. Here are some pump changes which will lower the NPSH3 or NPSHR of the pump.

1. Use a different pump with a lower NPSH3. This is mostly only an option if the evaluation is done during the system design and pump selection process. This may include a different end suction pump or use of a double suction pump. Since speed is a major factor (speed squared) effecting suction energy and thus NPSH and cavitation potential, choosing a larger pump that runs at a lower speed can be a good option. At this point we are not changing the pump hydraulics or system design point, so a new pump is required.

Pump Type	Start High SE	Start Very High SE
Inducer Pump	320x10^6	480x10^6
Vertical Turbine Pump	200x10^6	300x10^6
End Suction Pump	160x10^6	240x10^6
Double Suction Pump	120x10^6	180x10^6
2 Vane Lift Pump	100x10^6	150x10^6

Suction Energy Level	NPSHmargin
Low	1.1-1.3
High	1.3-2.0
Very High	2.0-2.5

$$NPSHA = NPSHR * NPSHmargin$$

FIGURE 5.20 Suction energy levels for pump type and NPSHmargins.

2. Modify the existing pump to achieve a lower NPSHR, while still meeting the required design point hydraulics. This allows you to keep the same pump footprint and lower cost of the fix. This may include a different lower NPSH impeller (increase impeller eye area) if offered by the pump manufacturer. In some cases, a lower pump speed can be used (if a larger impeller is available to still meet hydraulic system needs). Of course, the impeller tip speed may remain unchanged at larger diameter and lower speed (changes offset each other). An inducer (small axial flow impeller attached to the eye of main impeller) may also be an option for the exiting pump. Install a larger eye impeller if available but the drawback is a larger eye impeller may be more vulnerable to recirculation issues, so an evaluation of the pump operating point is necessary to ensure the operating point is close to BEP. If it is a vertical turbine multistage pump, then lowering the first stage impeller my help solve an NPSH or cavitation issue.

Either way, to improve the NPSHR, a pump modification of some sort will be needed which will require some investment. Outside of changing the pump NPSHR, other pump design changes are to provide longer life in spite of a current cavitation condition, such as improving impeller or casing materials. These are listed in the pump selection section as to the hierarchy of materials and resistance to cavitation conditions.

Another impeller design feature is vane overlap shown in Figure 5.21. Vane overlap reduces suction recirculation and reduces suction energy. This will be at the mercy of the pump manufacturer designer, therefore not a lot of options in an existing pump.

Impeller coatings are sometimes looked at to improve the cavitation resistance. Epoxy coatings mixed with ceramics are often used but with limited success. Mechanical fatigue at the bond surface leads to failure from the high implosion pressures of cavitation, like the one shown in Figure 5.22, which was removed in less than 12 months and the coating was found coming off the impeller. Not only was it

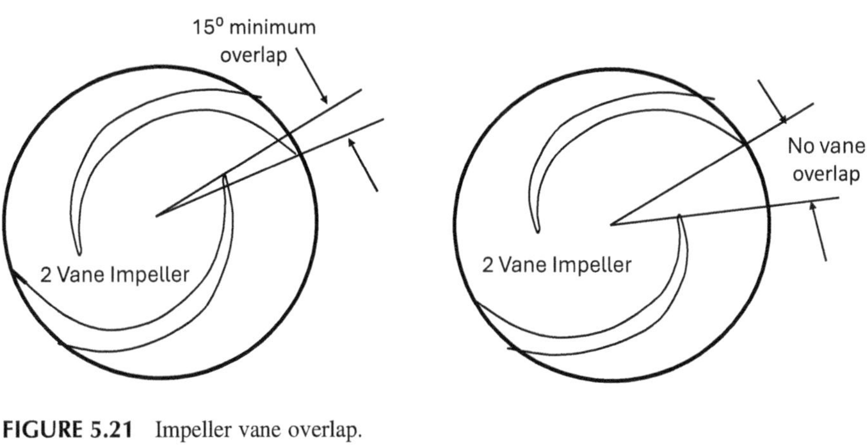

FIGURE 5.21 Impeller vane overlap.

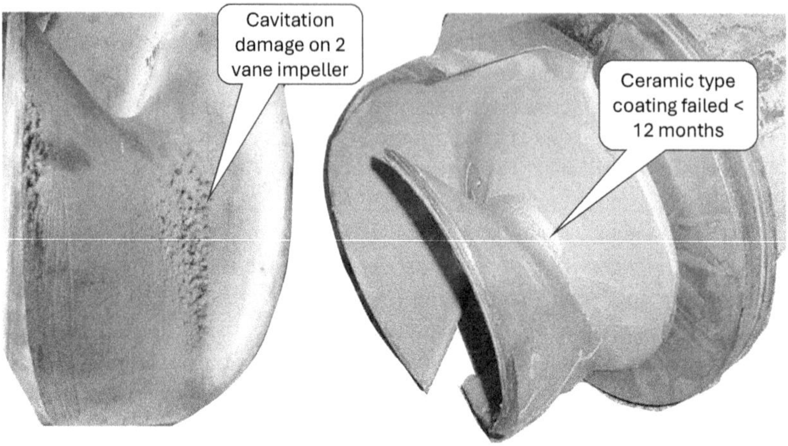

FIGURE 5.22 Failed ceramic coating on impeller from cavitation.

not effective for very long, now an impeller balance issue has been introduced on the pump impeller.

In high intensity cavitation conditions, a softer coating typically works better to dissipate the cavitation implosion energy. The elastomer-based coating may provide some benefit if the cavitation energy is below the elastic limits for the material; otherwise, a similar coating failure may result. In general, coatings are not good solutions but just band-aids for cavitation issues.

One final caution on using coatings is the thickness of the coating. Every mil of coating changes the dynamics and performance of the impeller hydraulics. Around the leading edge this can increase the NPSHR. Loss of vane area also reduces the pump capacity, affecting its hydraulic performance. Therefore, the larger the pump, the less impact coating has on pump performance.

So now let's consider the pump system side of this and the net positive suction head available (NPSHA). NPSHA is mostly a function of system or hydraulic design as well except for a couple of areas. NPSHA is an absolute pressure so if the tank is under a vacuum, the vacuum must be expressed in absolute head level. The NPSHA equation is shown below.

$$\text{NPSHA} = P1 + Hs + Hvp - Hf - Hv$$

where:

NPSHA = net positive suction head available
$P1$ = absolute pressure in suction tank in feet
Hs = static head in suction tank in feet
Hvp = vapor pressure of fluid in feet
Hf = friction head loss in suction piping to pump suction in feet
Hv = velocity head in feet

Looking at the physics of failure for NPSHA, there are five areas to effectively modify NPSHA – tank pressure (*P*1), tank static head (*Hs*), vapor pressure (*Hvp*), suction line friction head loss (*Hf*), and velocity head (*Hv*). So, what are the practical levers to modify NPSHA in the pump system?

1. Increase suction pressure to pump by increasing the system static head (*Hs*):
   - Raise the level of liquid in suction feed tank. This may be the easiest and most practical (low cost) thing that can be done if the tank is tall enough.
   - Lower level of pump. This is not easily accomplished and if investment is done to relocate pump, then a new pump with better NPSHR characteristics is probably a better choice.
   - Install a booster pump. This again requires an investment but will increase the system NPSHA and provide margin for the primary pump.
2. Decrease vapor pressure (*Hvp*): For a given system, the fluid itself is not going to be changed so if the process allows it to be cooled that is about the only lever available. Although, only a small amount of improvement can be attained with this variable. Cooling fluids may lower the vapor pressure, but lower temperatures usually mean larger cavitation bubbles. Cool water can many times be most damaging.
3. Increase absolute pressure over liquid (*P*1): If the tank has a pressurized air pad on it then this might be an easy effective solution for improving NPSHA. Be sure to check the tank and relief valve ratings and integrity before increasing the operating pressure. If it is an atmospheric tank, then this is not an option.
4. Reduce suction friction losses (*Hf*): This will also require a redesign but here are the options.
   - Increase suction piping size. Likely not worth pursuing as it will not generate a substantial reduction in friction loss unless there is an original system design issue.
   - Decrease fluid velocity by either larger pipe or lower flow in system. Modify system for lower flows is likely not an option since system must support the process.
   - Shortened suction pipe is not going to produce significant friction loss so a low ROI option.
   - Remove minor friction losses in suction line (extra valves, elbows, suction strainer, etc.). This is likely not an option with process support or will not produce a significant reduction to justify the investment.

An important way to affect the pump NPSHR is with pump design, which many times means a different pump. This pump may involve lower suction energy, lower speed, and operating point to the left of BEP among other things to affect the operating NPSH3 and eventually the NPSHR. There are other small things that some people do such as overfile the inlet vane tips to help a cavitation issue or underfile vane tips to increase pump capacity. The reminder on filing vane tips is to always leave enough vane thickness to hold the integrity of the impeller vane. Also, if you file vane tips, you are basically creating a new pump so there is not an exact pump curve to match your special pump now. In general, I don't like these types of modifications as you don't really know the pump performance and I'm not sure if it may provide significant benefit for most issues. However, there are times where some special modifications may be required to solve a problem.

While some of the options for NPSHA or NPSHR may seem like your hands are tied, moving one or two parameters slightly from the pump side (NPSHR) or system side (NPSHA) can add up to a good situation for the pump and system to provide cavitation-free operation.

Any obstruction on the pump suction side can suddenly create cavitation, even when the pump and system are properly designed. Things such as a foreign object in the suction line or buildup in the suction or failed valve could create a large pressure drop into the impeller and cavitation may result.

One such situation observed many times in some pump systems is buildup in the bottom of the tank around the suction nozzle inlet. This is particularly problematic on sump or wet well systems where the tank can't be easily drained and cleaned out. A sump typically contains fluid, as it serves as a low-point drain typically for a sewer system. Drain systems always collect some solids that sink to the bottom of a sump so blocking the pump suction is almost inevitable at some point.

Figure 5.23 shows how the buildup can accumulate in a sump and block off the pump suction. For a sump, the pump is already in a tough suction lift condition so additional blockage is even more threatening to pump performance and reliability. It is usually a good idea to vacuum out sump every two to three years, depending on the solids and contamination level in process stream.

In extreme cases, a pump cavitation problem may only have one solution and that is to install a better pump for the application where the NPSHA > NPSHR. The next case study will look at a chronic cavitation issue and the solution that resulted from the failure investigation.

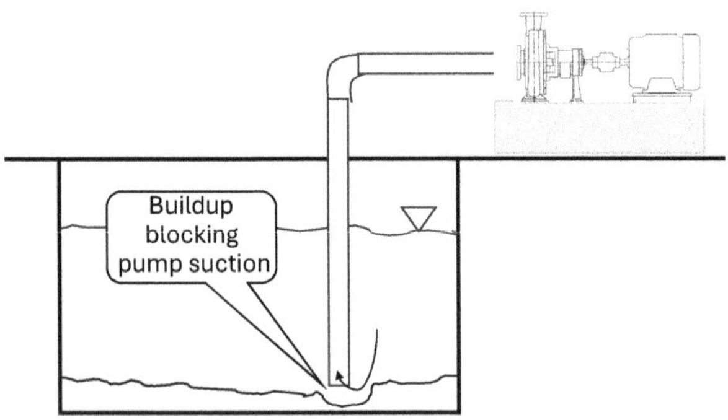

FIGURE 5.23 Sump buildup effect on pump suction.

CASE STUDY 14 CAVITATION DAMAGE AND SOLUTION

A wire pit white water pump was operating to transfer reclaimed white water from the forming section of paper machine to a white-water chest, save-all chest, and a dilution destination. It was a 12 × 10 × 17 pump originally installed with a 13″ impeller and design point of 6,000 gpm at 79 ft and NPSH3 = 20 ft. After a paper machine speed up, the pump was modified with a design point of 7,000 gpm at 102 ft and 79% eff with NPSH3 = 23 ft. The pump then had a 15.5″ diameter and four vane impellers. The pump was driven by a 300 HP, 1800 rpm fixed speed motor. Operating at 1800 rpm, the NSS = 14,086 is very high for a pump-specific speed of 3,100.

The pump had a MTBF ~7 months ever since it was started up ∼26 years earlier. The failures have been numerous with various levels of failure as described below. The annual maintenance cost for failures was around $50,000 and at times the downtime cost was around $30,000.

Level 1. (pre-failure conditions) – high vibration, high vane pass vibration
Level 2. hidden failures – impeller pitting, wear plate damage, shaft deflection
Level 3. minor failures – sealing (packing failure), shaft wear damage
Level 4. functional failure – bearing failure (unplanned failure)

Initial failure analysis focused on returning the pump to precision standards from assembly, alignment, and balance. While there were several opportunities, the smoking gun for the poor reliability was not found in the maintenance impact area. There were some items to address and some interesting observations in which there was not much that could feasibly be corrected with the existing pump.

The pump was precision aligned many times within standards of better than 0.002″ offset and 0.0006″/in of angularity with limits for 1800 rpm without any significant change in pump vibration or reliability. While misalignment was not a major issue, an evaluation later showed that there was a significant thermal growth misalignment from the hot motor. The large, open frame motor ran very hot in the 190–200°F range. The resultant thermal growth of the motor and pump machine set had the motor growing ~0.005″ more than the pump. Thermal growth adjustments and hot alignment were conducted on the pump, but no significant change in pump reliability was observed.

There were never any bearing installation issues observed nor any lubrication-related deficiencies of the pump bearing assembly. The bearing failure modes were typically either bearings damaged from excessive vibration over time or water leaking from the packing sealing arrangement getting water into the power end. The bearings were in an oil flinger design, so the bearings sat outside the oil level when down. This probably further decreased reliability when water entered the bearing housing.

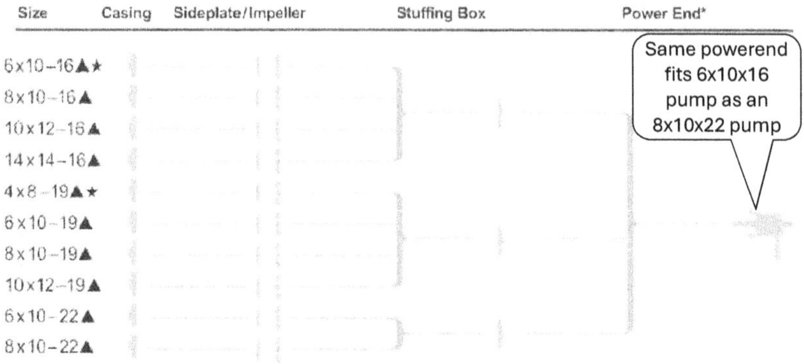

FIGURE 5.24 Example pump component interchangeability chart for pump-by-pump size.

One observation was that the pump's power end was the smallest available for the range of pump sizes it fit. Also, the pump had the smallest power end available, so any design margins for that power end (bearings, shaft, etc.) were pushed to the maximum. A sample chart is shown in Figure 5.24 so the same power end fits an $8 \times 10 \times 22$ pump as a $6 \times 10 \times 16$ pump but of course the same pump (same bearings, shaft stiffness, etc.) on different sizes does not see the same bearing loads, shaft loads, and resulting shaft deflection. This means bearing life is inherently shorter than a large power end on a smaller pump.

Pump vibration is either mechanical or hydraulic. For mechanical vibration, there were many areas to evaluate. First the structure or foundation could be a contributor to pump vibration. The foundation was found to be too narrow for the pump size. The foundation mass was also found to be low mass at around 2.2:1 ratio with that of the pump instead of 5:1. At times some casing bolts might be found loose. This was partly due to the high running vibration and also low bolt preload from installation. Bolt condition (corrosion) reduced bolt preload on installation as thread friction ate up the torque. Torque procedures were also not closely followed on many of the large casing bolts.

Pump unbalance was another concern with a few minor opportunities. The key length was 1.5″ too short, which created 3.14 oz-in of unbalance and well outside G1 balance. Key phase on the spacer shaft was also incorrect. Shaft runout was less than 0.001″ although shaft deflection under operating loads was more than an estimated 0.004″ at 5500 gpm.

After the initial rounds of pump failure analysis and number of actions were completed to see if any impact could be made. Pump assemblies were changed with precision assemblies from a balance standpoint (G1 balanced rotors), precision alignment with key length and key phase correct, new shaft couplings, new pump casing and suction wear plate, new impeller with proper impeller clearances, and even a new motor. The same performance, pump vibration, and pump reliability were seen after all this precision work. Maintenance teams were frustrated to say the least, but maintenance was trying to solve a problem that didn't have maintenance as a root cause.

Initial pump hydraulic analysis did not reveal a root cause either. The pump operated between 5,500–7,000 gpm which is in the preferred operating range (POR) for this pump as shown in Figure 5.25. Also, the NPSHA for the system was 38 ft, while the curve NPSHR or NPSH3 was 23 ft. So NPSHA > NPSHR cavitation initially did not look like an issue. Other problems like air entrainment and submergence also looked like non-issues for the application as both were designed and operated within acceptable ranges. It became even more important to understand the physics of failure (POF) of the components that were failing on this pump.

The pump exhibited very high vibration all the time, even after a complete rebuild. While there are numerous failure modes on the pump, the one big constant is the very high vibration. It appeared that the severe pump vibration was linked to all the other failure modes. If the high vibration could be

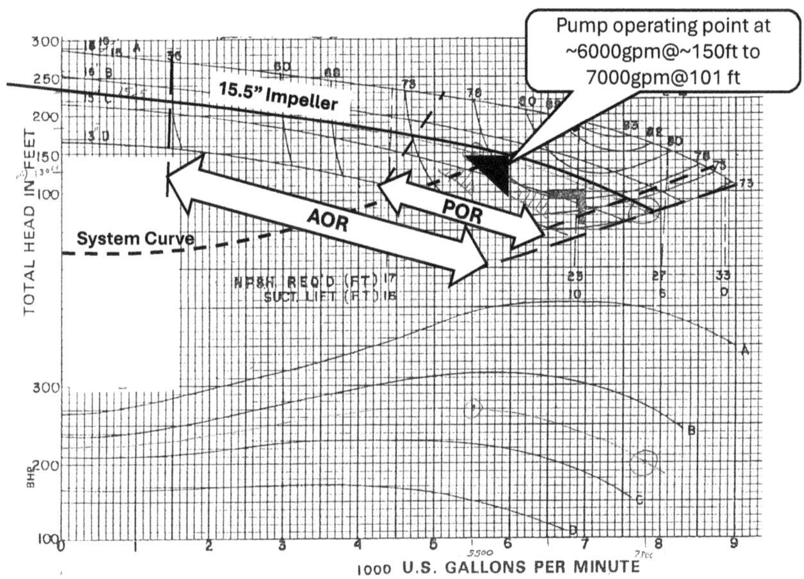

FIGURE 5.25 Pump performance curve with operating point.

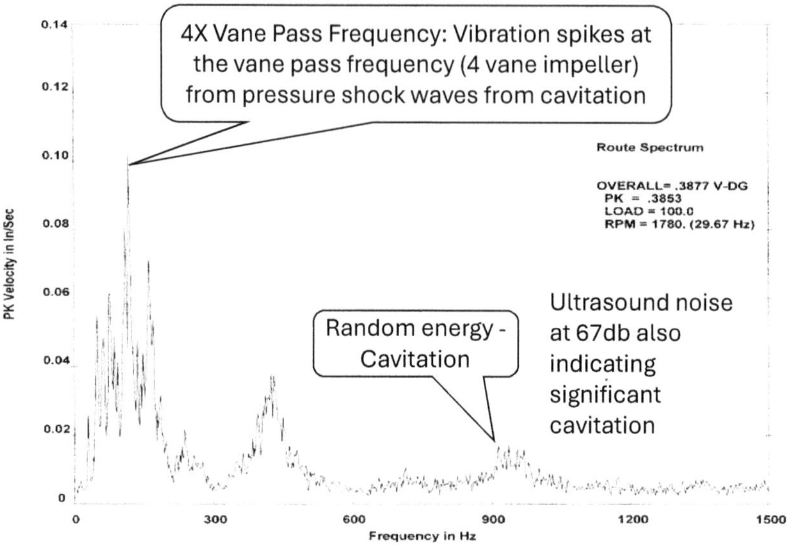

FIGURE 5.26 Typical pump vibration velocity spectrum showing cavitation and vane pass issues.

resolved, then the pump reliability may also be resolved. So high pump vibration was the initial focus of the more in-depth failure analysis. The pump vibration signature is shown in Figure 5.26. High vane pass vibration and cavitation vibration was evident in the velocity spectrum. The overall vibration trends on this pump had vibration anywhere from 0.4–0.7 in/s. The ultrasonic noise on the pump suction was also very high at 67 db as the cavitation noise was clearly audible.

From a physics of failure standpoint, the impeller pitting damage was a significant clue to the root cause of pump failures. The impeller was a short life component with respect to cavitation pitting damage as shown in Figure 5.27. The impeller pitting was in the suction side of the impeller vanes, which was consistent with suction cavitation damage. The impeller was made of stainless steel which has good cavitation resistance, but it was no match for this application. Cavitation is very damaging with some references noting implosion pressures of 10,000 psi on the impeller surface.

While there are various methods to determine cavitation potential and NPSHmargin, suction energy (SE) was used on this application. As discussed previously, many applications will require an NPSHmargin to operate without significant cavitation damage. From the evidence on this pump from the impeller damage, vibration data, and ultrasonic data, the pump is experiencing significant damaging cavitation.

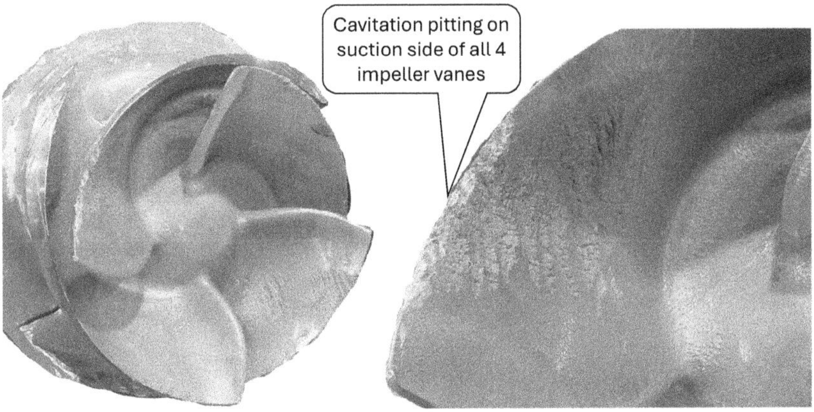

FIGURE 5.27 Pump impeller cavitation pitting on suction side of impeller vanes.

So, applying the suction energy equation, the SE ~274×10^6 which make this a very high SE pump that requires a 2.5 NPSHmargin. Applying this margin to NPSH3, yields an NPSHR = 58 ft. With an NPSHA = 38 ft, now NPSHR > NPSHA and cavitation is a significant issue which agrees with the evidence found on the failing pump impeller and vibration. The cavitation implosions were occurring on each impeller vane which drive the vane pass vibration very high.

The solution now is to solve the insufficient NPSHA issue or lower the NPSHR. Just to be even, we need to erase 20 ft of NPSH gap. There are not enough variables to get more than 3–4 ft more on NPSHA and that was from operating the tank level higher. Now that leaves NPSHR. No alternative impeller was available to meet the required hydraulic conditions, leaving the selection of a new pump with lower NPSH, lower SE, and the hydraulic envelope that meets the process demand. The pump curve in Figure 5.28 shows the pump that was selected.

One of the key action items to solve the issue resulted in a new pump being installed. Some of the pump operating parameters were targeted to select a more ideal pump for the application to avoid the previous pump reliability issues. A summary of the two pumps with data after the new pump was up and operating is shown in Figure 5.29. While the NSS only dropped slightly, the NS was lower by 33% and the SE was cut in half as well as the pump speed. The HP was reduced and shaft stiffness improved. A TEFC motor replaced the open frame motor and lowered the motor temperature and significantly reduced the thermal growth of the motor. The pump bearing housing temperature dropped by 30°F as well as ultrasonic cavitation noise reduced by ~40% and non-existent after the pump upgrade.

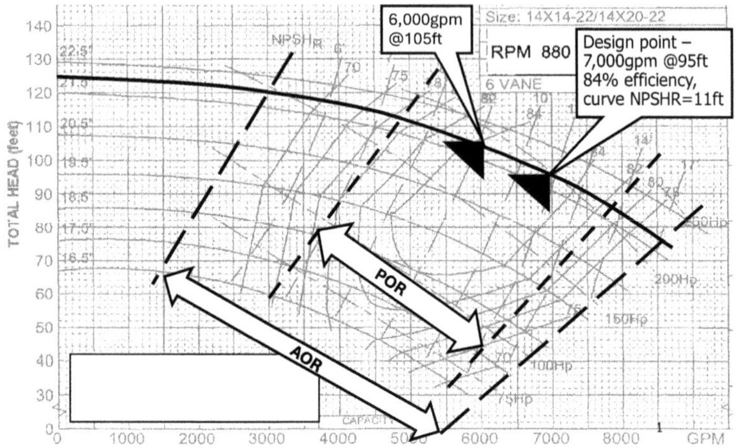

FIGURE 5.28 New pump curve with operating points.

Pump	Impeller size/vanes	Speed rpm	Eff.%	BHP	Shaft Stiffness	Motor Temp,F	Thermal Misalign.
12x10x17 PWO	15.5/4	1800	79	254	39	190	0.005
14x14x22 3175L	21.875/6	900	84	187	7	115	0.001

Pump	Key (lbs) Unbalance	Vibration in/s	Suction Specific Speed	Suction Energy x10^6	NPSHR	UT Noise,db	Bearing Temp.F
12x10x17 PWO	18	0.7	14086	274	58	67	155
14x14x22 3175L	0	0.08	12467	141	11	42	125

FIGURE 5.29 Summary of the pump design parameters.

Some other improvements on the new pump included a dynamic seal instead of packing on the pump. This included no seal water usage (~3 gpm to sewer previously) and no routine maintenance to adjust or change out. The pump foundation was also improved to get more mass and stability from the previous design. The casing had a center discharge instead of the tangential discharge. The pump power end was larger and had a bearing end support on the drive end (previous power end was small and had no support and was cantilevered from casing).

The previous pump had a failure rate ~1.33/year (Reliability = 0.26). The new pump operated for almost eight years without any failures other than a few oil changes and coupling inspections. The pump required no major maintenance for those eight years until it was shut down following the permanent shut down of the machine. A 26-year-old pump problem had been resolved. A problem which had been an issue since it was installed is a good indicator that there is likely a design problem. This was a good reminder that a design problem will not be corrected by even the best maintenance team. A pump design problem requires a redesign of the pump or system or both.

Another interesting observation when looking at both the original pump curve and the new pump curve is that the operating point for both is within the POR and near BEP. The cavitation issue was related to the suction energy and NPSH and not the operating point of the pump. The opposite of this will be looked at thoroughly in Chapter 10, showing the impact of pump operating point on overall pump reliability.

Case Study 14 Learning – Importance of pump design and NPSHmargin on pump cavitation.

In evaluating the NPSH for a pump, some special system conditions may exist when the suction tank is under a vacuum (less than atmospheric pressure). Since NPSHA is an absolute pressure, a much lower NPSHA will result when the source tank pressure is below the atmospheric pressure. For example, look at a typical paper machine uhle box vacuum separator pump system shown in Figure 5.30. A uhle box is a vacuum box that the press felt runs across and the vacuum sucks the water from the felt. This water comes from the sheet as water is pressed into the felt and the sheet dryness is increased. The vacuum separator on the back of the machine separates the water from the vacuum (air flow). The removed water is then processed out of the system in a pump system.

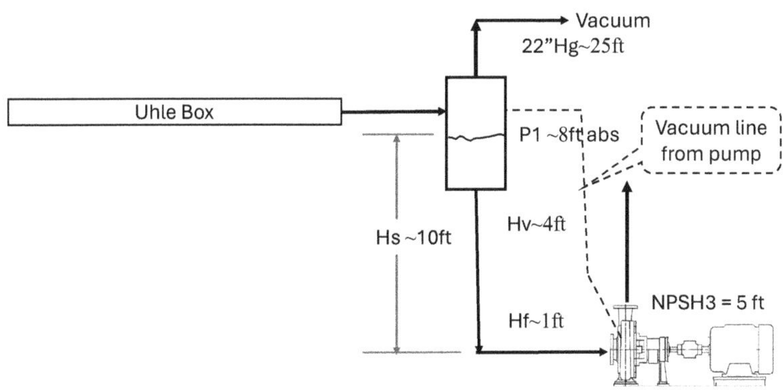

FIGURE 5.30 Low static head paper machine uhle box vacuum separator pump system.

For this pump system, a low NPSH3 pump would be desired (NPSH3 = 5 ft). The NPSHA from the separator system operating with 22″ Hg in Figure 5.30 would have an NPSHA ~ 13 ft. With an NPSHmargin = 2, the NPSHR ~ 10 ft. So NPSHA (13 ft) > NPSHR (10 ft), therefore no significant damaging cavitation will likely result, and the pump should operate properly. A higher NPSH pump may have issues in this application. An NPSH3 of just 8 ft would create the NPSHA < NPSHR and cavitation could impact pump reliability and performance.

For a basement uhle box there is not much static head to add NPSHA as shown in Figure 5.30. A top felt section may have an additional 20–25 ft of static head into pump suction. The NPSHA may be 37 ft. However, for the actual pressure at the pump suction nozzle is ~15–16 ft below atmospheric pressure. This effectively means that the pump must have this much lift capacity to operate properly. The specified pump has a pump lift ability of ~25 ft so it is sufficient for the application.

A couple of other pump observations for this application. Since the suction line is below atmospheric pressure, any leak on the suction side of the pump (at a flange or shaft seal) will inject air into the pump suction. This can kill the pump performance as the air collects at the impeller eye. To avoid this, sealing must be properly sized and maintained whether it is a mechanical seal or packing. Another option is to install a small vent line from the pump suction nozzle near the impeller eye back to the vacuum source to allow any entrained air to escape.

Another pump system setup on a vacuum box system may look like that in Figure 5.31, where the separator seal leg goes to a seal tank. This can be done if there is enough static head greater than the vacuum level. The pump in this system is only pumping out of an atmospheric open seal tank so the pump system side is somewhat straightforward. A key item in the system is to size the seal tank volume to handle the vacuum seal leg volume. On startup, the vacuum will draw water from the seal tank up to the required level (25 ft for 22″ Hg vacuum); therefore, there must be

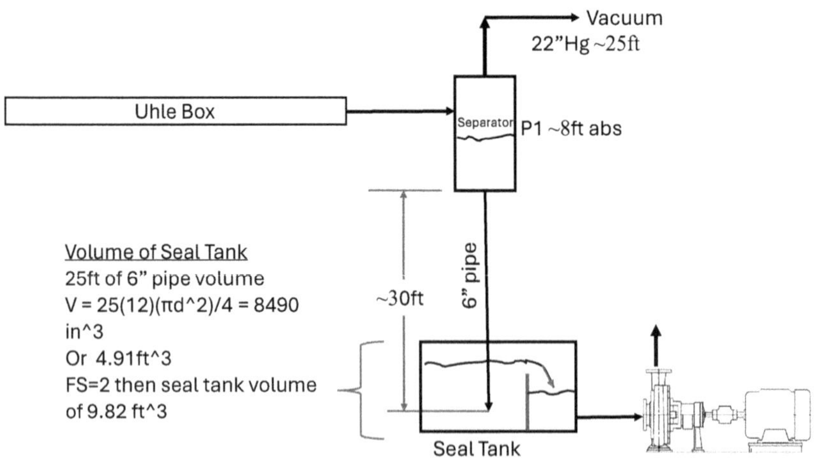

FIGURE 5.31 Alternate paper machine uhle box vacuum separator pump system (high static head) with seal tank.

sufficient volume of water in the seal tank to prevent the tank from emptying and becoming open to atmosphere.

At the operating vacuum level, the level would be 25 ft of pipe so if the piping was complete vertical run that would be 4.9 ft^3. With a $SF = 2$ then build a seal tank with 9.82 ft^3 volume. It is also advisable to not have long horizontal runs on this vacuum leg piping but to keep it as vertical as possible.

Other pump systems which may incorporate a tank under a vacuum might be a condensate receiver system like on a dryer system or operating in a vacuum (vacuum condenser) on a turbine. In these systems, it is important to apply and understand gauge pressure and absolute pressure in all pump calculations.

Note

1 1992, Nineth International Pump Users Symposium, Texas A&M University, Design Considerations & Application Guidelines for Pumping Liquids with Entrained Air Gas Using Open Impeller Centrifugal Pumps, Charles Cappellino, Daniel Roll, George Wilson, Goulds Pumps, Inc.

6 Pump and System Performance

Pumps and systems when they are a good match fit together like a hand in a glove or as Forrest Gump would say, like peas and carrots. When they are not a good match, it is like a bad marriage where there are always problems or conflicts such as poor pump reliability, piping issues, vibration, high energy usage, premature component failure, and the list goes on. Selecting the right pump is as much about the system as it is the pump.

When sizing the pump for a simple system, a single design point (flow) is chosen where it is believed the pump will operate most of the time. The system curve is the flow resistance of the system versus flow in the system. The same calculations are repeated for different flow rates within the system as shown in the pump selection chapter. This will represent how the system friction head changes with flow. Since the friction head is a function of fluid flow (specifically fluid velocity squared), the system curve is non-linear. The operating point of a pump will be the intersection of the pump curve and the system curve as shown in Figure 6.1.

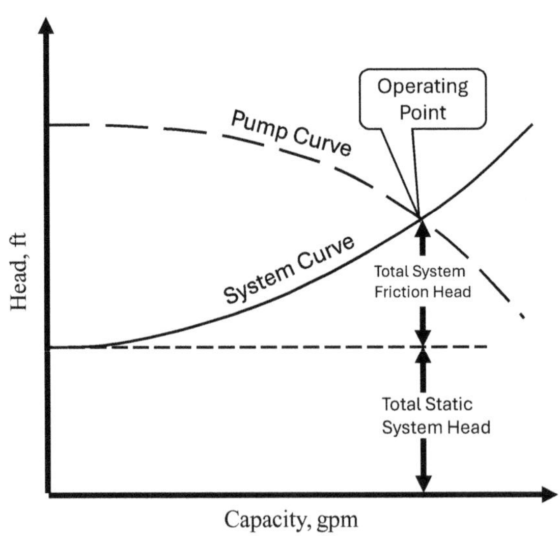

FIGURE 6.1 Pump and system curve.

DOI: 10.1201/9781003729525-6

PUMP CURVES

A pump curve shows how the pump will perform in a given set of system conditions such as pump flow versus flow resistance or head. The only pump curve that truly matters is the one specific to the pump installed, which is the actual pump impeller diameter and the operating speed. Pumps produce flow based on how much system resistance is put on the pump. The piping system is what determines the resistance on the pump. Essentially, pumps produce flow while the system produces the pressure. The pump flow depends on the system, so the system is the major factor determining pump performance.

A pump performance curve has lots of information on it as shown in Figure 6.2. Performance is graphed as flow versus head. Some of the items on the pump curve are pump size, model, speed, efficiency, NPSH3, impeller size, BHP, min flow line, eye area, impeller pattern number, curve #, impeller drawing #, max sphere and # impeller vanes.

The larger number on the pump size is the inlet nozzle size. The smaller number is the discharge nozzle size. Some manufacturers may list the suction size first or the discharge size first. The last number is typically the largest impeller size for the pump size. Eye area refers to the impeller eye area and can be used to determine the impeller eye diameter if desired. Impeller pattern number can be an important number as it is usually cast on the backside of the impeller. This will allow you to match the impeller with the specific curve in the field and to ensure the correct impeller is on the pump assembly. Many incorrect impellers can fit onto the shafts,

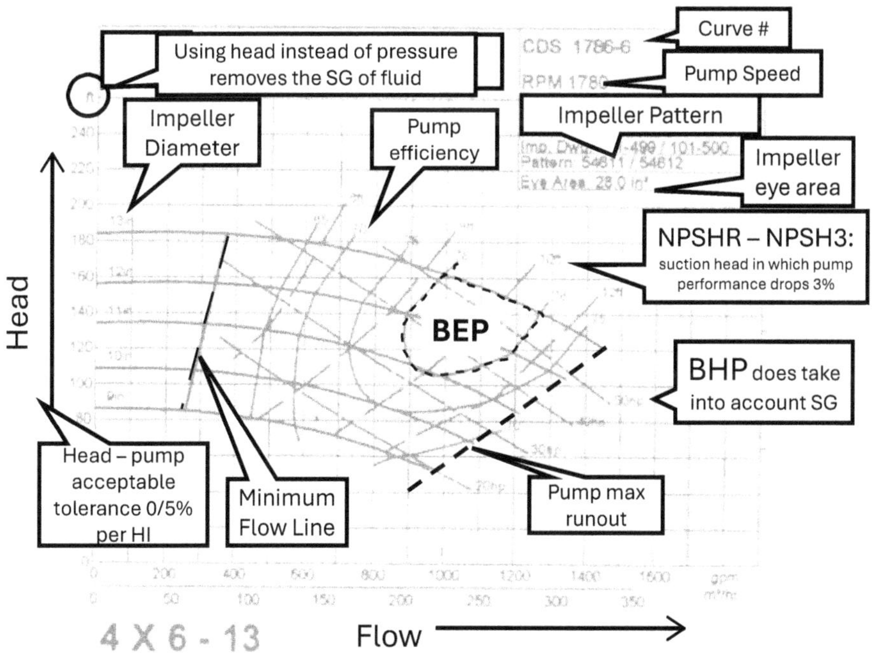

FIGURE 6.2 Typical pump performance curve.

but still be the wrong choice for the pump. The maximum sphere is the largest solids that can pass through the pump. Some curves may also have suction lift lines. One atmosphere is ~33 ft. The sum of NPSH and suction lift is 33 ft. So if the NPSH is 20 ft then the suction lift is 13 ft and vice versa. As pump flow increases, the NPSH increases, and the suction lift decreases as seen on the pump curves.

For a multistage pump, many pump curves may only show a curve per impeller stage. Therefore, this is similar to several pumps in series so the head on a single stage curve is multiplied by # of stages for total head, TDH. The BHP is also per stage so the same multiplication by # of stages for total BHP. Some pump curves will show performance for all stages so check with pump manufacturer preferences. For a self-priming pump, a priming curve may be given for a specific impeller diameter. This just shows how long it should take for the pump to prime. A general rule of thumb sometimes used is ~25 seconds for every foot of lift.

Some pumps will have two impeller diameters listed on each curve. This is an impeller where there is a major and minor diameter. The large diameter is the impeller diameter on the suction side of the impeller. The small diameter is on the rear of the impeller. Sometimes only a single diameter is given for a taper cut impeller, and this is the average diameter between the major and minor diameter. Other single diameter impeller notations are just a single diameter for a straight cut impeller.

VISCOUS FLUIDS PUMPING

Typical pump curves are based on water at 60–70°F with SG = 1. If viscous fluids are being pumped, then corrected pump curves must be produced, as head and BHP will both increase and flow, while efficiency will decrease. If the kinematic viscosity gets above 20–30 cst, then a viscosity correction is generally needed.

One method involves getting correction factors for the viscous fluid to determine the water pump curve flow, head, and efficiency for pump sizing. The correction factors may be taken from a viscous correction factor curve like the one in Figure 6.3. Viscosity in this curve is in either centistokes or SSU with the relationship between the two estimated with the equation below. SSU is Saybolt Seconds Universal.

$$\text{SSU} = \text{Centistokes} \times 4.64$$

An example from the viscosity correction using Figure 6.3 curves could be modeled as follows.

1. Choose viscous flow and head of the pump system, *Qvis* and *Hvis*. Start with *Qvis* and move up to the *Hvis*. Say *Qvis* = 1,500 gpm at 100 ft.
2. After head is attained, move along head curve until fluid viscosity is met (for example, use 220 cst).
3. Move vertical up to correction factors – *CQ*, *CH*.

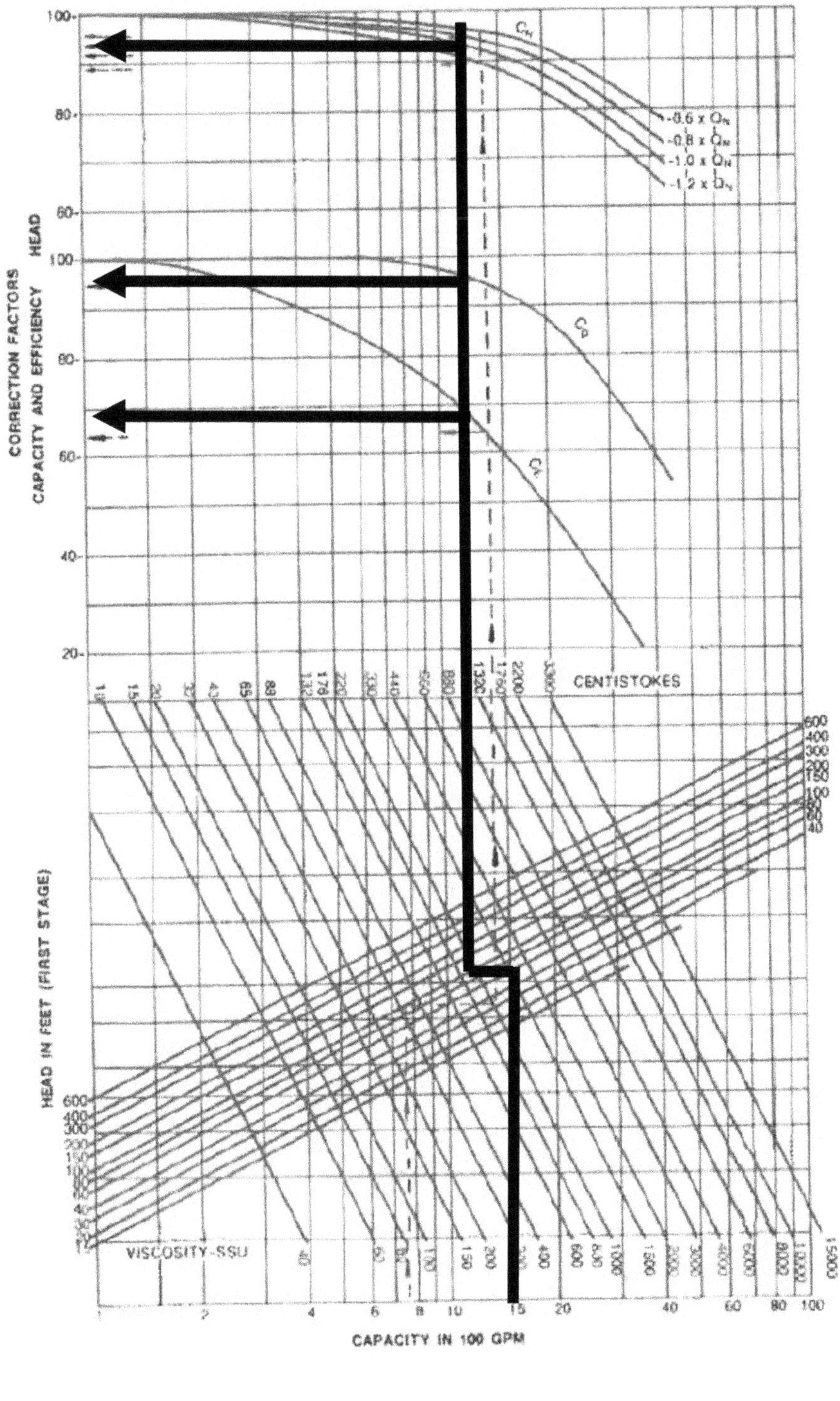

FIGURE 6.3 Viscous fluid correction factors for viscous pumping.

4. Solve for water flow and head.

$$Qw = Qvis/CQ$$

$$Hw = Hvis/CH$$

$$Evis = Ew*CE$$

where:
$Qvis$ = flow for viscous fluid
CQ = viscous flow factor
Qw = flow for water
$Hvis$ = head for viscous fluid
Hw = head for water
CH = viscous head factor
$Evis$ = efficiency for viscous fluid
Ew = efficiency for water
CE = viscous efficiency factor

5. Select a pump with the corrections for water flow (Qw) and head (Hw).
6. Get water efficiency for design conditions, Ew. Make correction for viscous efficiency using efficiency correction factor.
7. Calculate viscous BHP using viscous parameters – flow, head, and efficiency.

A newer method to produce new viscous pump curves is found in ANSI/HI 9.6.7-2021. Whichever methods are employed, just know there can be error with almost any method when dealing with viscous fluids. The chart in Figure 6.3 is only for standard pump designs with normal operating ranges with open or closed impellers. If NPSH issues are involved, there may also be additional error.

Pumping sludge can have other challenges around air entrainment as sludge may ferment releasing gas creating variations in consistency. In general, sludge containing up to 8% solids can be handled by centrifugal pumps. For higher solid concentrations, a closer look at NPSH requirements becomes important. When solids exceed above 8–10%, positive displacement pumps (lobe pumps are good choice) are more suitable. At higher solids, the sludge also tends to channel the water flow. Sludge velocities should be kept above 4–5 ft/s to avoid sedimentation in the line and grease buildup. Sludge can also have similar adjustments for head and power with the following relationships.

$$H\text{sludge} = H\text{water}\ (1 - TS)$$

$$P\text{sludge} = P\text{water}\ (1 + TS)$$

where:

Hsludge = head for sludge, ft
Psludge = power for sludge, HP
Hwater = head for water, ft
Pwater = power for water
TS = total solids $< 8\%$

Another pump curve issue that occasionally arises is curve droop. This can cause unstable pump performance and operating conditions if the pump operates in the droop region. The pump flow can oscillate between two operating heads as flow swings. This can cause a surge on the pump and system. Startup of a pump with curve droop can also be tricky if the system already has a pressure that operates in the droop region. On startup, you can't always avoid this droop region. Starting up the system with a droop curve is best with the discharge valve open so the droop region might be avoided. Generally, I want to operate a pump with curve droop outside the droop region with a head below the dead head pressure to avoid any unstable pump operation.

CASE STUDY 15 PUMP CURVE DROOP EFFECT

A small boiler feed water pump system was experiencing some random vibration in what appeared to be a surge-type condition. The pump system had fixed-speed pumps with a pressure control valve. The pump ran a small package boiler with a steam rate of 81,000 pph at 300 psig. The pump was an eight-stage pump that had an original design point of 180 gpm at 860 ft as shown in Figure 6.4. When the boiler load was above 150 gpm, there was little issue with getting into the droop area of the curve which was at a head of ~910 ft.

There were periods of boiler load where the load was lower, and the resulting head was 924 ft which put the pump in the droop area of the curve (see Figure 6.4). When this occurred, the pump had the potential to surge. Surge was observed many times by seeing some water standing on the steel baseplate vibrating, as the pump moved across the curve droop area.

Some other process changes later increased the load to the boiler system which helped get the pump further away from the droop portion of the curve as

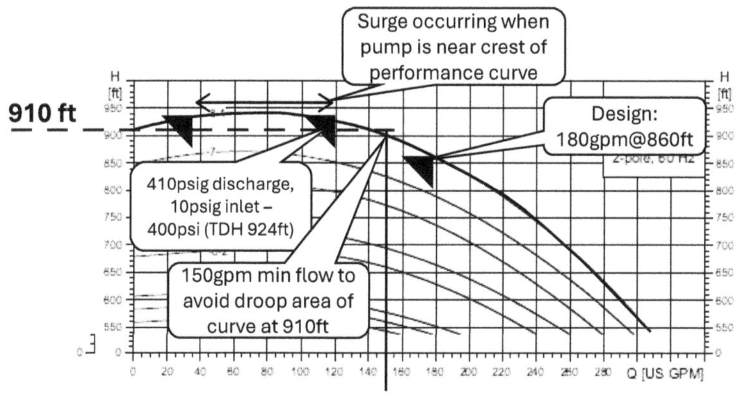

FIGURE 6.4 Original pump curve and operating points with curve droop.

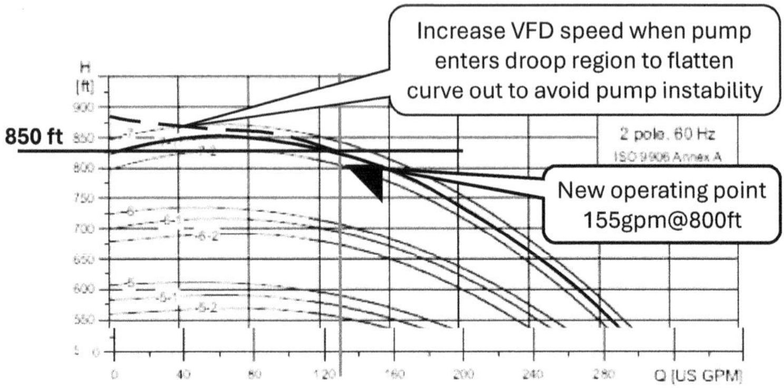

FIGURE 6.5 Correcting curve droop effect with a VFD.

shown in Figure 6.4. Since the original pump was operating at a lower operating point, the control valve was operating in a heavily throttled position at 15–40% open. Therefore, the pump was replaced with a seven-stage pump which allowed the control valve to operate in a more open position of 50–60%. This change resulted in improving the buffer of about 50 ft of head away from the droop portion of the curve.

The pump also had a system NPSHA ~7 ft while the NPSHR ~9 ft (including a 2 ft pump manufacturer's 2 ft margin). This put the NPSHA < NPSHR so the cavitation was likely also affecting pump performance and reliability. This led to changing the pump to a low NPSH pump, which had a smaller (low NPSH) first-stage impeller. The low NPSH pump had an NPSHR ~5 ft, so it was an improved NPSH condition after the pump change in regard to cavitation.

If the pump had been designed with a VFD, there is a drive/controls modification which could have been manipulated when the flow approached 150 gpm, the VFD would increase speed to ramp up the lower flow portion of the curve to remove the droop effect as shown in Figure 6.5. This may require an overspeed depending on where the VFD is operating. Motor BHP should not be an issue since the flow is also drastically reduced in that region but it is always good to verify.

The design of the pump and system led to the pump performance and reliability issues. Modifications to the pump design operating point and a new pump would address this issue. The pump's original design did not match the actual operating point conditions.

Case Study 15 Learning – Do not operate a pump within the curve droop region for the normal operating point for stable pump performance.

SYSTEM CURVES

The system curve basically has two components – static head and friction head. The system curve starts at the total system static head when flow is zero. This is a positive number most of the time when pumping to higher elevations. But this will be negative if pumping to lower elevation, where friction head is the primary resistance (negative static head). The friction head varies with the flow through the system which includes not only piping but valves and any other process equipment like filters and screens. The other process equipment can have its own pressure drop characteristics as a function of flowrate. This will have to come from the manufacturer's data. A quick system curve can be drawn with only two points – the system static head and a flow friction head datapoint. The pump TDH showing static head and friction head is given below.

$$\text{TDH} = Hs + fQ^2$$

where:

TDH = system total dynamic head
Hs = total system static head
f = friction factor
Q = flowrate

For a fixed speed pump with a throttling valve in the system, the pump curve is fixed, and the system curve changes as the system friction loss increases or decreases depending on the valve percentage open. The system curve then bends and moves up and down the constant fixed speed pump curve as shown in Figure 6.6a. With a variable speed curve, the pump curve varies with speed, but the system curve is constant. The variable pump curve moves up and down the system curve as the pump speed changes as shown in Figure 6.6b.

While most pumps operate singular in pump system, pumps can also be configured in series or parallel in process systems. Series pumping is done to increase the system pressure at a desired flow. Unlike a component in a system that has a pressure drop, a pump adds energy to boost pressure so two in series will increase the system pressure. The two pumps in series don't have to be the same exact pump, but it is always a best practice to have the same size pumps especially from a flow standpoint. The two pumps operating in series do not experience the same pressures. But they will basically pass the same flow since they are arranged in series (unless flow is bled off the first pump). If the pumps are not identical (flow capacity), then each pump will operate at a different point on its curve.

If the second pump in series is pulling a higher flowrate, a suction pressure deficiency could result under the right conditions (enough pump size mismatch occurs). While both pumps can be evaluated together from a system standpoint, they must also be evaluated individually from an individual pump standpoint. It is always good to start the first pump before starting the second pump, ensuring the second pump has liquid available before being started and before shutdown, the pumps should be stopped in the reverse order.

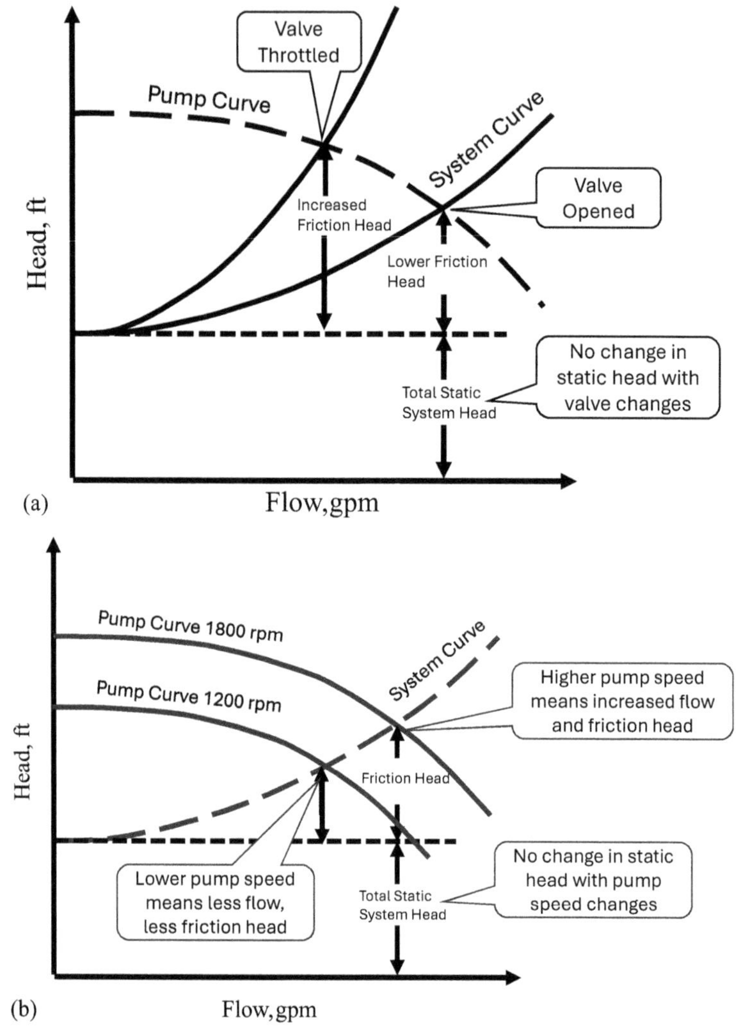

FIGURE 6.6 (a) Fixed speed pump and variable system curve. (b) Fixed system curve and variable pump curve.

Pumps can also be configured in a parallel arrangement. Pumps with installed spares are configured this way but they do not typically operate together in parallel. True parallel pumps operate at the same time in the same piping system. Many think that by having parallel pumps they can easily double the flow, but this is not true due to the exponential nature of the system curve. The flow will always be something less than 2X the flow of one pump. Many times, it may only be 1.5X as much. It really depends on the system design.

How a pump and system interact with each other is determined by the slope (shape) of both the pump curve and the system curve. A flat pump curve will maintain a near-constant head over the flow range. For a pump system where a steady system pressure is desired, a pump with a flat curve has an advantage. This would be good for a pump system where the pump is supplying fluid to a distribution header, such as a dilution water system. As different end uses vary flow demand, the header remains fairly constant for control loops to control the process. A pump with a flat curve can be a drawback in a system where small changes in a system head can drastically change flow output. If having stable flow is desired, then a flat pump curve is not what the process needs.

A steep head rise pump curve has a large increase in head over the flow range. This type of pump would not be ideal for a distribution-type pump system but would be great for a pump system where it would take a larger change in head to change the flowrate. This would offer good resistance to flow changes when the system changes. Figure 6.7 shows how different pump curve shapes may be affected by various system curve changes.

Pumps and systems change over time and are not static in their operating points. What initially may have been a reliable pump and operating point can shift over time. Here are few of the things that can change over time in a pump system.

- Increase system friction (reducing pump flow): See pump and system curve affect in Figure 6.8.
- Decrease in pipe diameter due to corrosion, abrasion, or sclerosis.
- Increase in pipe wall surface roughness due to corrosion, abrasion, or sclerosis which affects resistance to flow (friction head) as shown in Figure 6.9. Smaller pipe inside diameter also increases fluid velocity, which also increases friction head.
- Buildup in a system component (valve, piping) causing blockage and increasing pressure drop.

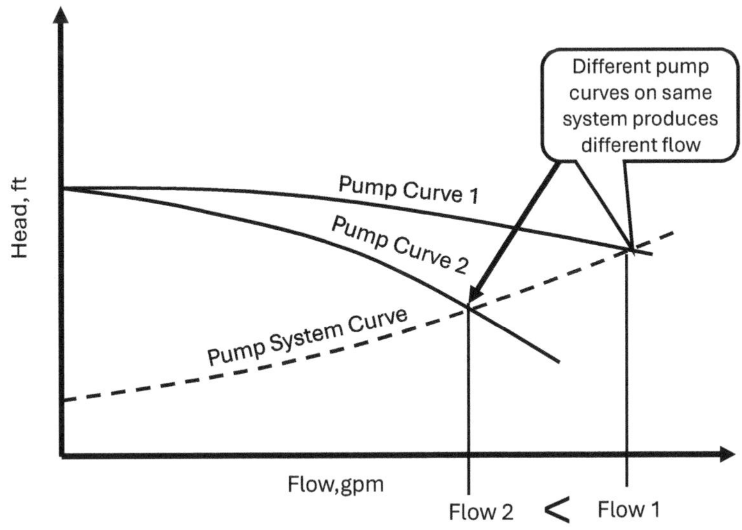

FIGURE 6.7 Pump curve shapes and system curve affects.

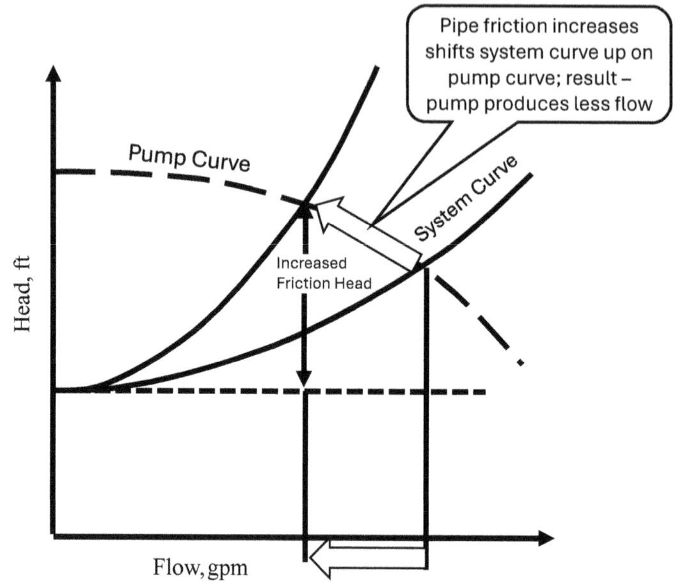

FIGURE 6.8 System curve increasing as system deteriorates.

FIGURE 6.9 Surface roughness of pipe over time changing.

- Decrease in system friction (increasing pump flow): See pump and system curve affect in Figure 6.10.
- Decrease in capacity requirements due to leaks, control schemes, etc.
- Change TDH due to static head changes (tank levels).
- Wearing of piping components – valves.
- Wearing of pump components – impeller, casing, wear rings, and causing internal leakage/recirculation.

The system curve will change whenever the system condition changes. For instance, increased system friction would look like Figure 6.10 for any deterioration of the

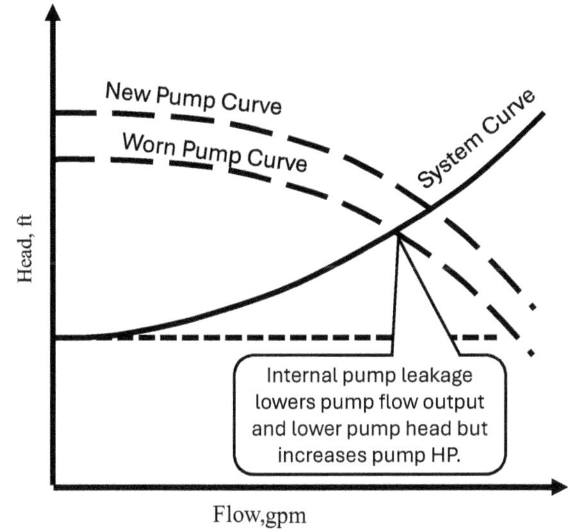

FIGURE 6.10 Pump curve decreases as system changes (leakage, etc.).

system where the friction head increases. As the system head increases, the pump curve moves down the system curve. As a result, the pump will produce less flow as the system head has increased.

Pump system curves may take on various shapes like pump curves. A system curve has two main components – static head and friction head. The system curve shape and interaction with the pump curve also greatly affects how the pump will or can respond in the system. A static head-dominated system curve will be very flat and look like that in Figure 6.11. A high static head system will require the pump to generate most of its head before any flow is generated.

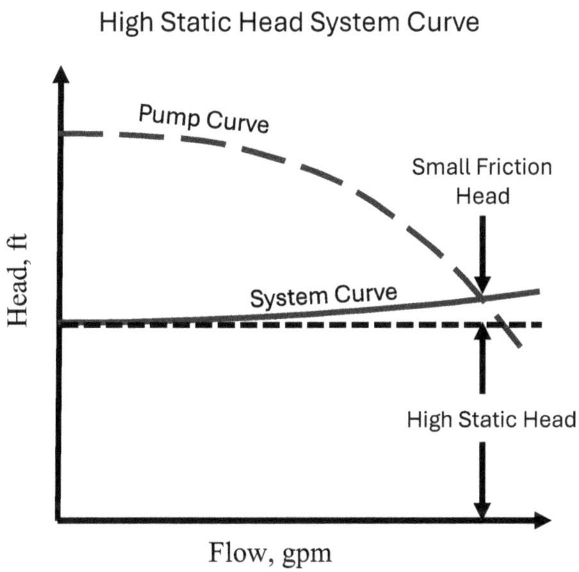

FIGURE 6.11 Static head dominant system curve.

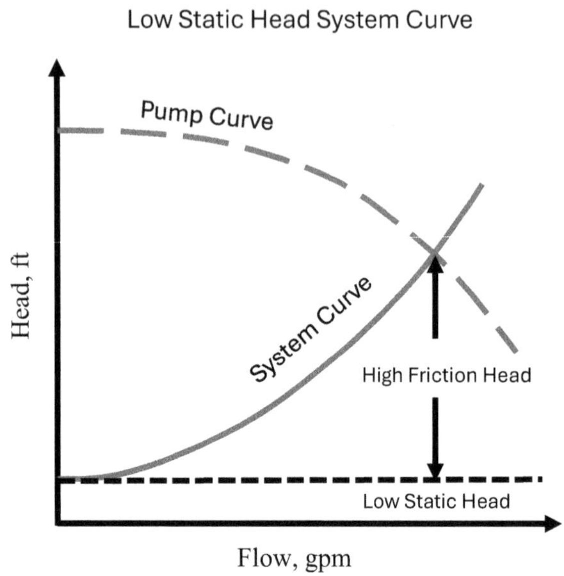

FIGURE 6.12 Friction head dominant system curve.

Figure 6.12 shows a friction head-dominated system curve where the system curve is very steep on the pump curve. Contrary to a high static head system curve, a pump operating in a high friction head system (low static head) will begin to generate flow very quickly. A pump operating in a high friction head system may want to startup with the discharge valve throttled to reduce the amount of time the pump is operating at very low head. This could have the pump potentially operating outside the AOR for a period of time until the system fills up, so throttling the valve on startup can eliminate or reduce this time.

CASE STUDY 16 PUMP PROCESS CHANGES IMPACT ON PUMP SYSTEM

Another pump system change that can happen over time is when a process or a system change shifts the operating point of the pump. The pump may no longer be a good match for the new system or process flow change. This change can also greatly affect pump reliability. Since pump systems are live systems and experience many changes over the life of a plant, pump systems cannot be assumed to still be operating at the original design conditions. I would venture to assume that after ten years as much as 30% of pump systems have seen some changes. Over a period of 30–40 years, as much as 50% of a plant's pump systems may be impacted by process or system changes.

In a papermill, a blend chest stock pump had experienced significant changes from its original design over many generations that had reaped the results of such changes. Generations later, new operational and maintenance

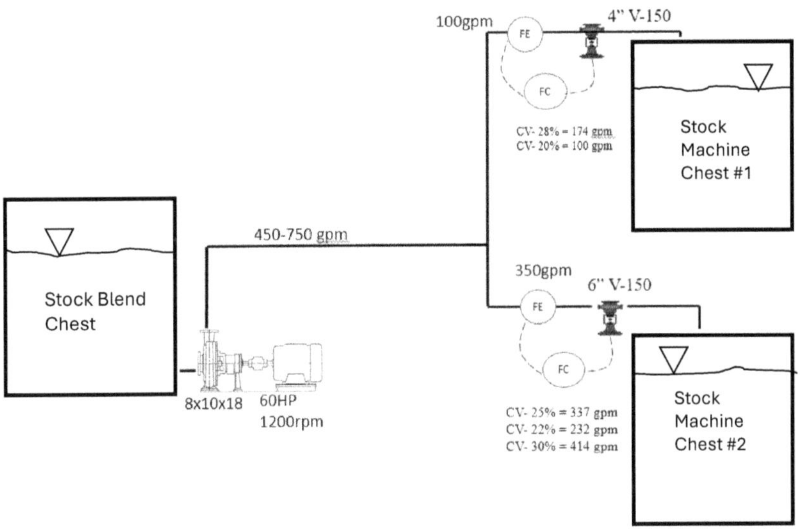

FIGURE 6.13 Process diagram of stock pump system.

staffs had little knowledge of the major shift in the process and pump system. The existing process diagram consisted of a pump pulling from the blend chest and supplied two different machine chests as shown in Figure 6.13. Each machine chest controlled the flow by a single flow control loop. The control valves were operating very closed off with a normal operating range of 20–30% open. The pump was an 8 × 10 × 18 end suction stock pump driven by a 60 HP, 1,200 rpm direct drive motor.

The pump had been experiencing poor reliability with high maintenance on the pump and piping system. High vibration was also present much of the time. The original design point for this pump was 2,200 gpm at 80 ft. The original pump operated at BEP with 80% efficiency and required 56 BHP. With process changes, the pump system now had a very different operating point of 415 gpm at 115 ft, as shown in the pump curve in Figure 6.14. The pump was operating outside of AOR and very inefficient. Recirculation cavitation was an issue which drove pump vibration and poor reliability. These changes also led to high shaft radial loads, leading to shaft deflection and poor shaft sealing.

Several options existed for this pump system to improve its operating conditions for the pump and system as listed below.

1. Install a smaller impeller on the existing pump.
2. Install a VFD on the existing pump.
3. Install a recirculation line with either flow or pressure control with the existing pump.
4. Install a smaller new pump at new design operating point.

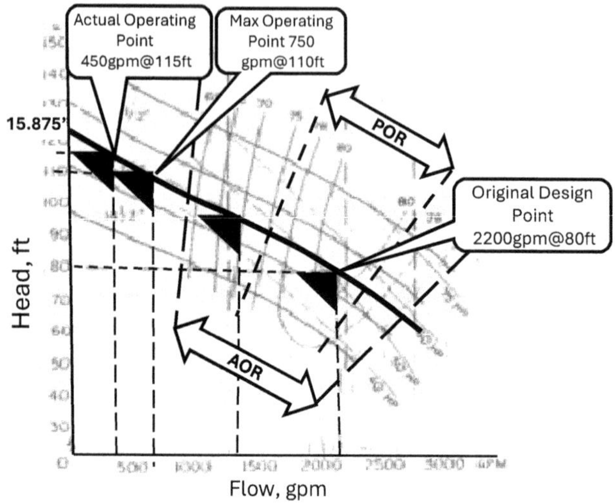

FIGURE 6.14 Original pump curve with various operating points.

Option 1 with the smaller impeller did not really change the operating point on the pump as it still operates outside the AOR for the pump, as can be seen in Figure 6.14. It does reduce head and radial load a little which saves a small amount of energy, but it doesn't address the major issues with the pump system. Option 2 only lowers the head but does not solve all the issues. Since there are two destination points and separate controls, a VFD would be on header pressure control and likely would not be able to reduce head enough. A VFD on pressure control does nothing to help oversizing from a flow standpoint. Both these options also may improve the control valve opening more since the head is reduced.

Option 3 increases the system flow and reduces system head so that the pump will operate within the AOR and even POR if desired. In doing this, much of the pump system reliability issues are resolved but it comes at an energy price as most of the flow produced by the pump is wasted in recirculation. Also, this option adds additional piping, control valve, and control loop to maintain; therefore, from a LCC standpoint, it is not very attractive.

Option 4 to install a new, smaller pump is the best overall option as it not only moves the operating point to within the pump AOR, but is also energy efficient in its impact as shown on the pump curve in Figure 6.15.

A summary of these four options is shown in Figure 6.16, which has some of the hydraulic comparisons and energy usage. The difference in the original design, current operation, and then all the options can be clearly seen. Under the current process conditions, it is clear that option 4 with a new pump properly sized to deliver a reliable pump system while supporting process demands.

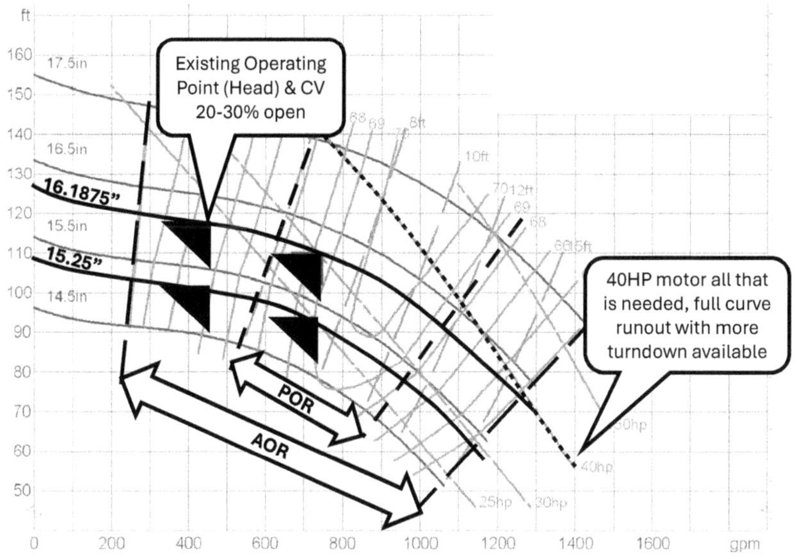

FIGURE 6.15 New, smaller pump performance curve.

Condition	Flow, gpm	Head, ft	Eff, %	BHP	OP Within AOR (POR)	CV % Open	Annual Energy Cost, $	Energy Savings, $
Original 8x10x18, 3175	2200	80	80	56	Y (Y)	NA	20K	NA
Existing 8x10x18	450	115	35	38	N (N)	20-30	14K	Ref
1) Existing Pump Smaller Impeller	450	97	35	32	N (N)	40-50	11K	3K
2) New Pump 4x10x18 3175	450	97	60	19	Y (N)	40-50	7K	7K
3) VFD Existing Pump	450	97	35	32	N (N)	40-50	11K	3K
	450	85	35	28		60-68	10K	4K
4) Existing Pump Recirculation Control	1250	97	70	45	Y (N)	40-50	16K	-2K
	1600	90	76	49	Y (Y)	55-65	18K	-4K

Smaller impeller accomplishes about same results as VFD

Energy inefficient

FIGURE 6.16 Summary of pump system options to improve pump reliability.

This pump system reliability issue had pump and system design as the root cause category. Maintenance had spent lots of effort to try and make the pump system reliable with no success. This is another good reminder that having a pump system audit every five to ten years is not a bad idea to evaluate pump systems as process changes occur in plant operations.

Case Study 16 Learning – Always verify pump operating points and never depend on original design points. Modify pump systems for the pump and control valves to operate within the AOR.

SERIES AND PARALLEL PUMPING

Series pumping is when two pumps are pumping in the same system where the discharge of the first pump feeds the suction of the second pump. To make the series pump curve, just add each pump head at each flow rate. If they are the same pump, then it is basically doubled head (TDH) for each flow for the pump curve. Due to the system curve, the operating point will be slightly different from just double head, as the head and flow may also change depending on the system. Figure 6.17 shows an example of series pump curves with the same pumps.

For pumps operating in series, check the maximum inlet pressure for the second pump that the seal arrangement can handle that pressure. Also, the other pump components such as casing need to be verified for the final system pressure ratings. If there is seal water for these pumps, verify there is enough pressure for the second pump to overcome stuffing box pressures. Pumps don't need to startup dry or without proper suction pressure, so typically, start up the first pump in series first and then the second pump.

Parallel pumping is when two pumps are running parallel in the same piping system. This is done for several reasons but basically two pumps share the pumping load for a system. This is different than having two pumps piped up to the same system and only needing one pump and the other is an installed spare pump. Having an installed spare is one of the easiest ways to increase the pump system reliability but it can be an expensive reliability. True parallel pump systems require both pumps to operate to maintain the process and they should be engineered for that scope. An installed pump and a spare may be operated in parallel, but the pump system should be evaluated before doing so or poor reliability could result.

A common myth about parallel pumping is that the flow will double when you add the second parallel pump to the system operation. This is not accurate as the friction head on the system curve increases by the square of fluid velocity so this

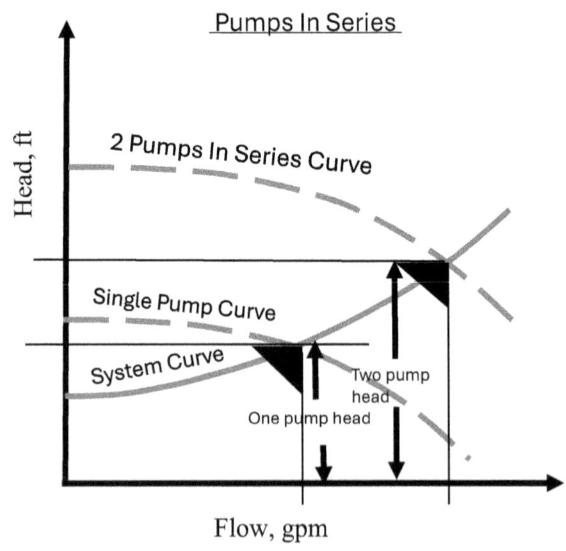

FIGURE 6.17 Example series pump performance curves in system curve.

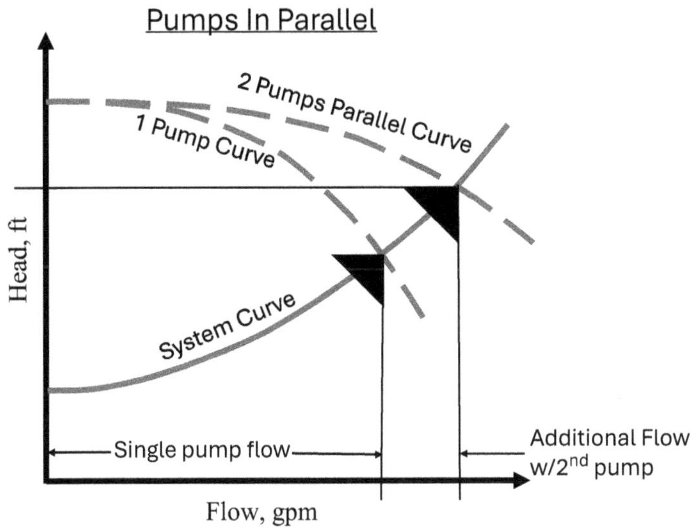

FIGURE 6.18 Example parallel pump performance curves with system curve.

non-linear response and intersection with the parallel pump curve will result in a reduced flow from adding the second pump as shown in Figure 6.18. The pump curves are basically the flows added up at the same head but when the system curve is added in, there will be less than 2X flow when the system curve intersects the two-pump parallel curve. How much more flow with the second pump added depends on the slope of the pump and system curves. A static dominant or low-friction head system curve is much closer to 2X flow with a second pump in parallel. A friction dominant system curve is just slightly more than one pump when second pump is added in parallel.

If the two pumps are the same, they will basically share the pumping total flow (load). For example, if one pump can pump 700 gpm and a second pump adds another 300 gpm, then the combined system flow is 1,000 gpm for both pumps. Therefore, in this case, each pump in parallel will produce ~500 gpm on the system. There is usually some small difference in system piping for one of the pumps, so technically, there will be a very small difference in each pump loading, but they will be essentially the same if they are the same pump. Of course, pump wear, clearances, and other hydraulic efficiency factors also make the load sharing different too.

It is best if pumps are the same pumps, but they don't have to be exact. The pumps should have similar shutoff head and specific speeds. It will mean that the load share for the two pumps will unevenly split unlike that of the same-size pumps. If these pumps are on an ASD, then pumps should be run at the same speed to even out system load (typically be the same speeds). When staring up parallel pumps, start the weaker pump first. Do not run one pump too long before starting the second pump, as one pump may run out on the end of its curve. When properly designed, most parallel pump systems are designed to operate both pumps in the system for even loading within the AOR for each pump.

CASE STUDY 17 SINGLE AND PARALLEL PUMP SYSTEM COMPARISON

A sump pump system was equipped with two self-priming pumps that were piped up in parallel as shown in Figure 6.19. The pumps were a 6×6 pump operating with the minimum size 11″ diameter impeller. The driver was a 10 HP, 1,200 rpm fixed-speed motor. The pump operation was a batch-type process where the liquid reached a high level to pump it down and shutting off once the level reached a low point. Upon reaching a high level, the pump would start and pump the sump down. Normal operation was for one pump to carry the normal process load which happened much of the time.

During times of high demand, when the sump reached a high-high level, the second pump would start to assist and in pumping down the sump. With both pumps running on one piping system, the pump system was operating in parallel pump mode. There were times where even the second pump could not maintain the sump level and the sump would overflow the basement U-drain system. The need and issue were to figure out how to best increase the pump capacity of the sump pump system for high demand conditions.

The pumps had an original design point of 700 gpm at 29 ft. A system curve was calculated for the piping, fittings, valves etc. in the system, which with a single pump operating would produce 700 gpm. A parallel pump curve was generated and shown with the system curve as shown in Figure 6.20. With the second pump coming on the total flow was ~1,000 gpm. With two exact same size pumps in parallel, the pumps individually would share the system load flowrate of 1,000 gpm, so each pump was pumping ~500 gpm. Due to increasing system curve friction, the pump had an incremental flow increase of ~300 gpm instead of just doubling the flow.

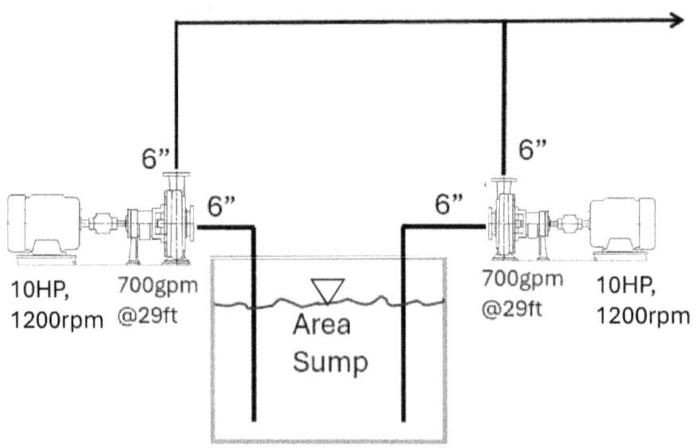

FIGURE 6.19 Process flowsheet for original sump pump system.

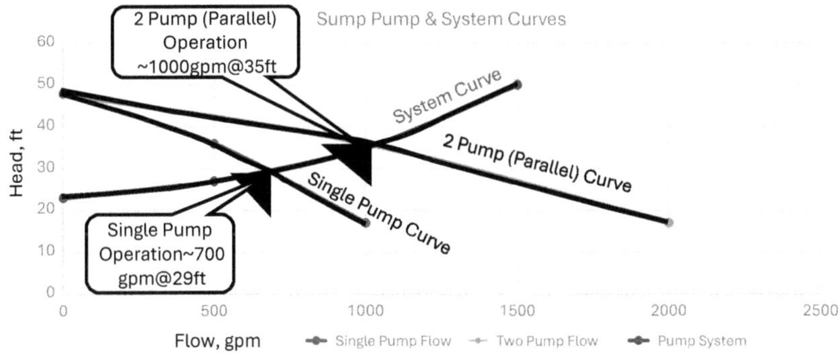

FIGURE 6.20 Single and parallel pump curves with system curve.

Several options were evaluated to determine the best path forward. Installing a brand new sump pump system was not even considered as there was insufficient capital to investigate that option. The options had to utilize the current base equipment. A summary of those options is shown in Figure 6.21. The option to just install a larger impeller on the existing pump and system presented some mixed results. When operating with one pump in the system, the pump produced an additional 150 gpm and a good system efficiency producing 73.8 gpm per HP. However, when operating two pumps in the system there was only an additional 100 gpm, when compared to the original pump system and the system efficiency dropped to 63.9 gpm per HP.

Pumping Conditions	Total Flow, gpm	TDH, ft	Pump Eff.%	BHP	System Eff. (gpm/HP)	Comments
Existing 1 Pump Operation	700	29	58	8.8	79.5	Overall system efficiency pretty good
Existing 2 Pump Operation (Parallel)	1000	35	58	15.2	65.8	Only +300gpm by adding 2nd pump in parallel operation (poor system eff).
Option 1 – 11.5" Imp, 15HP motor 1 Pump Operation	850	31	58	11.5	73.8	Significant project upgrade costs on this – motor, baseplate, cable, starter, etc.
Option 1 – 11.5" Imp, 15HP motor 2 Pump Operation (Parallel)	1100	36	58	17.2	63.9	Significant project upgrade costs on this – motor, baseplate, cable, starter, etc. Worst system efficiency. Poor long term investment. Only +100gpm total from existing system.
Option 2 – Separate Piping System (2 pumps)	1400	29	58	17.2	81.4	+700gpm more flow compared to existing 1 pump operation (+400gpm from existing 2 pump operation). Most energy efficient.
Option 3 – Add Vaughn Chopper Pump (Separate)	700	35	60	10.3	68	By adding 3rd pump, total system flow potential 1700-2100gpm, also chopper will handle trash.
Larger sump pumps will decrease normal pump times so pump cycles on/off well beyond reliable pump motor conditions						

FIGURE 6.21 Options for pump improvement.

While this option required only a larger pump impeller, it required a larger pump motor (15 HP from 10 HP original). This meant more modifications to the baseplate and electrical feed system with a larger starter and cable, so the project cost was fairly high, especially for the small incremental pump system performance improvement that resulted.

The best option ended up being to install another 6″ pipe system for the second pump so that both pumps operated as single pump systems. The option of installing a larger pipe system for the existing parallel pump system was not evaluated as the cost for that option would be more than just installing an additional 6″ pipe system for the second pump. The only project cost was piping with no pump or motor modifications. Option 2 created an additional 400 gpm from original system and another 300 gpm from option 1 (upgrading the pump with larger impeller and motor). The pump efficiency was actually the same for 700 gpm operation or 500 gpm operation, but the total flow increased. As a result, the system efficiency for this option was improved by 81.4 gpm/HP versus the original condition where only 65.8 gpm/HP was achieved. So, the system got more flow to manage the high-demand situations and also got more energy efficient in the process.

The solution to maximizing the pump capacity on this system was a pump system redesign from a parallel pump pipe system to a single line pump system. There was even another option not listed here which was to add a third pump to the system, but this was too costly with available budget. Like any pump system optimization, the options must fit the company's business plan and financial conditions so this can be a major factor in solution options.

Case Study 17 Learning – When solving pump system issues on parallel pump systems, evaluate the impact of the system and system curve on pump performance when evaluating the possible solutions.

Pumps in parallel can present some unique challenges in pump and process systems. Can any two pumps be placed in parallel and function in a pump system? This next case study will show how in an extreme case what can happen when adding two unmatched pumps in parallel.

As with every pump and pump system problem, one of the first challenges is to try and figure out which element is the root cause of poor pump system performance or reliability – is it the pump itself or something in the system? The cause and effect of each can contribute to poor pump performance so this requires patience and discipline to work through the problem. This next case study will explore this very dilemma.

CASE STUDY 18 UNMATCHED PARALLEL PUMP SYSTEM

A water pump system had two pumps operating in parallel. The two pumps were very differently sized and styled pumps. The original #2 pump was an $8 \times 10 \times 12$ end suction pump but was later upgraded to a larger #2 pump $8 \times 10 \times 19$ standard end suction pump sized for 3,500 gpm at 265 ft. The #2 pump had a 250 HP, 1,800 rpm direct drive motor. An upgrade was done on the #2 pump many years earlier from the original pump which was 2,600 gpm at 110 ft. The smaller #1 pump was a $6 \times 6 \times 18$ recessed impeller end suction pump sized for 1,000 gpm at 90 ft. The pumps operate in parallel in one piping system. They were also working off individual variable speed drives but controlled by one speed controller so that each pump must operate at the same speed. The basic process flow for the existing pump system is shown in Figure 6.22.

The #1 pump was experiencing poor reliability and during high periods of demand flow, when the pumps went to maximum speed, the two pumps could only put out ~2,700 gpm. If #1 pump was operating alone, it could pump 1,900 gpm by itself. Therefore, adding the second, larger pump increased the total flow by only about 800 gpm. While this would handle some high demand events, other times the demand flow required was as high as 3,500 gpm. When the pumps were unable to keep up with the flow, the rejects tank overflowed, flooding the U-drain system throughout the entire basement.

The original pump system would pump around 2,800 gpm when the pumps were running in parallel. Interestingly, the pump upgrade to #2 pump didn't really produce any more pump flow than the previous two parallel pumps. The two pumps were similar in size and operating point so the pump system did as much as it could with both pumps sharing some load, but at times of high

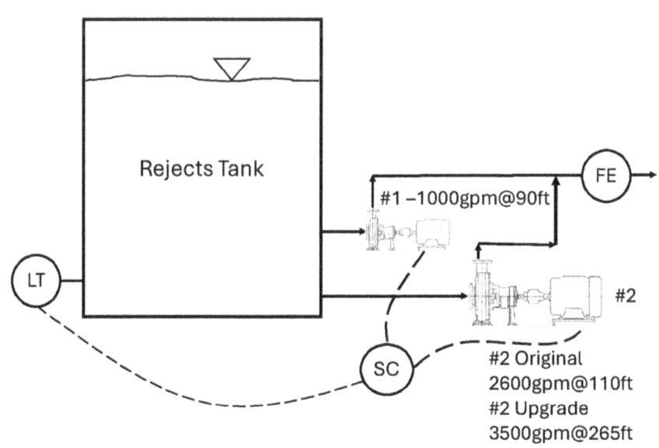

FIGURE 6.22 Process flow sheet for rejects system.

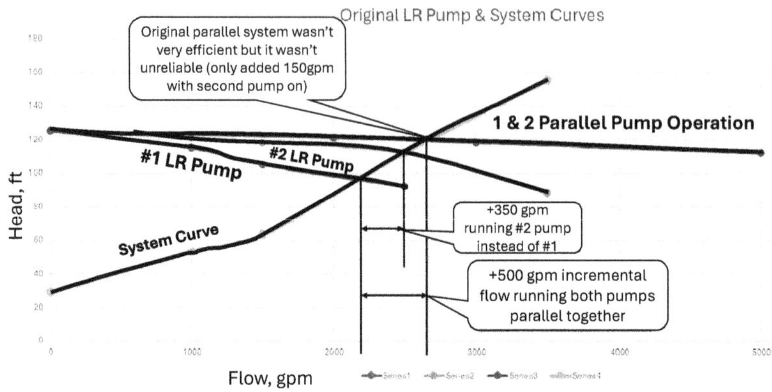

FIGURE 6.23 Original single and parallel pump system curves.

demand flow the original pumps were not enough. The upgrade enlarged the #2 pump to much larger than #1 pump but the speed control remained the same for both pump VFDs, so the pumps had to run the same speed which presented problems in parallel operation.

With #2 pump, there were some issues with air entrainment and pump wear on impeller and wear plate that was negatively affecting performance but there were issues operating the pumps together getting increased flow. A closer look at the pump and system curves on the original and upgrade began to tell the story. The original pump parallel system design curves are shown in Figure 6.23. The original pumps were matched pretty good, so they shared a load in a reasonable manner. Due to the very steep system curve, the second pump only added about 500 gpm to the total flow. During high demand conditions, this was not enough to prevent the tank from overflowing.

The first upgrade did not change the #1 pump but the #2 pump was changed to a larger pump. The larger pump was sized to match a similar rejects system pump in the same area. The problem was that the other rejects system handled high solids and had a much higher system head. The new #2 pump when paired with the original #1 pump in parallel created a mismatched parallel pump system. This became evident by some operating data of the two pumps as single pump systems and in parallel.

The #1 pump had a full load amp (FLA = 149) and when operating alone it was pulling 136 amps while pumping ~1,900 gpm by the flowmeter. When the #2 pump was run alone, the motor amps were 217 (FLA 290) while pumping ~2,700 gpm. The interesting thing was when both pumps were run in parallel at the same time, the amps dropped significantly on #1 pump and amps went up on #2 pump, but the total flow didn't really change. A summary of the conditions is shown in Figure 6.24.

Condition	Speed, %, rpm	Total Flow, gpm (curve head, ft)	#1 Pump, amps (FLA=149)	#2 Pump Amps (FLA=291)	BHP (WHP)	Actual Eff. %
#1 Pump On #2 Pump Off	#1 - 100, 1180	1900 (81)	136	0	118 (39)	33
#1 Pump On #2 Pump On	#1 - 100, 1180 #2 - 100, 1780	2700 (143)	60	235	#1 - 34 #2 - 194 (98)	51
#1 Pump Off #2 Pump On	#2 - 100, 1780	2800 (143)	0	217	179 (101)	56

FIGURE 6.24 Motor and pump conditions on singular and parallel operation.

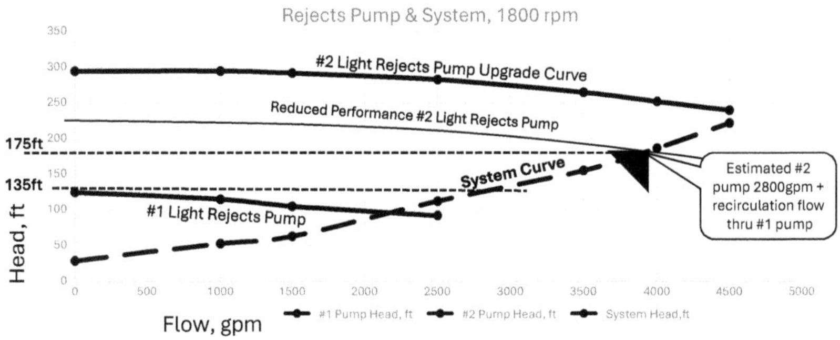

FIGURE 6.25 #1 & #2 pump performance curves with system curve.

With the #2 pump upgrade with the #1 pump operating in parallel, the system pressure was higher than the dead head pressure of #1 pump. Therefore, #1 pump was not contributing anything to the system flow. The #2 pump overpowered the #1 pump, as shown in Figure 6.25, with the pump curves and system curves. Even with #2 pump has wear and decreased performance, it still overpowers the smaller #1 pump in parallel with its operating head (175 ft) > (135 ft) dead head of #1 pump. When both pumps were running in parallel, the #2 pump overpowered #1 pump and it actually was backflowing through #1 pump. This is why the motor load went down on #1 pump and up on #2 pump. Remember the #1 pump is a recessed impeller pump, therefore, the casing was very open at the back connecting to the tank suction line as shown in Figure 6.26.

The negative effect on the #1 pump was very inefficient and wasted energy, resulting in reliability issues. The pump was running in a process dead head condition which resulted in high shaft loads and deflection. The #1 pump had chronic mechanical seal failures as a result.

The root cause of this pump system reliability issues was the system design or actually the redesign. System design error with installing a large and small pump in parallel with the same VFD speed controls built in poor pump system

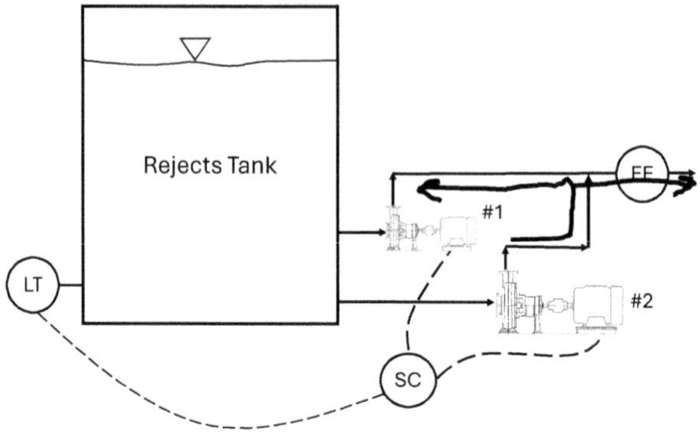

FIGURE 6.26 Recirculation flow through #1 pump.

reliability that no amount of maintenance or tweaked operation could fix. While the pumps could have been made to work better with separate speed controller, that would not have helped the total capacity shortfall that the pump system had. The correction to fix this system was to separate the pumps to operate on separate piping systems including their own VFD controls. This was feasible since there was an abandoned line on a system that was not being piped to the same destination. The rejects system total flow output went from ~2,800 gpm to a total of ~5,400 gpm with #1 pump at 1,900 gpm and #2 pump at 3,500 gpm. With this improvement, tank overflow events were essentially eliminated.

Case Study 18 Learning – Best practice is to use the same or very similar pumps in parallel pump systems. If different pumps are used, ensure that there is sufficient pump load on the smaller pump.

CASE STUDY 19 PUMP SYSTEM ISSUE OR PUMP ISSUE?

A screw press feed pump system consisted of a feed tank, pump, flow control loop (control valve and flowmeter), and screw press as shown in Figure 6.27. The pump was a 6 × 8 × 14 end suction pump with a 12.5″ diameter impeller that had a design point of 1,150 gpm at 62 ft. The pump was pumping paper stock at a consistency of 6%. The production rate would vary from 550 gpm to 1,000 gpm with the flow control valve throttling to regulate the production flow. The pump feeds the inlet end of a screw press to dewater the stock and increase consistency. The screw press is almost like a screw pump in series

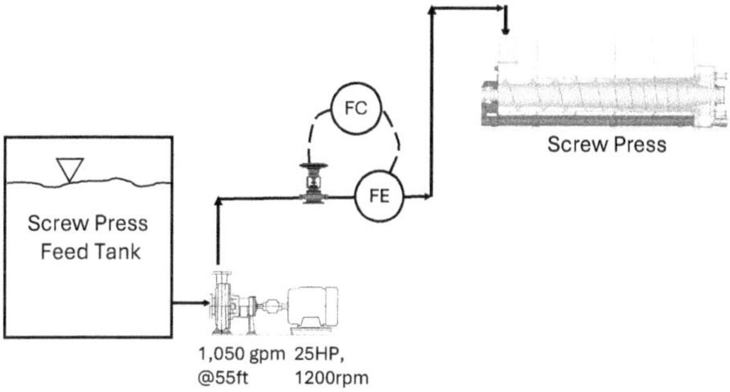

FIGURE 6.27 Process flow diagram for screw press feed system.

and when working well does not produce a pressure drop in the system but only conveys and dewaters the stock before the next stage of processing as it dumps by gravity into a tank.

The process began to have issues meeting the production rate through the feed pump system as the control valve went 100% open and the pump could only produce a maximum flow of 720 gpm. What is wrong with the pump and system? The first initial response by the area team was that the pump was worn out and at least the impeller clearance needed to be reset to its original target of 0.015″. During the next downtime opportunity, the impeller clearance was reset to target but the results did not change the pump performance. The impeller size was verified, and the pump wear was also inspected. The inspection revealed the pump to be in excellent condition so the pump being the root cause of pumping issues was ruled out.

Further observation revealed that the screw press was running high inlet pressure. Instead of operating at 0–1 psi inlet pressure, the screw press would run up to 6 psi inlet pressure at times and was now always operating between 4 and 6 psi. As the control valve opened to 100% trying to meet the higher flow setpoint, the system pressure backed up at the screw press inlet. So essentially the pressure drop shifted from the control valve to the screw press. The screw press now was a bottleneck, and the high pressure drop element in the pump system. That system pressure increases with the pump curve made a difference of 550 gpm to almost 1,050 gpm as shown on the pump and system curves in Figure 6.28.

How did we know this? Pump wear on this stock application has been proven to be very low in years of operation. Pump wear in low wear applications happens very slowly over time. The system pressure gradually increased over several months. Trends showed that the inlet pressure to the screw press was slowly rising. In these cases, proving whether it is the pump or the system can be difficult without some performance data.

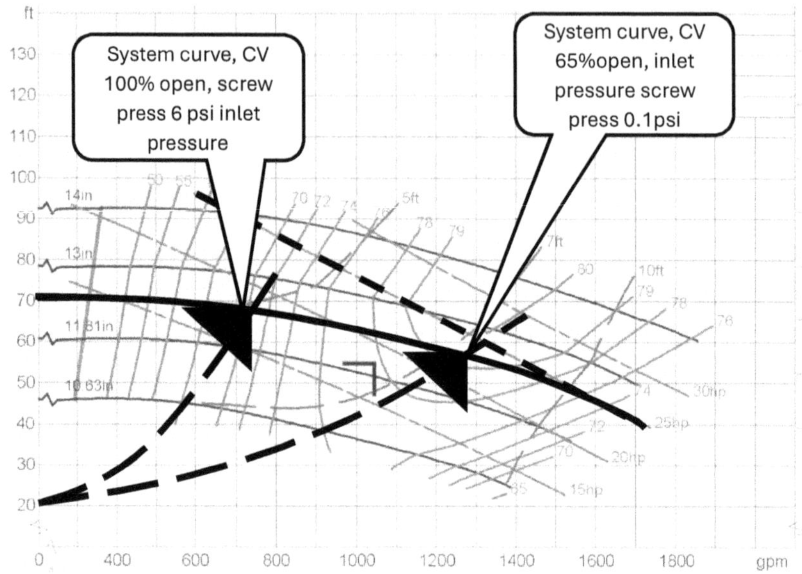

FIGURE 6.28 Original five-vane pump and system curves for optimal and problem condition.

Some data was taken on the pump to assess its hydraulic efficiency. There was a flowmeter in the process line so an actual process flow from the pump was available. From this the WHP was calculated at ~14.1 HP. The pump motor current and voltage were measured with an average of 29.6 amps and 468 volts. Since the pump motor was operating at nearly FLA of 32.5, estimates from motor data on efficiency 92.4% and power factor 0.78 were used to calculate actual BHP at ~23.2. Now knowing WHP and BHP, the actual efficiency was calculated at ~61%. The pump curve efficiency at 700 gpm was ~64%. So only 3% efficiency had been lost on pump's original condition. A 3% pump efficiency loss would only create about a 100 gpm and 3 ft of head difference in performance so the pump could not create the impact that the process was showing.

The root cause of the system issue ended up being the screw press was worn and partially plugged. The press was not able to process and dewater the stock at higher production rates, so the inlet pressure backed up when the control valve opened up. After the screw press was rebuilt and able to dewater the stock efficiently (new wipes, reset screw clearances), the screw press returned to normal operation as did the pump system. The normal condition was the control valve was now at around 60% open and the inlet pressure to the screw press was 0.1 psi while getting 750 gpm compared to before where 100% control valve only produced just over 700 gpm at 6 psi screw press inlet pressure.

Another system observation on this stock system at times was the effect of stock consistency on pump and system production. With all other variables held constant, the system curve effect between 6% and 3% consistency is about 600 gpm impact as shown in Figure 6.29. There had been time where the consistency has been higher than 6% even up to 6.5–6.8% consistency. This further hurts the ability of the pump to perform at higher production rates.

Another opportunity in the screw press pump system was a pump impeller change. A larger impeller wasn't much of an option as the pump had a 25 HP, 1,200 rpm motor so a larger impeller would not allow full curve runout. However, there was a different impeller and pump curve performance option that gave a little more benefit for this system.

The original five-vane pump curve shows that the head rise from 1,150 gpm to 550 gpm of 10 ft as shown in Figure 6.30. The three-vane head rise of 10 ft had a change from 1,100 gpm to 700 gpm as shown in Figure 6.31. Therefore, comparing the curve slope over 10 ft of system head change, the five-vane pump flow performance dropped by ~650 gpm, while the three-vane pump curve slope over 10 ft system head changed only dropped the pump flow performance by ~400 gpm. This means that the three-vane impeller yields another 250 gpm of pump system performance (flow) over the same system head change compared to the original five-vane pump curve.

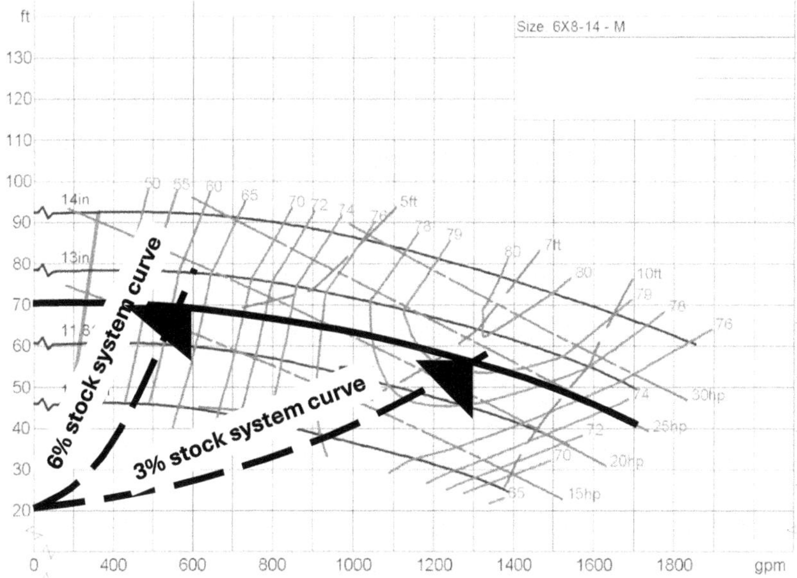

FIGURE 6.29 System curves for 6% and 4% consistency stock.

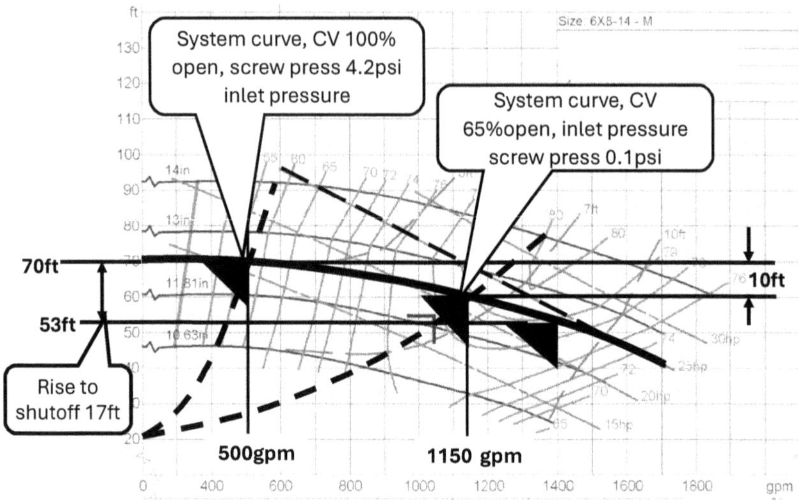

FIGURE 6.30 Five-vane impeller pump curve.

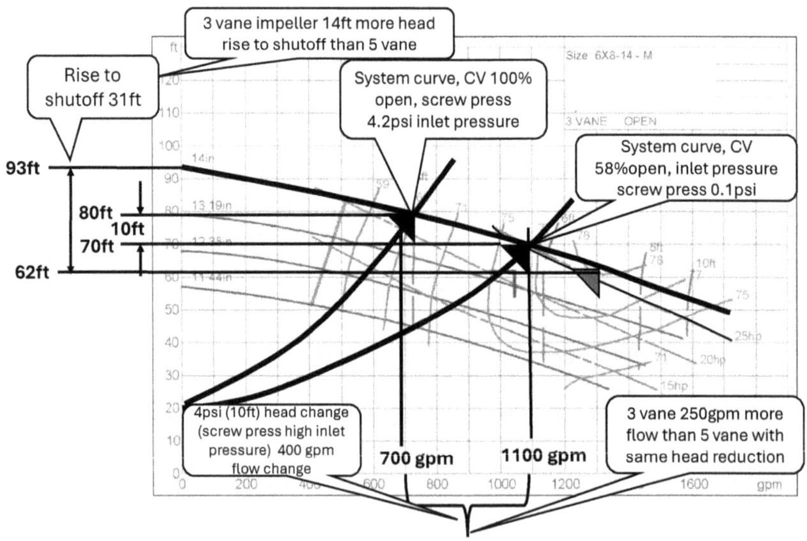

FIGURE 6.31 Three-vane impeller pump curve.

This pump system issue has many causes such as design, operations, and maintenance. The design element is using the three-vane impeller that provided an improved flow performance at the same head condition. The operations element is controlling the consistency in the system so that it doesn't go too high and limiting the pump output. The maintenance element is keeping the screw press maintained in good condition, so high inlet pressure does not result which pushes the system curve back up the pump curve-cutting capacity. The main root cause in this issue was a system problem, where a component in the pump system began producing too much pressure drop.

Case Study 19 Learning – Evaluate each variable friction head component when pumping issues arise. Also consider curve slope effects on pump capacity and process margin.

A similar situation arises when issues involve the pump or pump system and the motor itself. Is the motor faulty and causing the problem, or is something in the pump or pump system causing the issues? I've heard some say when the motor can't drive a load, it is a weak motor, and I've seen some change the motor only to find out that it wasn't the issue. This next case study will look at an issue with a motor in the pump system and how this was figured out.

CASE STUDY 20 PROCESS-INDUCED PUMP MOTOR OVERLOAD

A medium consistency (6–15% consistency) pump system had some intermittent motor overload issues. The pump system consisted of stock at 10–12% consistency feeding from a standpipe into the medium consistency pump that had a level control valve as shown in Figure 6.32. The pump was a 4×8×24 end suction medium consistency pump with a suction inducer with the impeller. The original design point was 400 tons per day at 12% consistency which had the pump with a design point of 555 gpm at 425 ft. The pump had a 250 HP, 1,800 fixed-speed motor as the driver.

Due to other factors, the process became unstable, and the stock consistency began to drop. The standpipe level began to increase, and the level control valve went from a normal position of 40% open to as high as 100% open at times. During these types of events, much of the time the motor would trip out on overload. Initially, this was not understood by the process operating personnel as to the root cause of motor overload. Most process pumps they operated did not trip out at all.

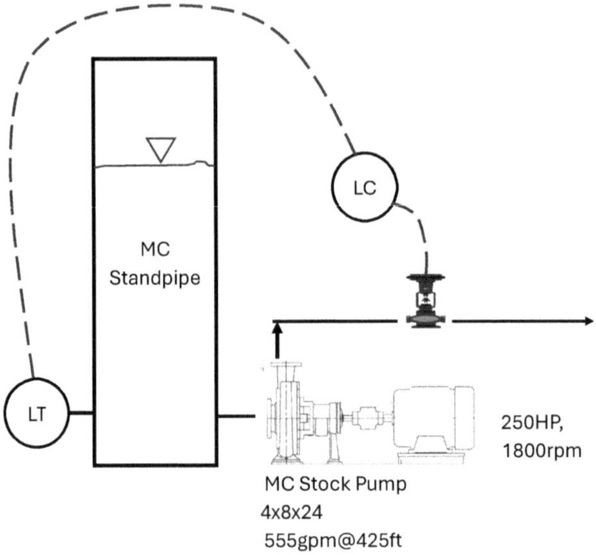

FIGURE 6.32 Process diagram for MC pump system.

The original design point of 555 gpm at 425 ft at 12% consistency stock would see a motor load ~165 HP, which is far from the driver motor HP of 250. During one of the upset conditions where the pump did not trip out, the motor load was checked and was ~223 HP. With the production rate constant from design, this would equate to a consistency of 6.7%. The same process calculations for fully loading the 250 HP motor would be a flow of 1,200 gpm with the consistency at 5.5% as shown in Figure 6.33. At 4% consistency, the flow increases to ~1,600 gpm, requiring 300 HP. The consistency was observed to be in the 4–5% range during some of the motor overload events.

With lower consistency, the flow must increase with the production rate (tons/day) constant. When the flow control valve opens and the TDH decreases, the system curve moves out on the pump curve and pump BHP increases. Higher consistency stock produces higher friction head than low consistency stock at the same flow so the system curve will be further to left on the pump curve.

It is also common for medium consistency pumps to not have full curve runout with the driver (motor) BHP. The medium consistency pump is not supposed to be pumping low consistency stock so in theory it should never be running out toward the end of the curve. The motor would be so large, mostly idle 99% of the time, and would not be feasible to design it to handle the full curve runout. In this particular pump, a 350 HP motor would be required for full curve runout. A lightly loaded motor is not desired, as efficiency and power factor both are reduced.

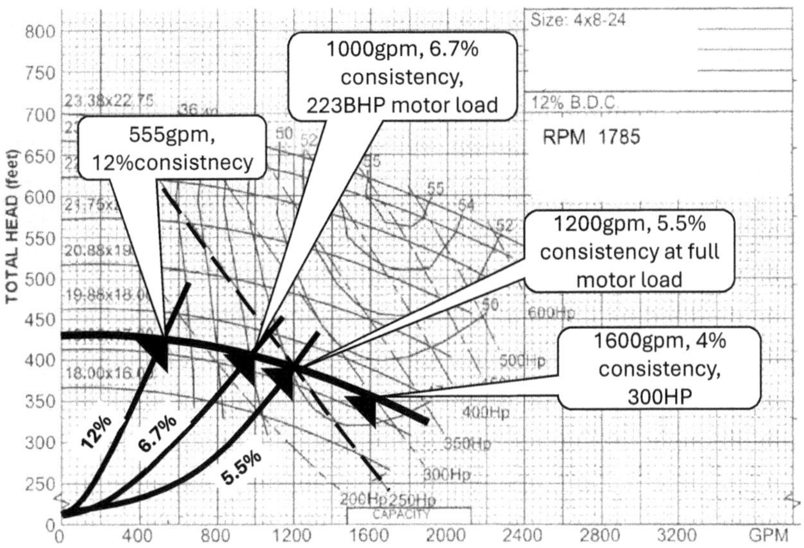

FIGURE 6.33 Pump and system curves at various stock consistency.

The root cause of pumping load issues in the case is operations by not controlling the stock consistency. In a more general pump application, not having full curve motor runout might be considered a design issue but not so much in a medium consistency process due to the very large motors for full curve runout. This was also a system issue and not a pump issue.

Case Study 20 Learning – Paper stock consistency can greatly affect not only the process flowrate but also the system curve friction losses that if there isn't enough drive HP to full curve runout, overload may result.

In the marriage of pump and system, any changes to either will produce a different pump performance and hence system reliability. Therefore, will installing a new pump or direct replacement pump (same dimensions, same design) going to affect pump system performance? This next case study will reveal a potential result if the pump user is not careful.

CASE STUDY 21 SAME PUMP, DIFFERENT CURVES?

A 6″ suction lift pump system was in need of a total pump replacement. The original pump had a 11.5″ diameter two-vane open impeller. This was a more expensive replacement than the typical rotating assembly replacement, as it also required a new casing. Alternatives were investigated by the area maintenance team including other pump manufacturers. An alternative pump from a different manufacturer was found to fit into the same footprint as the original. It was one of these spin-off companies that copied another pump manufacturers pump to capture market share. When the proposals for the original lift pump (pump A) and the alternative pump (pump B) were obtained, the alternative pump B was much cheaper than the original pump. Pump B also had the same mechanical dimensions as pump A and same performance as sales representative told it. It appeared to be an easy substitution with cost savings.

The original pump A already had reliability issues and was operating well outside of the AOR. The pump was operating at 1,800 rpm and only had a system head of ~35 ft which left the pump operating off the end of the pump curve. The original pump had a MTBF of about nine months in its current condition, operating outside the AOR. While operating, pump B had a MTBF of about 50% compared to the original pump A.

Looking at the pump performance curves further revealed an interesting difference in the two pumps that were direct replacements. Pump B had a different hydraulic performance curve than pump A. The two pump curves are shown in Figure 6.34, where the differences are easily seen. Pump A had a low flow AOR limit of 325 gpm at 110 ft of head, while pump B had a low flow

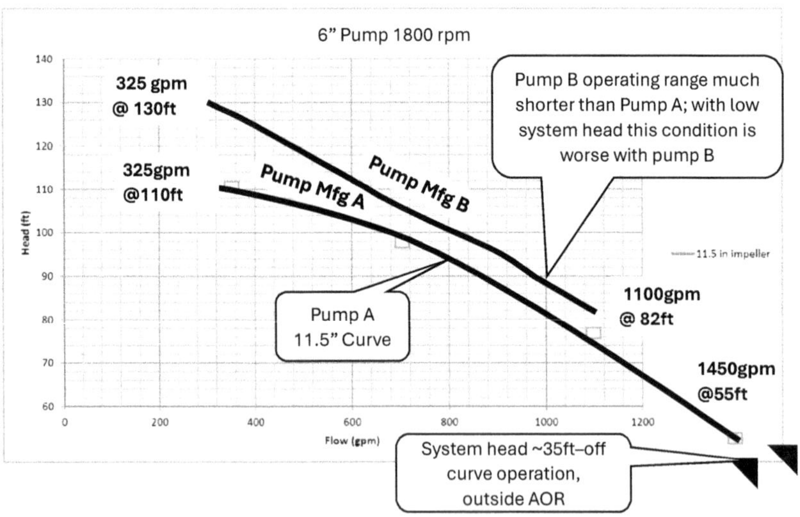

FIGURE 6.34 Pumps A and B performance curves.

AOR limit of 325 gpm at 130 ft. This was not a big issue since the pump did not operate in this part of the system anyway. The issue with the pump design operating point was that it was beyond the AOR, high flow limit. Pump A had a high flow AOR limit of 1,450 gpm at 55 ft, while pump B ran out of AOR at 1,100 gpm at 82 ft. While pump A operated outside its AOR at a system head of 35 ft, pump B was much further outside its AOR at the same head.

Another issue was discovered after a few short runs of pump B that a hydraulic error had been made when the replacement pump was purchased. Pump B was purchased with a maximum size impeller of 12.375″ which further made the operating condition worse since it operated outside the AOR and didn't need any extra head. The hydraulic performance and reliability impact was unknown to the area team that sought to replace a pump with a cheaper replacement that fit perfectly into place. Initially, pump B replacement was considered a success but over the next 1.5 years very poor pump reliability resulted. Later pump A was purchased and eventually replaced the worst-performing pump B.

The root cause of this pump issue is a design problem having two different pumps. Although the footprint is the same, consideration to the hydraulic performance being different was not evaluated. Another factor was that maintenance decided to substitute the pump to save initial cost, without performing a good evaluation of the substitute pump. This ended up costing more money in the long run.

Case Study 21 Learning – Always verify pump performance curves whenever any pump change is made, substituted, or pump is replaced with a different pump even if the pump physically is a perfect fit.

PUMP PERFORMANCE STANDARDS

There are several performance standards by the Hydraulic Institute on pump performance for pump manufactured when compared to their performance curves. They will range from 1B, 2B, or 3B. Different industries adhere to tighter tolerances on performance. For example, 3B for general industry pumps will be ±9% on flow and ±7% on head. The oil and gas industry may require a 1B, which is much tighter at ±5% on flow and ±3% on head. Pulp and paper have a default acceptance performance grade of 2B, which is ±8% on flow and ±5% on head.

Pump manufacturers may also apply different performance standards depending on the type of pump. For example, a large process pump or ANSI end suction pump may be manufactured with a 1B standard, while an axial flow or slurry pump may only be at 2B standard.

If there is a critical process where tight adherence to performance is needed, then a special request or performance test may be desired at the time of pump purchase.

FIGURE 6.35 Pump casing casting defect.

Aside from performance standards which inevitably will come from manufacturing tolerances that when assembled will result in performance error. A variety of quality or dimensional errors can occur in pump components, particularly in cast items, as shown in Figure 6.35. Here the pump casing had a significant flaw in the suction nozzle where the suction nozzle was not true on the casing. This flaw would not only likely affect the pump vibration from excessive pipe strain (potential leaks too) but also may affect the flow into the impeller eye and certainly decrease the pump efficiency and hydraulic performance.

The lesson here is to exercise caution when using a cheaper pump or a substitute part that is supposed to be the same part as the original pump manufacturer. Lower performance and poor quality can erase any purchase price savings, and this cost will be on an annual basis not just one time like initial purchase price. Good pump manufacturers will also sample test thousands of pumps each year to ensure their manufacturing process stays within the 1B, 2B, and 3B performance test quality. If 2–3% of efficiency loss between pumps A and B on 300 pumps was noticed, it would be thousands of dollars lost annually in energy loss.

For business models or applications where performance is critical, certified pump performance curves can be generated from most pump suppliers. Normally, only a small percentage of pumps are tested to verify they fall within the acceptable tolerance range. However, if any end user wants to verify that the pump they are getting meets the performance criteria it was selected for then that specific pump purchased can have performance test performed, generating a certified performance curve. This is not something that is really needed or can be justified from a financial standpoint for most business models or applications, but can be useful in some situations.

The pump curve shape can affect many different aspects of pump system operation and potential pump system optimizations. This next case study has a process system issue and is in need of a modification. The curve shape and slope offered another solution that wasn't really seen on the initial investigation.

CASE STUDY 22 PUMP CURVE SLOPE PUMP SYSTEM ADVANTAGES

A stock screen system was experiencing an imbalance in the process which led to the screen feed tank overflowing much of the time. There had been some process changes upstream which added more hydraulic load into the screen feed tank that left the screen and pump system inadequate for the extra process load. The screen equipment was already maxed out from taking any more hydraulic load, so increasing flow through the screen was not an option.

The screen feed system, shown in Figure 6.36, consisted of a simple feed tank and pump feeding the screen. The pump is a 4×4×12 recessed impeller end suction pump with a design point of 300 gpm at 122 ft. The pump was driven by a 75 HP, 1,800 rpm motor with a variable speed drive. The drive was controlled by flow control from the screen rejects.

The original design of the screen was with a 1.5% consistency, but the process changes had the consistency at 0.5% which further limited the screen capacity. The original screen maximum capacity was 400–500 gpm at 1.5% consistency, but the lower consistency 0.5% limited the capacity back to near 300 gpm. The screen needed to operate with a good feed pressure and manage differential pressure for proper rejects control.

Most pump curves have reduced head as flow is added to the pump or as the head drops from the system then the flow increases. Initially, the solution to the feed tank overflowing was to add another pump to the tank to pump out the

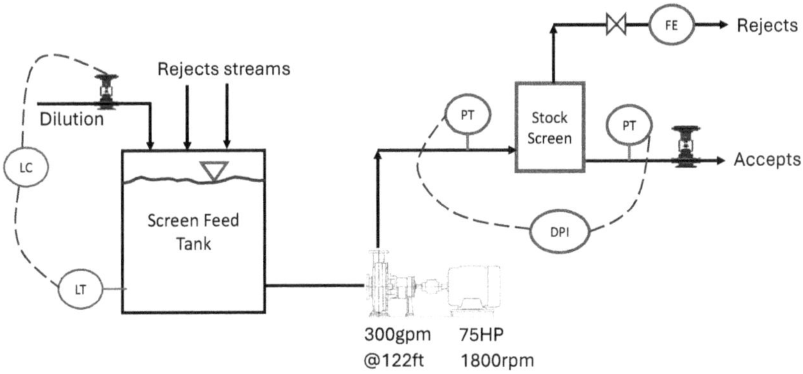

FIGURE 6.36 Process diagram on screen feed pump system.

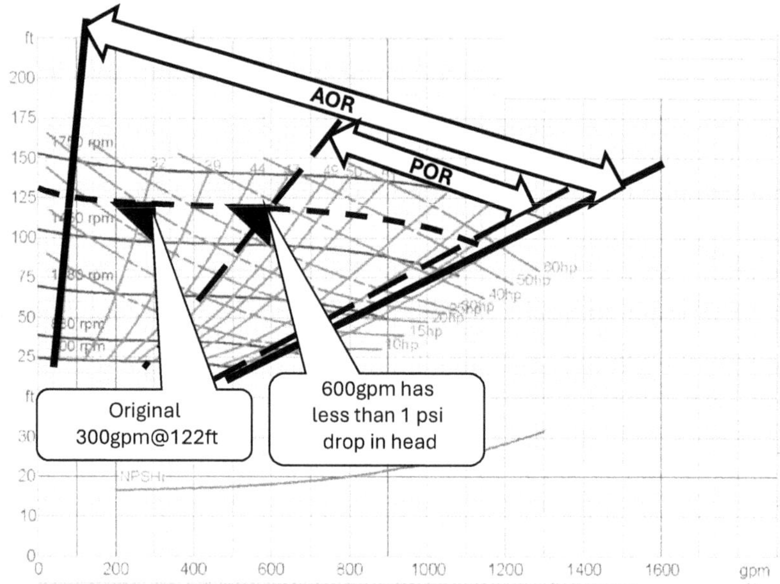

FIGURE 6.37 Pump performance curve with operating points.

extra process fluid to another tank. Losing feed pressure from the existing pump by adding hydraulic load was initially not desired; however, looking at the pump curve, in Figure 6.37, changed the system modification scope. The pump curve was very flat and increasing the flow by an additional 300 gpm resulted in less than 1 psi change in pump feed pressure to the screen.

The system modification scope now was to add a line and control valve to the existing pump so more process fluid can be sent somewhere else and prevent the feed tank from overflowing. Since there was an existing tank flow control valve that added makeup dilution water to the tank, the new control valve coming off the feed pump would also be a level control valve. The control would be a split range control where valve A would control in tank level when in the range of 0–50%, while valve B would control when the level is over 60%. This is shown in the modified process diagram in Figure 6.38.

The solution to this pumping capacity issue is a design problem with the pump system controls rather than the pump itself. The flat pump curve and current operating point presented some unique opportunities to resolve this process issue. Until a thorough evaluation of the pump hydraulics was completed, the best solution was not clear, so it is always good to have the right resources (pump system expertise) to evaluate pump system solutions.

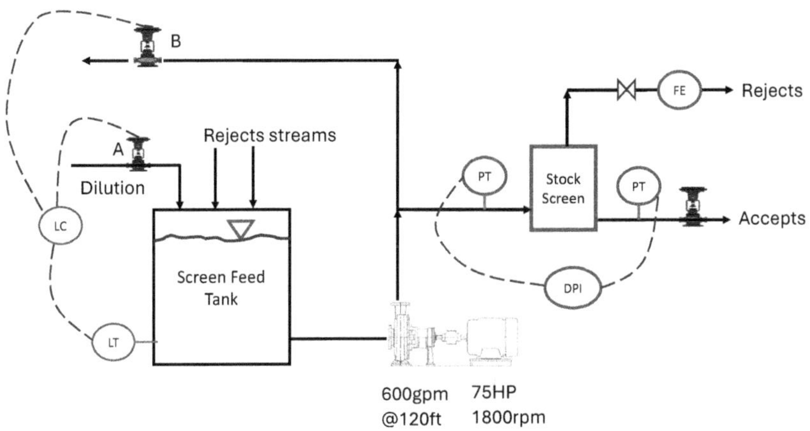

FIGURE 6.38 Modification to process control to manage tank level.

Case Study 22 Learning – Evaluate pump curve slope in response to the system to help solve some pump system issues.

PUMP CONTROL METHODS

There are four main pump system control schemes used in most pump and process systems – on/off (batch) control, inline throttle control, bypass/recirculation control, and variable speed control. Only a small number of systems will allow for on/off control where a tank level setpoint will typically kick the pump on and off as tank rises and falls. This type of on/off control is common in sump pump systems. There are also some systems that are on/off control where the pump is keeping a constant header pressure charge like that in a fire suppression system. A surge accumulator tank is typically employed in this type of pump on/off header pressure control setup.

Inline throttle control and bypass control are still used a lot in most process systems, but due to energy conservation efforts, adjustable speed drives are being used more frequently. Also, the advancements in adjustable speed drives, particularly for AC power, are much better and economically feasible than they were 40–50 years ago.

Probably the most common control method is a throttled valve arrangement. A control valve is installed inline in the system to regulate pressure, flow, or level within the process system. As with any throttling control valve, there is a corresponding pressure drop associated with the percentage of valve opening. This can be calculated/estimated by the following equation.

$$\Delta P = Q^2 SG/Cv^2$$

where:

ΔP = control valve differential pressure, psi
Q = flow, gpm
SG = fluid specific gravity
Cv = control valve flow coefficient

Typical pressure drops on control valves in pump systems is in the 3–7 psi range. Target-preferred control valve operating positions are 40–60% for best reliability and energy management but an allowable range of 30–70% is typical. Choose a control valve Cv with a greater operating travel range (% open) across the flow range desired as this makes the flow less sensitive to valve position and easier to tune. Best practice is to have 20 pipe diameters straight run before a control valve and 6 pipe diameters after. Even control valves can be affected by turbulence and non-linear flow.

In simple systems with a pump and single throttled control valve on the discharge, a straightforward improvement that can have a big impact not only on reliability but also pump energy usage is resizing the pump or control valve for the system. The pump modification is a much easier change than the control valve. If the pump was oversized for the system from a pressure (head) standpoint and the control valve was operating very throttled (closed less than 30%), there would be a need and opportunity to trim the impeller down to reduce the friction head in the system. This would also allow the control valve to operate in a more open position taking less pressure loss. There are just a few steps to execute this simple improvement.

1. Calculate the control valve ΔP (pressure drop) at the target flowrate from the control valve equation with the Cv at the CURRENT percentage open that the control valve is operating. This can be attained from the control valve manufacturer in a table or curve.
2. Calculate the control valve ΔP at the target flowrate from the control valve equation with the Cv at the NEW percentage open.
3. Take the difference in the CURRENT ΔP and NEW ΔP. Convert to feet of head.
4. Go to the pump curve at the target flow (or current flow) and locate the point on the curve. Subtract the head difference from step 3 at the same flow. Read what size impeller diameter is required. This will result in the control valve operating at the percentage open after the impeller is trimmed to the lower diameter. The head drop is basically transferred from the valve to the pump TDH.

What does this look like on an actual pump system? Control valve systems can be simple or complex. A simple system is one in which there is a single control valve and loop in the pump system. This was described above. A complex control system is one where there are multiple control valves or loops in many different piped destinations and control configurations fed by one pump. The next case study will show an example of a simple pump system.

CASE STUDY 23 OPTIMIZING A SIMPLE PUMP SYSTEM WITH CONTROL VALVES

A simple pump stock transfer system from a feed tank to a mix tank was originally designed with a $6 \times 8 \times 14$ end suction pump equipped with a 40 HP, 1,200 rpm fixed-speed motor. The original design point was 2,100 gpm at 50 ft. The transfer pump also had a flow control valve and flowmeter loop to control transfer to the mix tank as shown in Figure 6.39. The control valve was an 8″ V-ball with a magnetic tube flowmeter in the system. The stock system was on recycle fiber so there was considerable trash and contamination in the paper stock.

The pump system had several different wear failure modes such as the control valve and discharge piping wearing through, the pump suction nozzle wearing, pump impeller and wear ring wear, and pump casing wearing. The pump did not operate at the original design point but was operating between 760 and 1,000 gpm most of the time, which was not very close to the original design point as shown in Figure 6.40.

The current operating point was very close to the allowable operating range (AOR) low-flow end. This operating point contributed to the suction recirculation on the pump suction nozzle, which also increased the pump wear in those areas. With abrasives in the stock and low flow through the pump, the abrasives would spend more time going around the inlet part of the pump accelerating wear.

The 8″ flow control valve typically ran anywhere from 38% open to 20% open which is below the ideal control valve operating range. While there was

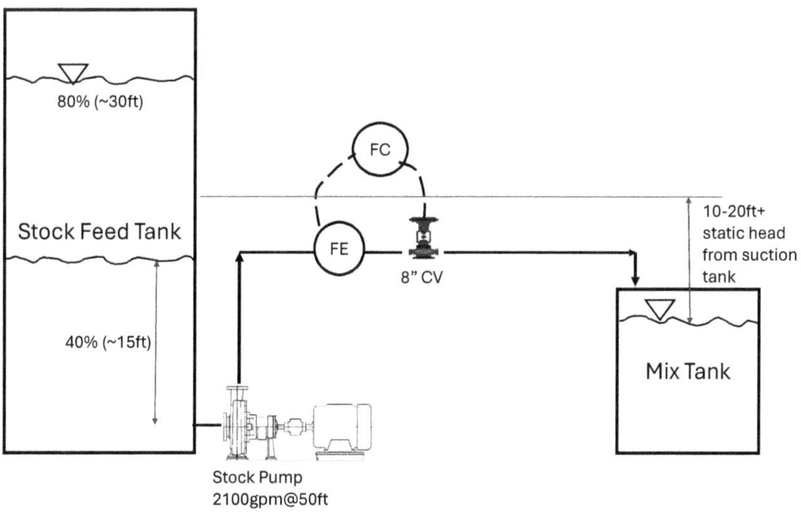

FIGURE 6.39 Process flow diagram for stock transfer pump system.

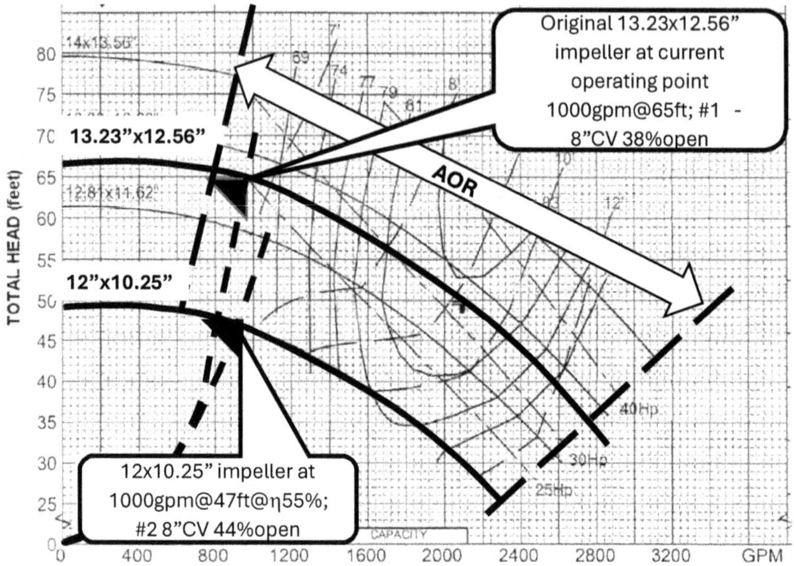

FIGURE 6.40 Pump performance curve with system curve and operating point.

nothing that could be done with the size of the pump from a flow capacity versus the process demand standpoint, there was some improvement that could be done with the pump system control. Since there is excess pump at the current impeller size design, the impeller can be trimmed down with the goal of getting the control valve in a more optimum operating range of 40–60% open.

At a flow of 1,000 gpm, the control valve is open ~38%, which correlates to a Cv of ~190 from the control valve manufacturer which calculates ~27 psi (62 ft) pressure drop. With the smallest size impeller ($12\times 10.25''$) for this pump, the head is lower by ~18 ft. Therefore, calculate the control valve Cv with a lower pressure drop of 44 ft (18 ft lower head from pump curve between the two impeller diameters at 1,000 gpm). The new Cv would be ~230 which would be a new control valve open at ~44%.

While there was only a small improvement in the control valve operating range, the pump impeller change did move the control valve over the 40% optimal control range. This also improved the control valve wear potential by having a higher open percentage. The pump also experienced some improvement by operating at a lower pressure which reduced the pump energy usage from 28 HP to 21 HP, so a savings of 7 HP which is ~$2,500 per year at 0.0055/Kwhr.

Note a slight improvement in the pump operating point, which means a little less suction recirculation as it is closer to BEP and further into the allowable operating range for the pump. Typically, this also leads to a decrease in pump efficiency due to large impeller cutwater gap when a very small impeller is used compared to a larger impeller.

This is a good example of showing that if a pump is oversized for a system and operating point that just trimming the impeller size will not change the fact that it is still oversized. There is no real way to efficiently solve an oversized pump outside of a new pump. Finding more process demand usage is about the only thing to make an oversized pump better fit a system. A recirculation loop or new pump would solve it, but both would be costly.

This pump system also shows how a low or negative static head can affect the pump system. The system curve starts out below the head curve in this case where the suction tank level is higher than the discharge static head. It shows that the static head difference by itself will produce about 350 gpm without the pump even running. Remember the pump is the total dynamic head that adds to the system. This was a high friction head pump system.

A VFD with the reduced impeller size was also evaluated to improve the pump operating point. The VFD would run ~700 rpm and pump motor load ~11 HP (the original was 28 HP). While this was not executed at the time, it was later installed to realize the full potential to make the pump system more efficient and eliminate the throttled control valve erosion. The control valve had to be left in the system as a shut-off valve since the static head was high enough to produce flow with the pump off. The control valve had to shut when the pump was down due to the high static head.

Basic pump and system design was the root cause of pump reliability issues in this simple system. The system reliability was improved by trimming the impeller, which not only enhanced pump reliability but also system and control valve reliability.

Case Study 23 Learning – Always match pump sizing with control valve operating range in pump systems.

A complex pump system from a controls standpoint is one that has more than one destination or multiple control devices in the system but is fed from one pump. Since the centrifugal pump performance is not linear and pump performance is directly proportional to system head, the control valves and pumps must be optimized together especially in complex systems. Optimizing complex pump systems (those with multiple control valves) can involve improving the pump itself or the control components, and many times may require both. It really depends on the details of the system design and what issues exist.

The end goal for optimizing the pump and system for complex control systems is that the pump and control valves both be operating within their respective AOR and, if possible, POR. The AOR portion should be a goal that can and should be achieved. The POR may not always be met for every system design. There are now pump system software products available to model complex systems which once modeled can be an efficient way to optimize complex systems. If these tools are not available, this can still be done in a manual type of fashion. Keep in mind that the process flow for each destination control point is not changing but only the control valve operating positions and pressures. A few key steps can be followed when evaluating a complex system.

1. Evaluate the pump and control valves at each operating condition and determine their operating points at both high and low limits to establish their operating range. Many complex systems will have a range of operating points due to the grades or process conditions within the plant. This will need to be known to determine where each component is operating which will set the path forward for optimization. This will be the flow and head operating point ranges for the pump and control valve percentage open range.
2. Decide on what components need changing – the pump, which control valves or both. Which components are operating outside the AOR part most of the time. Unless there is a bypass control on the pump, if the pump is oversized (from a volume standpoint) there may not be much that can be optimized. Analyzing an existing system and optimizing does not change the process demand itself. We are only modifying the pump and control systems to match each other, and the process demands as they are currently defined. If the control valves are out of range, then evaluate each control valve and note if the valve is too small or too large for the application.
3. If all control valves were operating too closed ($< 30\%$ open) then the pump impeller could be turned down to a smaller diameter and allow all the control valves to operate at a higher percentage open. If the system had pressure control of some sort, then the header pressure setpoint could be lowered to accomplish the same effect. Using the control valve Cv for the current percentage open and target percentage open at the same process flowrate (which is unchanged), calculate the difference in the pressure drop. Subtract the pressure head difference from the pump head at the current impeller diameter on the curve. The new pump head on the curve will show what size impeller diameter is needed. Also, check the other control valves with the new lower head and see how much they change in percentage open. If all these conditions are satisfied, then an energy savings can easily be calculated for the new impeller size and lower system head that has been achieved.
4. If one of the control valves is operating too open ($> 70\%$ open) and the rest are within the AOR, then only that control valve station may need to be resized. Therefore, using the control valve performance tables, use the current control valve flow coefficient (Cv) and go to the smaller valve to match that flow coefficient until the percentage open is within the AOR. If the control valve was operating too close, the same process would be done except going up in valve size to get the valve to open up into the AOR using the flow coefficient in the performance table.

For complex pump systems with multiple control valves, it is important to remember that all the other friction loss components are fixed. It is the control valves that are variable friction losses that can be modified. The next case study will walk through a complex pump system and its optimization.

CASE STUDY 24 OPTIMIZING COMPLEX PUMP SYSTEM WITH CONTROL VALVES

Many times, it is not advisable to have a VFD on complex pump systems especially with header pressure control because the header pressure can seldom be reduced due to some individual control valves being out of range. In this example, a VFD was used on the original design. The current operation: however, left the VFD operating at 100% speed to try and meet the process demand of all the operating points.

The pump system was a part of dilution system used for pulp cleaning as shown in Figure 6.41. Some of the destination points had flow control valve loops, while others had manual valve flow control. Most of the process destination points had plenty of control valve operating range left; however, there were three that had the control valves 75–100% open (noted and circled in Figure 6.41). The pump was an 8 × 10 × 15 end suction pump with a design point of 2,000 gpm at 154 ft. The pump had a 125 HP, 1,800 rpm motor with a VFD.

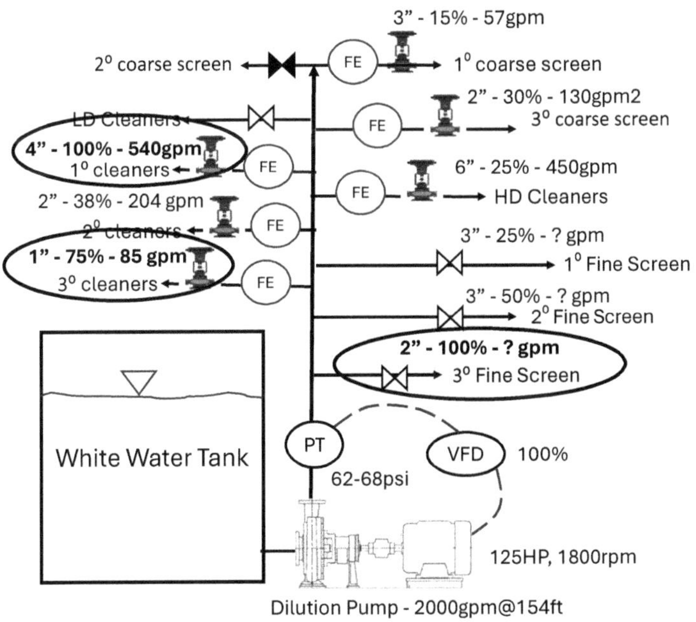

FIGURE 6.41 Process flow diagram dilution pump system.

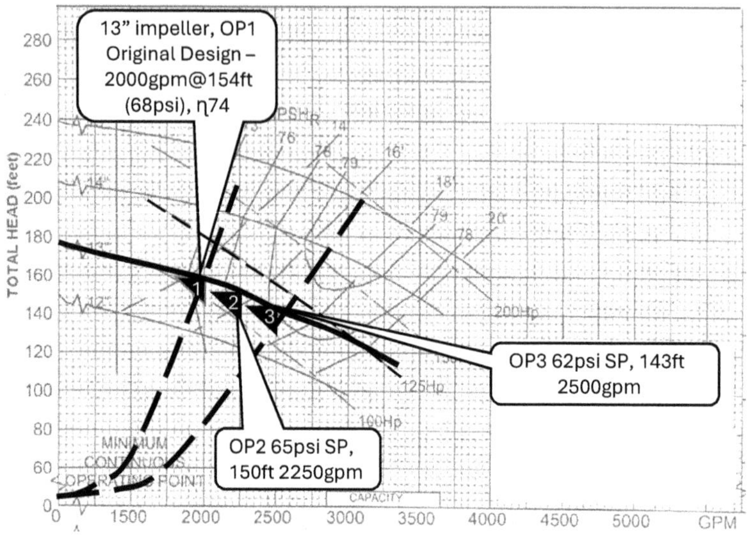

FIGURE 6.42 Original pump curve, system curve, and operating points.

The pump currently did not have enough capacity to feed enough water to three of the process destinations in the current system design. While the pump had a VFD, it was already at 100% speed. The pump operated between 62 and 68 psi header pressure. The static head relative to the pressure transmitter in the system was basically zero, so the pump TDH was 143–154 ft which correlated to a flow of 2,000–2,500 gpm. The original pump curve at 100% speed with operating points is shown in Figure 6.42.

With the 125 HP motor not leaving much room for modifying the current pump, existing pump modifications are limited. Over-speeding the pump or installing a slightly larger impeller (13.125″) might help a little but it will not be enough to give the entire system proper turndown and operating range required for reliable operation. Affinity laws quickly show only about a 3 ft increase in head and 3.5 HP. Any increase, however, would not give full curve runout with the existing 125 HP motor.

So, what are the remaining options to correct the imbalance in the pump system? There were three options considered.

1. Install a larger pump for the current system.
2. Upgrade the existing pump hydraulic capacity.
3. Upgrade existing controls on the three maxed-out systems.

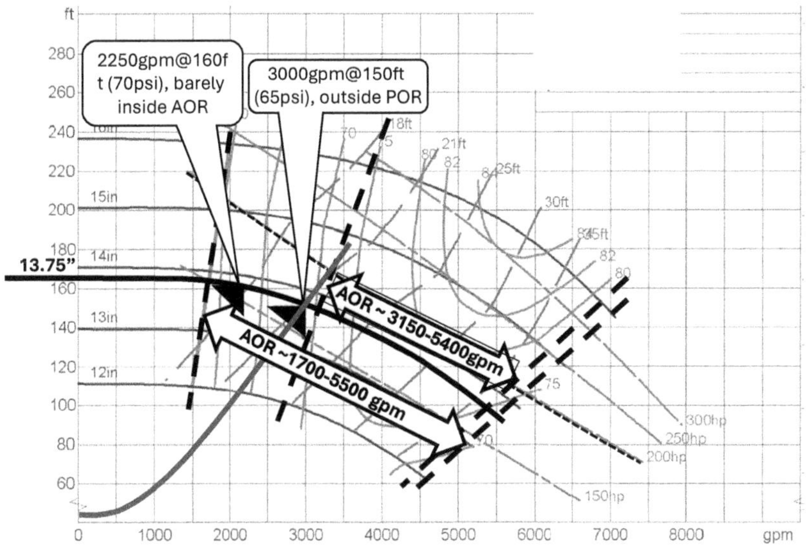

FIGURE 6.43 Option 1, larger pump curve.

For option 1 (larger pump), the existing system would require a pump to supply not only the flow required but a higher header pressure so that the three destination control valves can control and distribute the necessary flow from the new, larger pump. The new pump specified was an $8 \times 10 \times 16$ end suction pump with a 13.75″ impeller. At 3,000 gpm at 150 ft (65 psi header pressure), the pump operates just outside the preferred operating range, as shown in Figure 6.43. An upgrade to a 200 HP, 1,800 rpm motor with a new VFD is also required. Additional modifications include the pump motor baseplate and piping modifications, which would also be necessary, so this was a costly fix.

The second option was to install a 13.5″ impeller on the existing pump with a motor upgrade to 150 HP, 1,800 rpm. It would also require a larger VFD. Since the pump is not changing, there would be no piping modifications required. The existing motor baseplate was large enough to install a 150 HP motor frame with slight modifications to the foot attachment locations. The modified pump curve is shown in Figure 6.44. This upgrade would increase the total flow potential by 500 gpm at the same header pressure. While this might be an improvement over the existing condition, the problem control valves would likely still not be anywhere near a good control range.

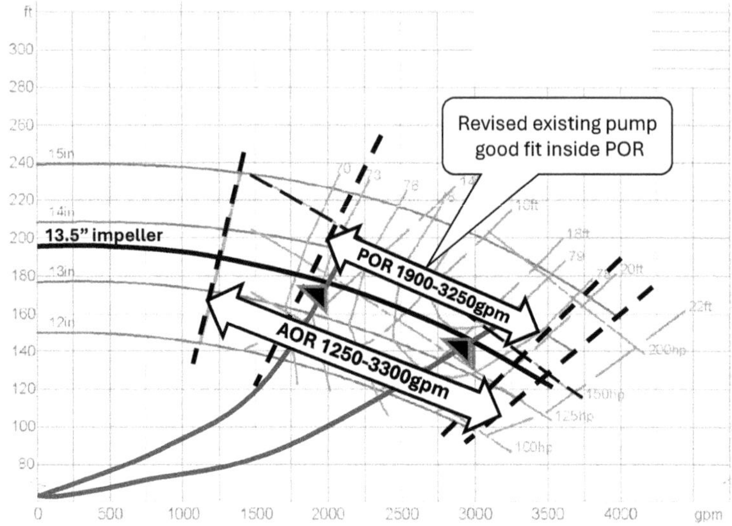

FIGURE 6.44 Option 2, modified pump curve with larger 13.5″ impeller.

The third option includes modifying the three destination points where the control valves are maxed out so that an upsized control valve station will operate in the optimal control range while providing necessary flow to the destination process points. The first area to evaluate was the primary cleaners. The existing control valve station had a 4″ control valve and flowmeter which operated at 100% open. The system modifications here included upgrading the control valve, flowmeter, and related piping to 6″ size as shown in Figure 6.45.

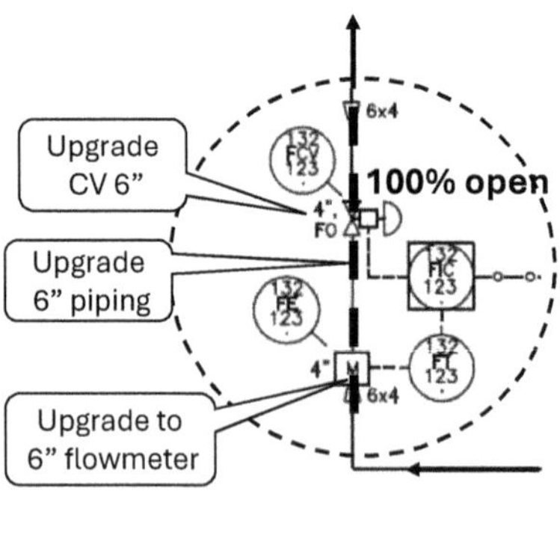

FIGURE 6.45 Primary cleaners control system upgrade.

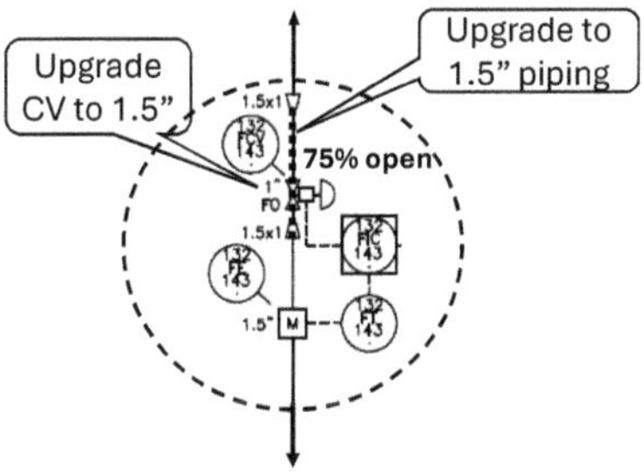

FIGURE 6.46 Tertiary cleaners control system upgrade.

Another system note is that with the existing flowrate range (550–650 gpm) in the 4″ system, the fluid velocities were 15–16.5 ft/s which is very high. For the 6″ system upgrade, the fluid velocities are lowered to 7 ft/s at 650 gpm. So much higher pressure drops, and system pressure would be required without an upgrade to the system to ever get target process flows to the primary cleaners.

The tertiary cleaners was another destination which needed correcting. While the tertiary cleaners were not as bad as the primary cleaners, the control valve was operating at 75% open. Since we are looking at operating at a lower system pressure as more flow is added to the pump, a lower header pressure will be used potentially in the future. So this system likely needed a modification as well so the 1″ control valve and associated piping was to be upgraded to 1.5″ as shown in Figure 6.46. At the current flow and head conditions and using the control valve Cv flow data table, the 1.5″ control valve would be operating around 55% open.

The tertiary fine screen system was the other destination that was maxed out at 100% open. This system needed a new 3″ piping feed and 3″ valve upgrade, as shown in Figure 6.47. The piping also had to be upsized on that section of the piping system.

Evaluating all three options, a clear picture of the best choice was more obvious, as shown in Figure 6.48. For comparisons, there were only a few different conditions evaluated since the variables of pump speed, header pressure, and control valves offered too many conditions. A flow range of 2,250–3,000 gpm and header pressures of 44 psi and 65 psi were used. Options 2 and 3 with the existing pump (impeller and motor) and system control upgrades gave a 39 HP or $15,000 annual energy savings compared to no energy savings for option 1 and 3 (larger pump and system controls upgrade). At the higher flow condition, the larger pump option costs an additional $9,000. It should also be noted that the project scope and cost were higher on the new, larger pump option and required a larger outage for installation.

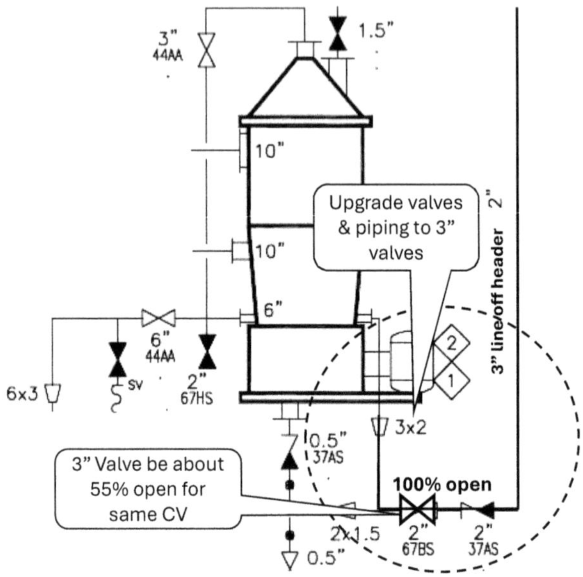

FIGURE 6.47 Tertiary fine screen system upgrade.

Condition	Flow, gpm	Head, ft	Pump Eff, %	BHP	Header Pressure, psi	Pump Speed, rpm (%)	Annual Energy Cost,$
Existing Pump & CV	2250	150	76	112 (122A, 477V)	65	1780 (100%)	40,265
2&3 – Larger impeller & CV	2250	100	78	73	44	1400 (78%)	26,237
2&3 – Larger impeller & CV	3000	150	79	144	65	1780 (100%)	51,756
3&1 Larger pump & CV	2250	100	50	114	44	1400 (78%)	40,974
3&1 Larger pump & CV	3000	150	68	167	65	1800 (100%)	60,000

15K energy savings

FIGURE 6.48 Summary of pump and system options considered.

The root cause of pump system issues was system design with the pump matching up with all the system end point controls. Specifically, there were several end points where the controls were not adequately sized for the need. There wasn't a large reliability issue with the pump, but the process conditions could not be met with the existing pump and system design, so this was a pump system performance issue.

Case Study 24 Learning – Evaluate the pump and system components to optimize both together for ideal operating range of the pump and control elements.

Another pump system parameter that can affect pump performance and reliability is static head. Especially when the static head has a large operating range. This can occur when there are large tanks involved, and tank levels vary in the operation. Steady state systems where the static head remains constant are much easier to control. The next case study will show where the variable static head can have a significant impact to pump operation.

CASE STUDY 25 STATIC HEAD EFFECTS ON PUMP SYSTEM PERFORMANCE

A broke stock transfer pump system consisted of a large stock tank, pump, and pressure control valve, as shown in the process diagram in Figure 6.49. The stock tank fed four different destinations with broke stock, and each varied its demand, so the system output flow had a large swing or turndown. The pressure control loop was to create a steady flow load for the pump and maintain header pressure for the destination control valve loops. The pump was a 6×8×22 end suction pump with a 20″ diameter impeller and a design point of 1,800 gpm at 165 ft. The pump had been experiencing poor reliability and short life on the rotating assembly as well as severe pitting damage on the discharge nozzle of casing.

The pump system has several variables which affect pump and system performance. The destination flow control variables, the tank suction head, and the header pressure recirculation control. The pump TDH may be expressed by the system parameters in the following equation.

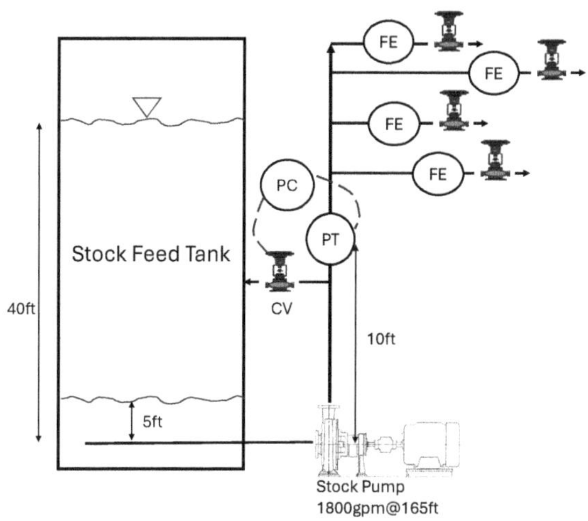

FIGURE 6.49 Process diagram for stock feed pump system.

$$\text{Pump TDH} = \text{PT} + 10 - \text{tank level}$$

The header pressure setpoint is controlled by the operator and can be moved to different positions. Broke process streams vary on the supply and demand side of the operation so the tank levels and header pressure can vary. The stock feed tank was a relatively tall tank at nearly 50 ft with tank levels operating as high as 40 ft and as low as 5 ft. For example, if the header pressure is set at 74 psi and the tank level is 5 feet, then the pump TDH is 175 ft. At 175 ft TDH, the pump flow is ~1,400 gpm.

Another condition was observed when the pump header pressure was operated at 56 psi and the tank level was 40 ft. The resulting pump TDH was ~90 ft and the corresponding flow ~3,200 gpm, as shown in Figure 6.50. This operating point was well outside the AOR and causing flow turbulence on the discharge side of the pump. With the variable demand and header pressure recirculation control, the pump operating point can cover a range, as shown in Figure 6.50.

While there can be a range for the header pressure setpoint, there must be a limit to the header pressure to keep the pump operating within the AOR. While the operator could calculate these manually, it is difficult to rely on repeatable operation so these could be programmed into the DCS controls for the header pressure setpoint. The pump TDH target is 135–175 ft, which keeps the pump

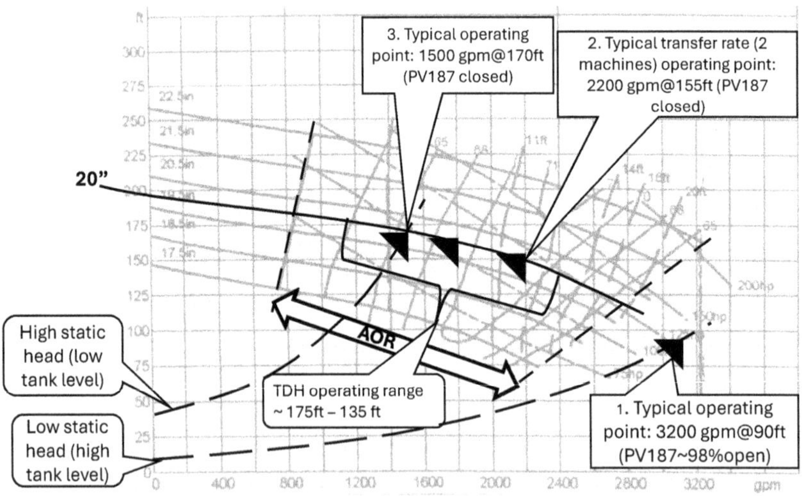

FIGURE 6.50 Pump and system curves with operating points.

in the AOR and near the POR with a flow range of 1,400–2,500 gpm. Therefore, looking at various conditions, the following limits were set up for the header pressure setpoint, depending on the tank level.

$$\text{Pump TDH} = \text{PT} + 10 - \text{Tank Level}$$

$$\text{PT} = \text{Pump TDH} - 10 + \text{Tank Level}$$

Example:

$$\text{PT} = 175 \text{ ft} - 10 + 5 \text{ ft} = 170 \text{ ft (74 psi)}$$

$$\text{PT} = 175 \text{ ft} - 10 + 40 \text{ ft} = 205 \text{ ft (89 psi)}$$

$$\text{PT} = 135 \text{ ft} - 10 + 5 \text{ ft} = 130 \text{ ft (56 psi)}$$

$$\text{PT} = 135 \text{ ft} - 10 + 40 \text{ ft} = 165 \text{ ft (71 psi)}$$

Therefore, for a programmed automatic controls algorithm in the DCS controls, the pressure transmitter limits could be set by the equations below. This would prevent the operator from setting a header pressure setpoint in a position which makes the pump perform outside its AOR.

$$\text{IA Equation: PT minimum setpoint (ft)} = 125 + \text{Tank Level}$$

$$\text{IA Equation: PT maximum setpoint (ft)} = 165 + \text{Tank level}$$

This pump system issue did not have anything in the pump design that would account for large swings in tank static head level differences. Some of this could be improved by some more advanced-level control programming, which could be considered system design. The pump system issues could be considered operational. Although, more knowledge by operations can also accomplish similar results to prevent pump operation outside of AOR. This would be a little more outside an operator's normal skill level to manage a complex pump system.

Case Study 25 Learning – Don't underestimate the effect of tank suction head on pump system performance and reliability.

Bypass or recirculation control also uses a control valve but recirculates the pump flow to control header pressure, process flow from pump, or tank level. The big difference here is the way the energy is wasted. Recirculation control results in greater energy waste, as the higher flow does not perform any real work in the process – just recirculated back to the source tank. In both throttle and bypass

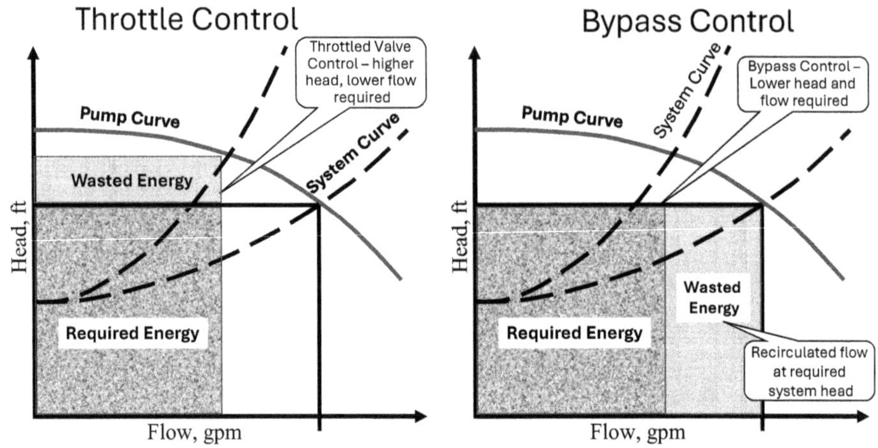

FIGURE 6.51 Pump control methods and system curves – required system energy and wasted energy.

control, the system curve is variable and moves up and down on the fixed pump curve, as shown in Figure 6.51.

The required pump system energy is shown, as along with the wasted pump system energy. The wasted energy for throttle control occurs from elevated pump head to meet a flow requirement. Wasted energy for bypass control occurs from elevated pump flow to meet a head requirement. Bypass control many times is on header pressure control and throttle control many times will be on flow control. Other control designs, such as level control, may also be designed using either control strategy.

The last, most efficient, and generally the most reliable system is the variable speed pump control. Also called an adjustable speed drive (ASD), it is mostly done by an electric motor-driven variable speed drive. While not very common, an ASD may also have a variable speed mechanical drive such as a turbine or variable speed belt drive. ASD, VSD, and even VFD get used interchangeably for the variable pump drive even though VFD is not the same. VFDs are probably 99% of ASD.

With an ASD, the pump curve is variable and moves up and down the system curve, as shown in Figure 6.52. ASD control like other pump system control may be header pressure, flow control, or level control. The system energy required flow is controlled by ASD, while the system head is dependent on the flow from the pump ASD which is directly proportional to the pump speed. ASD control is the optimum energy management control method for maximum pump system efficiency. Pump hydraulic efficiency is also usually maximized but can decrease at very low speeds.

An ASD is not always the solution to every pump problem or system but many times it can be the best solution when it is viable. An ASD can be a good option when the friction head is at least 50% of the total head in a system. If the total static head is most of the system head, then an ASD will still have to run a high speed to overcome total head, and the ASD benefit is minimized. The next case study will take a quick look at evaluating a pump system for VFD application.

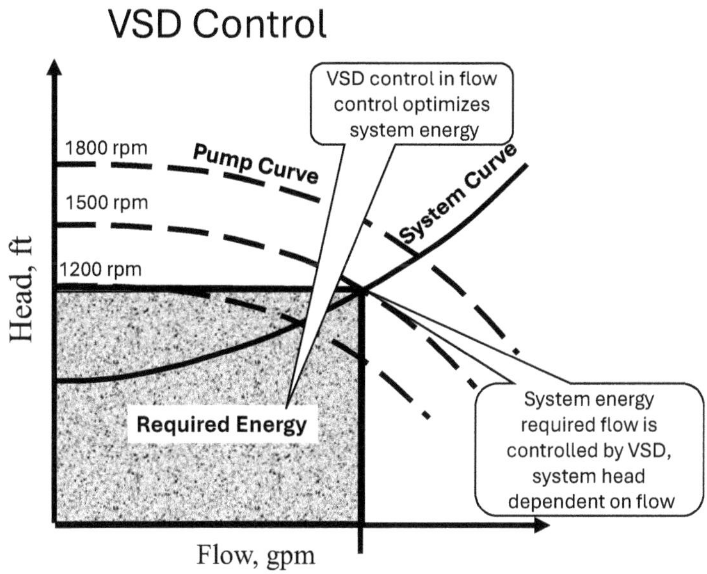

FIGURE 6.52 VSD pump and system curves showing required energy.

CASE STUDY 26 VFD PUMP SYSTEM EVALUATION

A disk filter feed pump system operated with a standard flow control loop and fixed speed pump, as shown in Figure 6.53. The pump is a 16 × 16 × 19 end suction pump with an 18.63″ × 16.19″ diameter impeller and an original design point of 8,830 gpm at 90 ft. The pump was originally designed with a fixed speed of 250 HP, 1,200 rpm motor. The flow control is a 16″ butterfly valve and a magnetic flowmeter. Typical operation had the control valve 47% open with a flow of 7,900 gpm. The range that the control valve operated was 40–55% open. At 40% valve open, the flow was ~6,400 gpm.

The original pump and system curve with the single control valve in the system is shown in Figure 6.54. As the control valve opens, the system friction head is reduced, and the flow increases, as seen with 7,900 gpm and the control valve at 47% open. When the control valve shuts to 40%, the flow was reduced to 6,400 gpm as the system friction head is increased. The 45 ft of total static head can be seen on the system curve, and the corresponding friction head of 55 ft at the 40% control valve position.

The static head in the system was ~45 ft, and the friction head was ~55 feet. The friction head to total head ratio was 55% which is greater than 50% (rule of thumb for VFD feasibility), so this pump system was a potential candidate for a VFD.

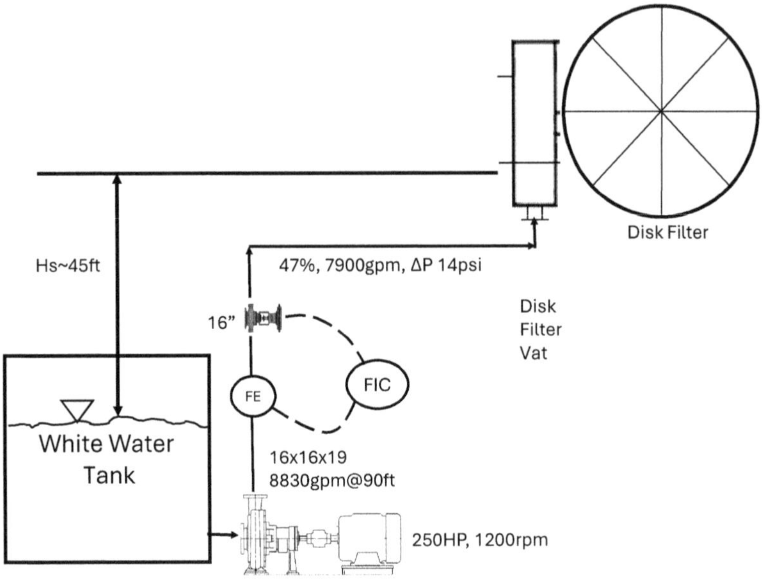

FIGURE 6.53 Process diagram for disk filter feed pump system.

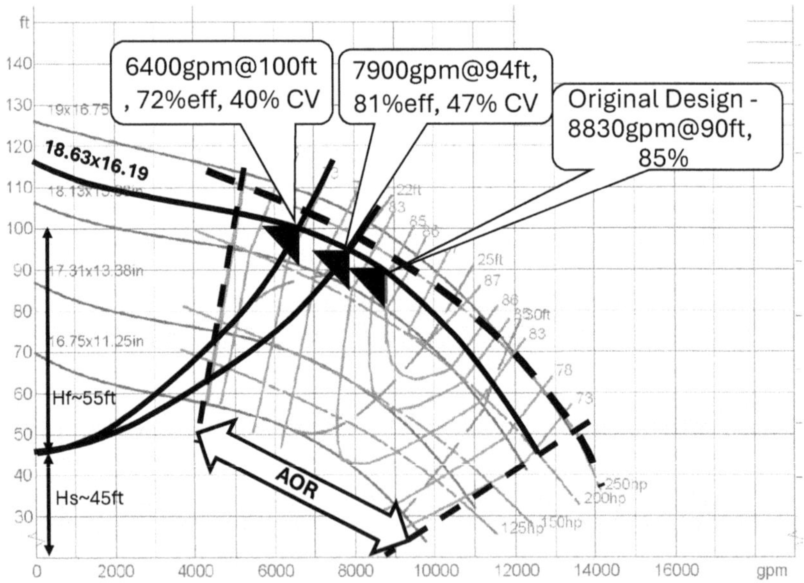

FIGURE 6.54 Original pump and system curves with operating points.

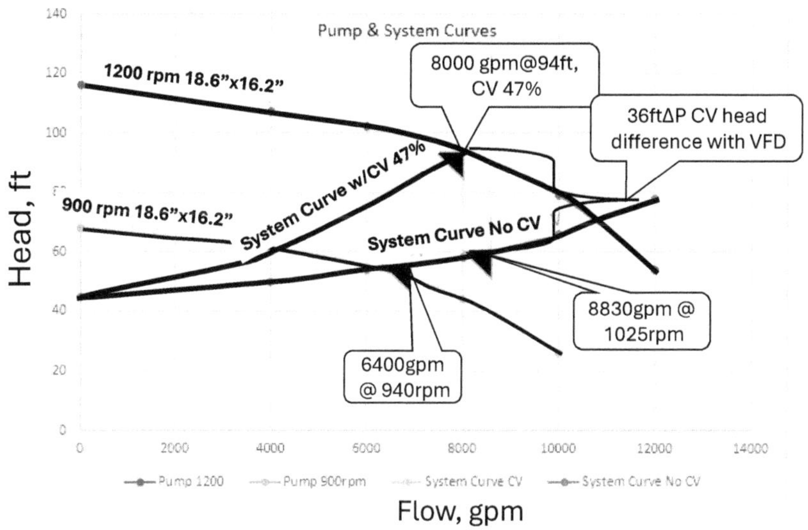

FIGURE 6.55 VFD pump curve and system curve with operating points.

With a VFD, the control valve is removed, and the system flow is controlled by the VFD. With the control valve removed from the system, there is only piping, fittings, and manual valves that contribute to friction head. The system static head remains unchanged. Figure 6.55 shows the pump and system curves and corresponding head values for the original control valve system and the VFD control system. Three operating conditions are shown on the curves for three flow conditions; 6,400 gpm and 7,900 gpm show a typical range of operating range for the pump system, and 8,830 gpm correspond to the original design condition of the pump system. The friction head difference is the main variance comparing the control valve system to the VFD system and ranges from 30 ft at the design flow of 8,830 gpm verses 46 ft friction head difference at the lower flow of 6,400 gpm.

The VFD operates in a range of 900–1,025 rpm for the flow range. Looking at the required BHP, a 200 HP VFD would satisfy the original design point of 8,830 gpm, but the existing operation never saw the flow go above 7,900 gpm. At 7,900 gpm at 58 ft, 87% efficiency yields a BHP ~133 so a 150 HP VFD should handle all of the current operating conditions. Using a 150 HP motor would certainly keep the drive much simpler avoiding a medium voltage drive. Figure 6.56 shows a summary of the original and VFD energy comparisons.

Condition	Flow, gpm	Head, ft	System Friction Head, ft	Pump Eff, %	BHP	Pump Speed, rpm (%)	Energy Savings, HP	Annual Energy Savings,$ ($.055/KWh)
Existing Original Pump & CV	8830	90	30	85	240	1180 (100%)	NA	NA
Existing Original Pump & CV	7900							
Existing Original Pump & CV	6400	100	55	72	228	1180 (100%)	NA	NA
VFD Pump	8830	60	15	87	156	1025 (87%)	84	30,000
VFD Pump	7900				133			
VFD Pump	6400	54	9	85	100	900 (76%)	128	46,000

FIGURE 6.56 Summary of VFD and fixed speed original design energy comparison.

Note from the table how the friction head is much lower, and the pump efficiency is higher with a VFD option. The energy savings for the VFD would be between $30,000 and $46,000 annually.

Case Study 26 Learning – High friction head pump systems are good candidates for a VFD with not only good energy savings but reliability benefits as well.

There are lots of different challenges with an ASD. Choosing the correct impeller size so that the pump speed will be at the right target speed to satisfy other mechanical and hydraulic requirements is important. It also needs to ensure the pump system has sufficient turndown capability to support the full range of process demand.

Some of the low-speed considerations might include the following. When pumping slurries or solids, low velocities can allow the solids to settle or plugging in the pump or piping. If the pump has a dynamic seal, then the expeller must produce enough head to seal off from the pressure in the stuffing box. With a variable torque machine like a pump, the torque of the motor must be checked against the torque required by the pump at the lower speed, or the motor may trip due to overload. A motor is a constant torque machine when speed is lower than the rated speed, but the pump is a variable torque machine. This next case study will look at the torque evaluation for the same disk filter feed pump in the previous case study.

CASE STUDY 27 VFD MOTOR TORQUE EVALUATION

Keeping with the same application as Case Study 26 on the disk filter feed pump system and VFD, another piece of the evaluation is to understand the torque balance between the pump and the motor operating at a lower speed. The pump is a variable torque device, while the motor has only the rated torque to deliver below the rated speed. Therefore, on the VFD as the speed decreases, the motor still delivers the same rated torque. The pump torque demand, however, at some point may be too much for the motor, so it is important to understand the pump torque required at its lower operating speed and compare to the motor torque rating.

In the disk filter example, two different motor sizes were considered – 150 HP and 200 HP (the original fixed-speed motor was 250 HP). The 200 HP rated motor torque at the nameplate rated conditions is 890 ft-lbs, while the 150 HP is 667 ft-lbs. The disk filter pump has a range of operation, so you'll want to evaluate the minimum and maximum expected operating conditions. The upper operating point is 7,900 gpm at 58 ft, which requires a 133 HP at 950 rpm. The torque for this condition is 735 ft-lbs. The lower operating point is 6,400 gpm at 54 ft which requires 100 HP at 900 rpm. The torque at this condition is 584 ft-lbs.

From the previous hydraulic and process evaluation, a 150 HP would cover the maximum flow condition; however, a 150 HP motor is not sufficient when looking at the available torque from the motor. The torque required at 7,900 gpm is 735 ft-lbs, while a 150 HP motor is only rated at 667 ft-lbs. So the only viable choice for the VFD motor to cover the range of flows for the process was the 200 HP VFD which has a torque rating of 890 ft-lbs.

Case Study 27 Learning – If operating a VFD or evaluating a VFD, always verify there is enough motor torque at that speed to operate the pump.

PUMP AND PIPING AFFINITY LAWS

While pump performance in a system is best done when the exact pump curve and system curve is detailed, pump performance can be estimated by using pump affinity laws. Affinity laws are just an estimate and in general accuracy is only good if impeller size change is less than 10%. Affinity laws also assume that there is no change in liquid exit angle. There are two sets of affinity laws – one with constant pump diameter and other with constant speed. Other pump parameters change when the one is held constant. If more than two of these variables change then affinity laws will not work. Note how much the BHP changes when speed or impeller diameter changes – cube of the change. This is good to remember anytime these small changes are made to the pump and that is to verify the BHP. The pump affinity law equations are shown below. Case Studies 24 and 52 used the affinity laws to compare pump performance under new operating conditions.

Constant Impeller Diameter:

$$\frac{Q1}{Q2} = \frac{N1}{N2} = \left(\frac{H1^{.5}}{H2^{.5}}\right)$$

$$\frac{\text{BHP1}}{\text{BHP2}} = \left(\frac{N1^3}{N2^3}\right)$$

Constant Speed:

$$\frac{Q1}{Q2} = \frac{D1}{D2} = \left(\frac{H1^{.5}}{H2^{.5}}\right)$$

$$\frac{\text{BHP1}}{\text{BHP2}} = \frac{D1^3}{D2^3}$$

$$\frac{\text{Eff1}}{\text{Eff2}} = \left(\frac{D2}{D1}\right)^2$$

$$\frac{\text{NPSH2}}{\text{NPSH1}} = \left(\frac{Q2}{Q1}\right)^2$$

Pipe and system affinity laws also exist and can be very valuable estimations for potential system changes. This happens when an existing pipe system is in place and a higher or lower flow is needed. The user may want to know what is the new friction head (total head result)? Since friction loss is the result of the square of the speed, this can be estimated. Another scenario could be that a flow may remain constant, but the pipe diameter may be changed. What is the resultant head in that system? Or, if only a fixed amount of head is available from the pump, and you want to determine the relationship between flow and pipe diameter when one of those variables is changed. These equations are shown below.

Constant Pipe Diameter:

$$\frac{hf2}{hf1} = \left(\frac{Q2}{Q1}\right)^2$$

Constant Flow:

$$\frac{hf2}{hf1} = \left(\frac{D1}{D2}\right)^{.5}$$

Constant Head:

$$\frac{Q2}{Q1} = \left(\frac{D2}{D1}\right)^{2.5}$$

A situation provides an example of how piping system affinity laws can be applied to analyze system performance. A system needs some additional flow, so the pump is to have an upgrade with a larger impeller. The existing flow is 3,000 gpm at a head of 115 ft. The increase in flow desired is 500 gpm, therefore, what increase in the system head will result from this additional flow and constant pipe diameter? Using the constant pipe diameter equation, the new system head estimate would be 156 ft. With this new head estimate, the pump engineer can now choose an impeller size from the curve at the new head for the system that will yield 500 gpm of more flow. It is always advisable to do the friction calculations before proceeding with a big change, but the affinity laws will give you an estimate upfront to determine its feasibility.

7 Pump Commissioning, Operation, and Troubleshooting

When a pump system has been designed and installed correctly, there is only one last hurdle to prevent good pump performance and reliability – startup operation. Breaking it down, there are three main stages: pre-operation (commissioning), actual startup, and normal operation. Errors can happen at any of these three stages that can cause pump operating or reliability issues. This chapter will cover just some of the areas where pump problems can occur during startup and commissioning activities.

COMMISSIONING (PRE-PUMP OPERATION)

I often think of commissioning as check-out activities (pre-operation). This is mostly focused on the pump system that is just being started up for the first time or maybe just had major maintenance activities. Although some will apply to anytime a system is started up.

The first thing to take note of is safety of the pump system. During commissioning activities, it can be common to run or flush a system and maybe shut back down for some adjustments or inspections. Make sure proper lockout has left the pump in the proper state. Are drains open or shut and valves are in correct positions for either run state or shutdown state? Before any startup, verify that guards are in place with proper gaps as per OSHA gaps. It is always good to stand at a safe distance during startup, as you never know what might happen the first time. Did someone leave a tool inside the pump? Could a flange blow out and spray harmful fluid on a bystander? There are a ton of things that could happen. Just good to be at a safe distance until it is clear that stable operation has occurred.

As part of checkout and signoff, the pump and driver should be properly aligned within target (see installation chapter for details). Have someone from the plant maintenance team sign off on alignment. Is thermal growth correction needed? If so, what are the plans for performing a hot alignment? Rotate pump shaft by hand (ensure lockout is still in place) to ensure it is free. This can also be done before the pump is coupled together as well.

If the motor has been changed or is a new installation, verify the shaft rotation before coupling to the pump. Has there been proper signoff of the pump impeller clearance setup? If the pump is connected to the motor, only do a quick rotation bump test. If the coupling is disconnected, it is not as critical. Do not want to dry run the pump on rotation test. If coupling is connected after rotation test, verify that the coupling gap and engagement is correct.

DOI: 10.1201/9781003729525-7

Have someone put a wrench on all fasteners to make sure they are tight. The torque wrench should have already been used on assembly, but this is just a last check to ensure nothing got missed. If the pump is run and shut back down later, perform a bolt tight wrench test again as some could have loosened.

Verify that sealing is set correctly. If the seal is packing, make sure the packing follower is snug but not overly tight, as excessive tightening can restrict shaft rotation, the packing will either get burned on startup, or it may even provide too much resistance and trip the motor. Turn the seal water on and verify it can get through stuffing box to the seal or packing. If the seal water hookup is a quench and drain setup, then set the stuff box pressure accordingly (detail in maintenance Chapter 8). Shortly after startup, adjust packing slowly and evenly after pump has run in for 30 minutes to an hour.

Another key check on the pump is lubrication. The pump will not run long or run well without proper lubrication. Understand the pump lube setup and the way the bearings are lubricated. Is the lubricant at the correct oil level in the pump (typically at the center of the bullseye sight glass)?

Lots of strange things can happen on an initial pump system startup, so be prepared for not only typical things but even the unusual happenings. The next several case studies are examples of actual events that happened on pump startups. Many of these have occurred many times over my career and illustrate how important executing basics are for a reliable pump startup.

CASE STUDY 28 NEW PUMP HIGH MOTOR LOAD

A submersible pump was replaced on a sanitary lift station with a new pump replacement. Upon startup the pump was drawing very high amp load and would trip out very quickly. The pump had a 6.5 HP motor and a full load amp rating of 7.1 amps. The pump motor was pulling over 50 amps on startup. What is wrong? The data doesn't even sound real. Some thought the system was backed up or clogged piping, but this didn't really make sense for a hydraulic pump engineer. That can't be it because the amp load actually goes down as you approach dead head condition. Some technicians were initially thinking that the pump had too small impeller diameter installed. That can't be it because smaller impeller than previous also means less amp draw or motor HP.

Has motor gone to ground? The motor was checked, and it was not to ground but upon further inspection it was observed that the motor wiring junction box showed to be configured to 230 V hookup; however, the application at the motor control box was wired up as 460 V (actually runs around 477 V). Higher voltage across a resistance and a higher amp draw will result just following the physics of failure of basic electrical theory.

There was an error in the supply of the new submersible pump motor assembly. The motor was sent back to the shop and wired for 460 V service and returned to the field for installation. Surprisingly, after many repeated startups, the motor was not damaged.

In this case, the spare pump/motor was designed or manufactured incorrectly. It was also missed later by the contractor that installed the unit. The problem was identified as maintenance was brought in to troubleshoot the installation. Since this was a replacement, it could be considered a maintenance issue as well and infant mortality-type failure by having the wrong part. In this failure, it was a pump problem not a system problem.

Case Study 28 Learning – Verify pump motor is correct for the application and connected correctly.

And the problems with this sanitary lift pump are not over. The next case study continues with another issue on the next installation where a new problem is revealed and later resolved. When you think or take some things for granted, they jump up and bite you.

CASE STUDY 29 NEW PUMP POOR PERFORMANCE (LOW FLOW/HEAD)

The same submersible pump from Case Study 28 was re-installed and wired up for the correct voltage. The startup of the pump ran well but this time the pump was not pumping enough. It was pumping very slow at an estimated 30% of target flowrate. The pump BHP was calculated at around 4.6 HP but the pump curve was very flat and it was hard to tell exactly where it was operating. What is wrong? A few thought the line is still plugged like we said the first time.

It was first suspected that it may have had the wrong size impeller – a smaller one. We never really verified when the submersible pump was sent back to the motor shop for rewiring the junction box, so the pump was really not inspected. The impeller diameter was verified by pump build specifications and nameplate on pump without disassembling the pump. It only came in two impeller sizes. We did not actually measure due to amount of work to be executed at the time.

Further inspection of the lift station, while the pump was running, showed that there was a lot of turbulence around the pump suction when the level dropped to a lower level. Almost like there was a blockage in the line and heavy recirculation was occurring. There were no closed valves in the discharge line and the inlet to the pump impeller eye has a tight mesh (near 0.5″) and the discharge line was a 4″ diameter. The possibility of line plugging seemed very low but possible.

The pump had a hinged flange connection for discharge as shown in Figure 7.1, which has on occasion not seated off properly on past pump changeouts when foreign debris from the pit got between the flanges. The discharge flange was inspected and found to be seated off correctly, so this was not the cause of poor pump performance and sump turbulence.

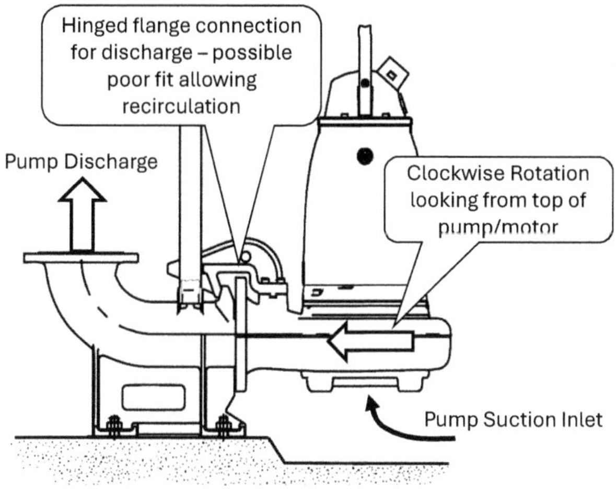

FIGURE 7.1 Submersible pump discharge connection.

The only thing that made any sense was the pump rotation was incorrect (backward). Poor performance and turbulence in the tank around the pump were both red flags to the pump rotation being incorrect. After verifying the pump was wired up backward, the pump rotation was corrected, and the pump began performing to curve performance.

An important learning is that some centrifugal pumps can still produce 25–40% of flow even with the wrong (backward) rotation. Never assume rotation on a new wired motor is correct without checking it. The submersible pump rotation inspection was not as easy as a standard pump where the shaft could be seen easily but rotation should be checked before lowering into the sump. This wasn't a typical ANSI pump with a screwed-on impeller but in those situations, the pump impeller may screw off the pump if operating backward. When it does, it will lock up the pump, trip out the motor, and cause some damage to pump impeller and casing.

This was an installation issue and considered a maintenance-related pump failure due to not verifying the pump rotation. This was also a pump issue and not a system issue.

Case Study 29 Learning – Verify pump rotation is correct; some pumps can produce 25–40% of flow even with wrong rotation. Always verify pump rotation on every installation, especially when motor rewiring is part of the scope.

After the pump system has successfully started up and the process is steady, take vibration data to establish baseline. This can be used for future troubleshooting and evaluate its initial condition. Also measure the pump bearing and motor temperatures. If you are pumping hot fluids, allow for some gradual warmup. Record normal pump conditions such as suction and discharge pressures, flow, and motor load. A good practice to take a screen shot of any process controls for baseline and future troubleshooting reference.

Monitor bearing temperatures but note 120°F feels hot to the touch. Many people may think the pump bearings are too hot just because you can't leave your hands on them, but they are not. A temperature measuring device must be used to verify temperatures – IR gives better snapshot of the whole equipment to see what is going on. For some applications, 170–180°F may even be normal for some pumping conditions. Anything under 160°F is not really a hot bearing but there are other considerations.

Keep in mind that bearing housing temperatures are typically 15–20°F cooler than the bearing inside but bearing housing outside temperature is generally all that can be easily measured and monitored easily. If using a non-contact IR or temperature gun, then make sure the surface emissivity is set correctly. If it is wrong, this can lead to large errors in the actual measured temperature. Dark, carbon steel-type surfaces will generally have an emissivity of 0.9–0.95. Any buildup on the pump bearing housing will also insulate and not give an accurate temperature; clean the surface off if necessary.

Temperature checks should be made close to the bearing locations and as close to the same location on each check. Operating temperatures can affect the lubricant selected for the application, so ensure the correct lubricant matches the operating temperature and type of bearings used. Higher temperatures may also require a higher clearance bearing such as a C3 clearance bearing instead of a standard clearance bearing.

Another factor to consider in bearing temperature monitoring is the temperature change from baseline to last measured. A change in temperature increase $>$ 50°F is a concern no matter what the absolute temperature measured. Monitoring frequency should be increased if this is observed and the root cause investigated. Small changes in bearing temperature are of little concern. A greased bearing may increase 25–30°F right after greasing but should come back down in a day or so.

At higher pump temperatures, operators may be tempted to put a water hose on the pump housing, but refrain from taking this action. Cooling on the outer bearing housing will reduce bearing clearance potentially even to the point of catastrophic failure. Also, water can infiltrate into the bearing housing cavity and the oil contaminating and initiating a water-induced bearing failure. A cooling fan might be the only appropriate cooling action. If housing cooling is needed and the pump is equipped with cooling, this could also be an option for bearing housings operating over 180–200°F.

It is always good to vent air out of the system, especially on a new system. A centrifugal pump with center discharge is basically self-venting when it starts up; however, a pump casing with a tangential discharge may trap some air in the top of the casing. It may be prudent to vent the top of the casing in these pumps at startup to ensure air is out of the casing. Double suction pumps should be vented at the highest point of the top casing to remove trapped air.

A self-priming pump primes itself but not on a new installation. The casing must be filled with fluid the first time, so it has the volume inside casing to act like a self-priming pump. This casing filling also includes if major pump maintenance has been performed, and the casing has been drained. If a self-priming pump is having a hard time to get started, look for leaks in the suction line. In cases where the pump is above the fluid and suction lift situation exists, the suction line is under a vacuum (below atmospheric pressure). A lift pump suction line under a vacuum will not show an external leak, it will suck air into the suction which will stall the pump and prevent it from priming. A leaking flange will not allow the pump to prime and start pumping.

Self-priming pumps also have a suction check valve usually near pump suction nozzle or just inside pump casing. If pump is having a longer prime time, this check valve could be broken. The check valve closes when pump shuts down and holds the liquid up in the suction pipe which shortens the priming time upon restart. These suction check valves normally do not last much longer than five years, so be ready to replace.

During checkout do not exceed motor starts per hour – larger motors and higher speed have fewer motor starts per hour. When starting or restarting a pump, pay attention if the system lacks a check valve or if the line is draining backward. Do not restart the pump or it can lead to a torsional shaft fracture, similar to the damage observed on the impeller in Figure 7.2.

For parallel pump operation, do not let the first started up pump run too long by itself. If the pump system is designed for both pumps to run, then they share the load so one pump may run off end of curve. If the pumps are not identical pumps in parallel, start the weaker pump first.

Always have a pump system engineer (or engineer on project) verify the pump operating point or performance to ensure that it is operating near its design condition

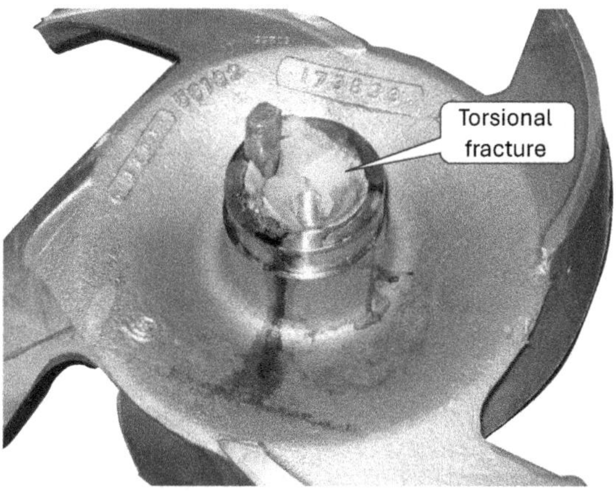

FIGURE 7.2 Shaft fracture from starting up pump that was turning backward.

before walking away at startup thinking everything is fine. This is the best time to have final signoff that the pump system is ready to go for a long reliable life. If there is an issue, there is momentum to engage the pump supplier, engineering, and project team to reconcile any installation or design issues. After this season passes, the mill will own any and all design misses and be far away from solving.

CASE STUDY 30 PUMP LUBRICATION INFANT MORTALITY FAILURE

The first failure involves a pump that was replaced during an outage. After startup, the pump ran into the second day and was found locked up and very hot after it had shut down the process. The pump was an API-style end suction pump with centerline mounts to negate thermal growth and had an enclosed impeller. This pump also had a very good bearing housing cooling fins. The pump was so hot that the paint was discolored and burnt and even peeling off in few areas. The lubrication arrangement was a flooded bearing sump (oil at the center of the most bottom roller). The pump operated at 3,600 rpm but had a good reliability history so not major system design issues to chase after.

So, what happened? It was found that the pump never had been filled with oil. The group that installed the pump thought the lube technician will come by later to add the oil. The lube technician thought the installers were adding the lubricant (this was actually the correct process). There are many ways to have the pumps lubricated but the best practice is that the pump installer has the responsibility to ensure that it is lubricated and ready to run when locks are removed. However, the failure was even more than that. The pump rotating assembly was prepared and placed at the job site, but no one noticed that it didn't even have an oil level sightglass installed. It was plugged. This was a mistake at the rebuild shop, but it should have been identified and corrected during job preparation prior to installation.

No oil was staged with the pump rotating assembly at the job. The alignment sheet has a section about lubrication on the machines being aligned. This was not completed or referenced on the installation. There was no operator, or maintenance rounds the following two days of operation to verify the oil level. If there were, they would have noticed there was no oil sight-level glass and that the pump was operating hot.

Pretty amazing that it ran for nearly two days with no oil. Couple of theories on why this was the case. One it may have had a small film of oil from assembly. Two, the design margin for an API-designed pump with respect to bearings, thermal growth, cooling fins, etc. allowing it to operate in more extreme conditions. The enclosed impeller and pump operating point resulted in low shaft radial loads on the bearings as the pump operated near BEP in its operation. The alignment was very good so reactive loads were also low.

This failure was clearly an installation issue and infant mortality failure, which is basic maintenance, as the root cause category. Although there were multiple opportunities to catch the issue before failure, such as operator basic care rounds at startup, but the root cause still goes back to maintenance not adding the proper lubricant on installation. This failure was obviously a pump failure and not a pump system issue.

Case Study 30 Learning – Verify pump has proper lubrication. Always check pump operating conditions after startup when signs of infant mortality may be evident.

So many pump failures might be caught and avoided by just doing basic checks after startup. If the pump, in the previous case study, had been inspected right after startup, the catastrophic failure could have been prevented. The next case study is one where some pump system checks were done by a reliability professional and likely avoided a pump reliability event. This is such a key piece (the last element) of a good pump startup and commissioning process that caught a startup issue and prevented failure.

CASE STUDY 31 NEW PUMP VIBRATING AND RUNNING HOT ON STARTUP

A project to replace an obsolete pump model was just completed with a new pump assembly in place within an existing system. The process was a floatation cell system as shown in Figure 7.3. The pump installation occurred on the run with one cell being bypassed so there was no real rush during pump installation. If one cell was not working, the chamber just overflowed into the next cell and the process was not interrupted. Point here is, process function is not easily recognized by an operator when one of five cells is not functioning properly, as the overall system can continue to appear normal.

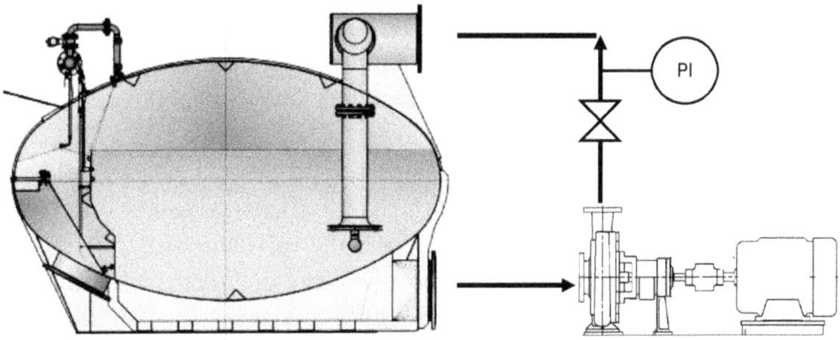

FIGURE 7.3 Process flow diagram floatation cell feed pump system.

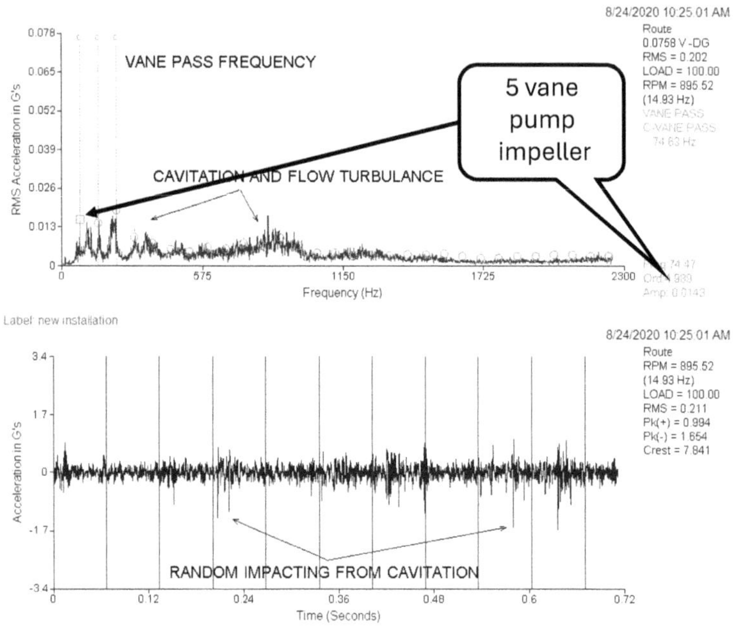

FIGURE 7.4 Vibration after startup of new pump installation.

The pump was a fairly large end suction pump with a size of 16 × 16–19 and an 18.13″ × 15.38″ diameter impeller. The driver was a 100 HP, 900 rpm fixed speed motor. Reliability checks conducted the next day after startup found the pump operating with high vibration and elevated temperature. The baseline vibration reading after startup is shown in Figure 7.4. Vane pass was elevated as well as a raised floor showing flow turbulence and cavitation at random frequencies.

Upon inspection of the pump operation, it was observed that the 16″ discharge valve was completely shut. The pump was running in a dead head condition. Somehow in the project commissioning and unlocking process, opening the discharge valve was not completed. The discharge valve was part of the lockout for the installation.

The valve was opened and set to properly run the process. The pump curve in Figure 7.5 shows the dead head condition and discharge pressures observed (28 psi) in the dead head condition. Also shown is the valve throttled in the properly adjusted position (80% open) with 15 psi discharge pressure (static head in this system is almost 0) and the pump TDH ~34.5 ft. The pump vibration dropped ~82% to normal levels once the pump was operating in the POR.

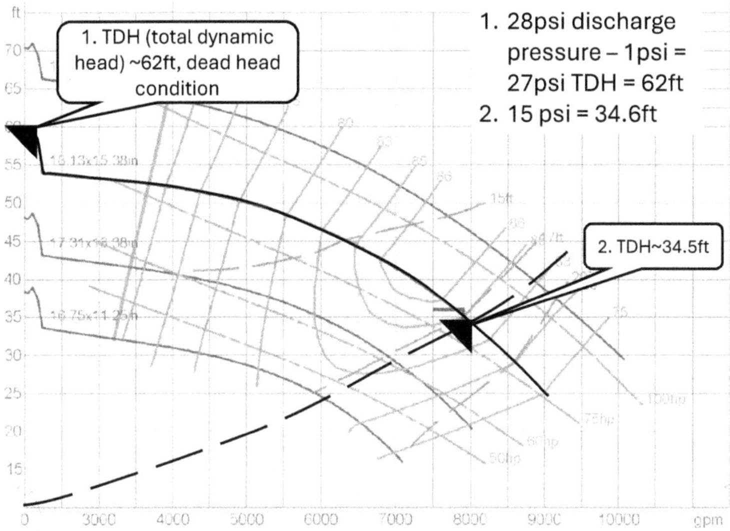

FIGURE 7.5 Pump operating point in dead head condition and after discharge valve opened.

The root cause of this pump issue is operations and failure to open valves correctly on startup. Having project and pump startup checklist is always a good practice. Fortunately, this was caught in the first 24 hours by reliability professionals before significant pump damage was done. This issue was a pump system issue and not a pump problem.

Case Study 31 Learning – Check valve position and pump operating points on pump systems after commissioning and startup.

It is advisable to start up a centrifugal pump with a throttled discharge valve. There are two circumstances where a pump system is empty upon startup: (1) when a pump system is started up for the first time (with discharge check valve) or (2) a system without a check valve (where fluid drains back out of piping). With the discharge line empty, there is only friction head against the pump. On startup, the pump will operate out on the pump curve until the friction head is developed as the pipe system gets filled up. If the pump does not have a driver with full curve runout, motor overload may be a possibility as well.

It is also advisable for low Ns pumps to have the discharge valve closed (fluid exists 90° from entry of impeller) and open the valve for high Ns pumps (axial flow pumps – fluid exits parallel to pump shaft). In any circumstance, the discharge valve should be opened shortly after startup to get the pump operating in a good operating point.

A simple checklist before starting up pump might include the following.

1. Pump and motor aligned.
2. All fasteners are tight on pump, foundation, motor, piping, etc.
3. Pump is lubricated.
4. Pump and motor rotate freely by hand.
5. Pump rotation is correct.
6. Seal water hooked up and on.
7. Suction valves are open, and tank is filled with liquid.
8. Discharge valve is in position for startup (closed, throttled, or open)
9. Prime pump if required, vent air on startup.
10. Verify pump operation after startup – pressure, flow, motor load.
11. Verify mechanical condition after startup – vibration, temperatures, etc.
12. Do not exceed motor starts per hour during commissioning process.

OPERATIONAL TIPS

While there is mostly a focus on pump design, installation, and maintenance, the operation of the pumps is also very important and operating groups have a major role to play. Some may be obvious and others not so much.

For pump systems with one pump and an installed spare, it is a best practice to rotate the two pumps on some frequency – maybe every other month. One approach that works well is to have the even pump runs even months and the odd pump runs odd months. The reason for this is the same reason to rotate shafts of spares in a storeroom – it reduces static failure modes of false brinelling and static corrosion. Installed spares have more vibration than storeroom spares so it is even more of a risk. This evens out the run times and wear for each pump. It also prevents any fluid in the pumps and lines from getting old, settling out or any other degradation.

For pumps supplied from a tank, operate tanks with level as high as possible or feasible. This not only can reduce energy consumption but improves suction conditions to reduce potential air entrainment and cavitation (higher tank level increases NPSHA). Since the suction head from a tank increases the system pressure, the pump will not have to work as hard to produce flow and head. If it were a VFD on pressure control, the VFD would run slower. Another benefit can be an increase in system capacity if needed.

CASE STUDY 32 TANK LEVEL EFFECT ON PUMP SYSTEM PERFORMANCE

A simple pump system transferred fluid from a small tank up one floor to another tank as shown in Figure 7.6. The pump needed some additional capacity. The pump did not have any flow control valves so it pumped whatever the system would allow it to pump. The pump was a fixed speed, 16 × 16–19 size end suction pump with a 200 HP, 1200 rpm motor. The pump was designed for 8,500 gpm at 70 ft and had a 17.68 × 14″ diameter impeller. At normal tank level of ~5 ft, the system had a static head ~30 ft. At the design point, this leaves 70 ft of friction head.

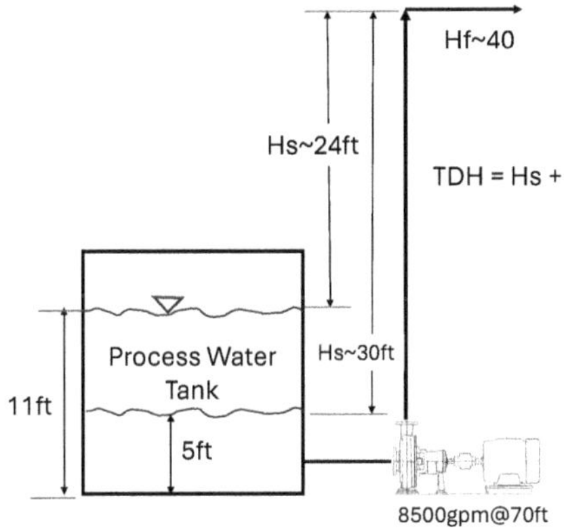

FIGURE 7.6 Process diagram for process water tank pump system.

There are several ways to get more flow from the pump. The obvious upgrade is a larger impeller if there is enough drive HP to accommodate the impeller upgrade. For this particular pump, a 17.9 × 14.69″ diameter impeller would add another 400 gpm and still have full curve runout as shown in Figure 7.7.

Another option to increase the pump flow is to increase the tank level in the supply tank and reduce the system static head. This pushes the system curve down and thus out on the pump curve. Pending on the slope of the pump and

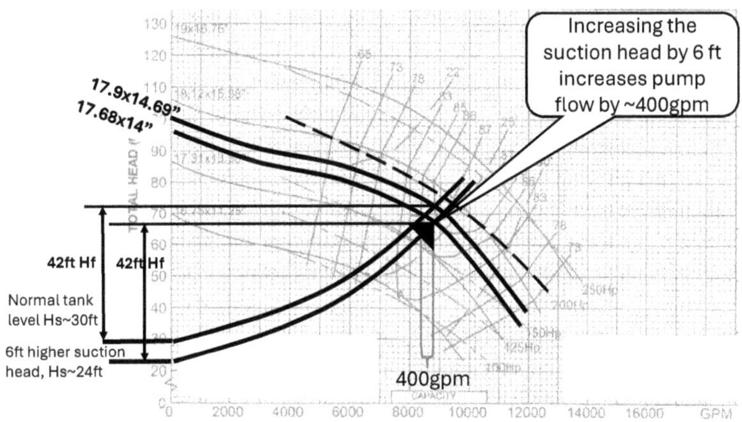

FIGURE 7.7 Pump and system curve with more tank suction head.

system curves, the result will vary. The system curve will also bend as a higher flowrate increases the friction head. In this example, the friction head at 400 gpm is around 42 ft (2 ft more than original flowrate).

Interestingly, this example shows that increasing the impeller size can produce the same increase in the pump flow as raising the tank level from 6 ft to 11 ft. The needed pump head can come from either increased tank static head or pump head but the tank head is more cost effective. The pump and system curves for the original, larger impeller and higher static head is shown in Figure 7.7.

Case Study 32 Learning – Consider increasing the tank level to increase pump performance without any pump modifications.

Understand the pump and system operational ranges (limits) and where it is operating in respect to those limits. Study and understand how the control system (control valves) reacts in partnership with the pump. This is especially true if a pump has header pressure control feeding many different destinations with control valves. Find the sweet spot for header pressure based on demand and controls for other system branches (see Case Studies 20 and 21).

Never modify a system without having a pump system engineer look over the changes to make sure pump and system can handle the changes. Even small changes can tip the scale and balance of pump performance and reliability. Even if the pump and system can handle the new demand or operating condition, the pump may be operating in a very unreliable new operating point. The negative effects of adding pump load may not show up for months. Pumps are designed for specific system conditions, so any changes create a new system design that is likely not as good as the original design. Some companies utilize a management of change process for changes such as this. This is a good idea to ensure that proper resources review proposed new process conditions.

TROUBLESHOOTING

Troubleshooting is an important part of pump system reliability. It is good to remember that troubleshooting is not RCFA. Troubleshooting is the process of finding the problem and correcting so normal operation can occur. RCFA is finding out the root cause that caused the interruption of process or failure of the component and then putting corrective actions in place to prevent future failures. Both are important in operating a pump system and both should be used. Good troubleshooting can save a lot of money, downtime, and frustration in the moment. Good RCFA can prevent future difficulty for years to come across many pump systems.

Troubleshooting is a critical task in operating and maintaining a pump system. The pump system contains the pump and the system. The first big hurdle to figure out when the process isn't working is to determine whether it is the pump or the system. A majority of pump system problems are typically the system and not the pump; however, in my years of experience, the pump is often blamed for poor performance. Probably because it is relatively easy to change, and it is someone else changing it. If the pump is the problem, then changing the pump will fix the issue but if it is not, then the organization has just wasted resources (time, money, and downtime) to execute an unnecessary activity, and you are still left with the issue.

One of the first things that must happen is identifying the pump that has the issue. Know the size, model, speed, and impeller size. There are around five ways to determine the size of the impeller installed in the pump. Only a few will confirm the impeller size, but all are good input to find out the diameter.

1. Get pump serial number from pump tag. Contact pump manufacturer and get original design specifications including pump curve, design point, and impeller diameter. This will not confirm which size is currently installed but it at least gives a starting point. Most original equipment manufacturers will not have changes if the impeller size was changed by the end user so proceed with some caution if this is the only source for size.
2. Look into the maintenance records to determine if the pump diameter has been checked or noted. Since the serial number data may not reflect the current condition, the maintenance records may indicate if an impeller change was made.
3. Remove pump rotating assembly and measure impeller diameter. This type of invasive investigative task may not be an option as the business may not support it, but sometimes an opportunity may be available and important enough to perform this activity. This will absolutely confirm the impeller so if accuracy is important, this will be the best to know for sure.
4. If the opportunity is available, run the pump for a short time under 0 flow condition (close discharge valve). Have a pressure gauge on the pump and read the 0 flow (dead head) condition. This will roughly tell you the impeller diameter by looking at the pump curve at the 0 flow (dead head condition) total dynamic head. Another estimate if you don't have the pump curve is the relationship below.

   $$SH \sim D^2$$

   where:
   SH = shutoff head (0 flow), ft
   D = impeller diameter, inches

5. Another way to get an estimate for impeller size involves the intersection of the pump curve HP line and the actual pump TDH. Pump TDH can be figured from discharge and suction pressure. Suction pressure gauges are generally not in place so tank level minus a small suction friction loss can get a good estimate for pump suction pressure. To estimate pump BHP,

motor data must be collected. Current and voltage for each phase of a three-phase motor and motor efficiency and power factor must also be estimated. This step will not only verifies the impeller diameter, but also helps to determine the pump operating point.

Once the pump information is known, including the impeller diameter and pump performance curve, the pump operating point must be determined. There are at least four ways to determine the pump operating point. The pump performance curve has flow, total dynamic head, impeller diameter, and motor BHP, which can be used to determine the operating point. Any two of these parameters will reveal an operating point. Below are at least four ways to get the operating point for the pump.

1. Pump total dynamic head, TDH. Get a suction pressure and discharge pressure reading. Static and dynamic head is included in these readings. Convert pressure readings to head. Also measure the distance from each of the gauges. The difference in pressure gauge distances can be significant in some pumps and systems, so this should not be overlooked. TDH is the total head the pump adds to the system at the pump centerline and can be calculated by the equation below; reference Figure 7.8 for a pump system with static head (static lift is different).

$$\text{TDH} = Hd - Hs + X$$

   where:
   TDH = total discharge head
   Hd = discharge head at pump centerline, ft
   Hs = suction head at pump centerline, ft
   X = distance between two gauges, ft
2. Motor load. For a three-phase powered motor, the brake horsepower is a function of current, voltage, motor efficiency, and power factor. Measure the current and voltage on all three phases (they should all be close or there is another potential issue). Motor nameplate voltage is slightly different from a plant voltage, so acquire an accurate measurement here. The power factor and

$$\begin{aligned} &Pump\ TDH \\ &= Total\ Discharge\ Head\ - Total\ Suction\ Head \\ &+ Head\ Difference\ of\ Pressure\ Gauge \end{aligned}$$

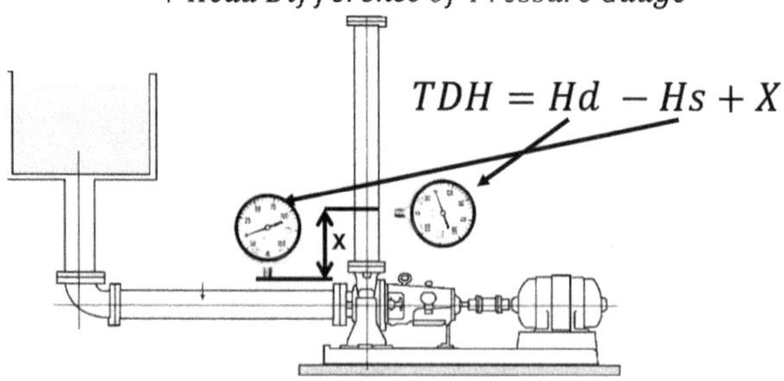

FIGURE 7.8 TDH corrections for pressure gauge locations.

motor efficiency will vary depending on the percentage of full motor load. Lower loaded motor and power factor will decrease. Full load to half load power factor may drop as much as 25%, but below 50% motor load, the power factor can drop further. If the motor is less than 50% loaded, the power factor can really affect accuracy of this calculation. Power factor can sometime be estimated from motor tables. Take an average of these and calculate the BPH by the equation below.

$$\text{BHPm} = I \times V \times \eta m \times \text{pf} \times \frac{1.732}{746}$$

where:
BHPm = brake horsepower at motor shaft
I = current, amps
V = voltage, volts
ηm = motor efficiency
pf = power factor

3. Have a flow measurement. This may come from either an inline flowmeter within the process or taking a non-invasive ultrasonic flow measurement outside the piping with a portable flow measuring instrument.
4. Mass balance on system with other known variables. There are many different process design methods and formulas that you could use. One example of this is some basic pulp and paper process mass balance calculations. These are used in dilution calculations. As a young engineer, a senior process engineer with Champion International, Rich Alger, taught me some of these principles. The following two equations can be used in paper stock applications.[1]

$$A + B = C$$

$$Ax + By = Cz$$

where:
A, B, C = process flows, gpm
x, y, z = stock consistency, % (whole #, not decimal)

There is other information than can be concluded during pump troubleshooting and collecting data on the pump system. Shutoff head does not confirm air entrainment condition, submergence issues, NPSH issues, or viscosity issues. If the pump is run dead head and the pump produces the head as expected, the following can be known.

1. Impeller diameter is correct per the curve.
2. The pump speed is correct.
3. The direction of rotation is correct.
4. Impeller clearances and efficiency is close to correct.
5. Pump fluid is correct for expected conditions.

The next case study shows an example of using mass balance approach to determine pump flowrate for actual conditions.

CASE STUDY 33 DETERMINING PROCESS FLOWRATE FROM MASS BALANCE

A sweetener pump supplying stock to the disk filter feed pump suction was being evaluated to determine the operating point as part of troubleshooting a reliability issue. There was some known process data being measured such as process water tank consistency, discharge consistency, sweetener pump consistency, and disk filter total flow. The unknown variables were sweetener flow and flow from couch pit which left two unknowns.

Using a mass balance of the small system as shown in Figure 7.9, we can develop the equations to solve for the flowrate from the sweetener pump. First, draw the system box boundary (dashed line). Flow in equals flow out for a simple mass balance. Complete the mass balance for the flow and flow/consistency as follows.

$$F3 = F1 + F2$$

$$F3C3 = F1C1 + F2C2$$

With two unknowns, two equations are required to solve the mass balance using the known variables.

$$F1 = F3 - F2$$

$$F3C3 = (F3 - F2)C1 + F2C2$$

$$F2 = \frac{F3C3 - F3C1}{C2 - C1}$$

Substituting in all the known/measured variables, solving for $F2$ yields 386 gpm. The pump had an original design point of 600 gpm at 60 ft. Now looking at the pump curve in Figure 7.10, we can know the current operating point. At 386 gpm, the pump was still operating in the AOR but to the left of BEP.

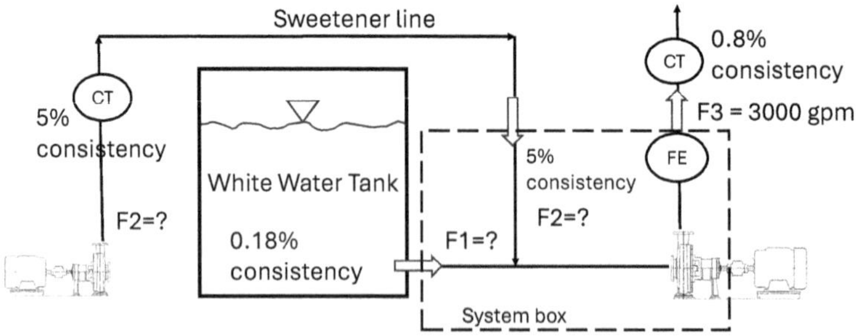

FIGURE 7.9 Process diagram using mass balance to find pump flow.

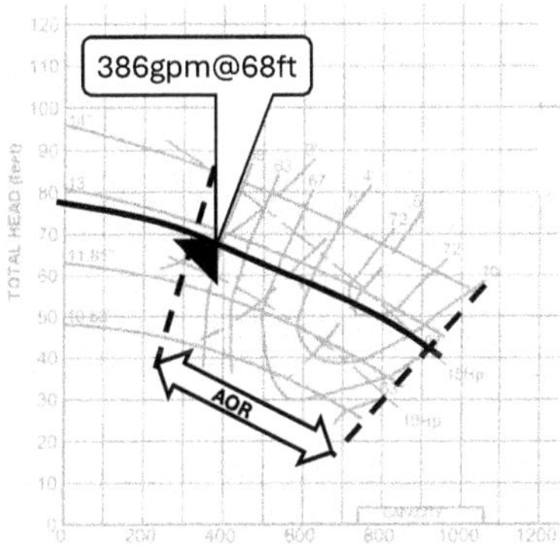

FIGURE 7.10 Pump curve with operating point from mass balance.

Case Study 33 Learning – Consider using mass balance approach when looking for system flows.

The next case study shows another example of using some pump operating data to troubleshoot and determine the pump operating point.

CASE STUDY 34 PUMP NOT PUMPING BUT HIGH MOTOR LOAD

A double suction overhung pump with a 13″ diameter enclosed impeller had a design point of 1,475 gpm at 790 ft. The pump was driven by a 450 fHP, 3,600 rpm motor. The pump began experiencing high motor load. The pump recently had maintenance performed.

Shortly after the maintenance work, the pump started experiencing motor overload trips. Maintenance went back into the pump to inspect. The pump was put back together and started up only to still have it tripping out. The area team thought there was an issue with the motor and the electrical engineer was brought into the discussion. Quickly the electrical engineer concluded nothing was wrong with the motor, but it was pulling 125 amps (full load amps was 100). He suspected a pump or pump system problem, so the mechanical/pump engineer was brought in to help troubleshoot further.

Using the equation for motor HP and knowing the impeller diameter, an estimated operating point for the pump can be calculated. The motor was pulling 125 amps and on a medium voltage feed with 2,300 volts. The motor also had a power factor of 0.87 fully loaded and a full load efficiency of 94%. The actual motor load was coming in ~545 HP, which was well beyond its rating and why the overload trip was occurring, but the question remained – what was the cause?

The pump curve with original design point and estimated operating point based on the motor load is shown in Figure 7.11. While the estimated operating point was at BEP, that was not where the pump was to operate in the system. The pump also did not have full curve runout with the current impeller diameter and 450 HP motor. The pump was showing a generated pressure (TDH) in the 650–675 ft range. So, both data sources showed the pump was well beyond the design operating point.

What was causing the pump motor overload issue? The pump system was checked, and all valves and components were in the correct positions as designed and normal operation. There had been no changes to the system, but all components were checked just to make sure.

Talking to the maintenance crews about events that had occurred previously. After the first trip events, maintenance was planned to try and determine what was wrong with the pump. Upon inspection, they found the wear rings on the pump were galled to the impeller and damaged. This was likely the initial cause of why the pump motor was tripping out – mechanical friction. The design operating point was only ~40 HP from the maximum motor rating (see the pump curve in Figure 7.11).

One of the pump wear rings was severely damaged and replacement wear rings were not readily available. The maintenance crew decided to leave one of them out and reassemble the pump. The resulting new condition added 700–800 gpm of flow to the pump via internal recirculation as shown in the

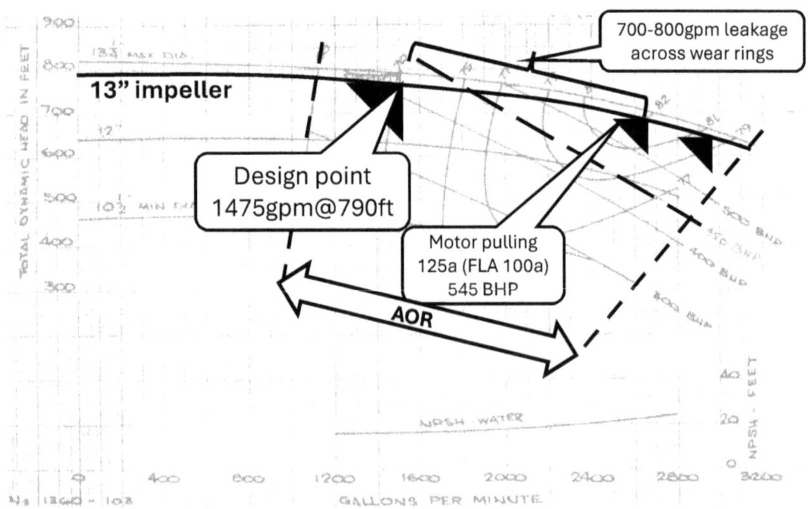

FIGURE 7.11 Pump curve with design and operating points.

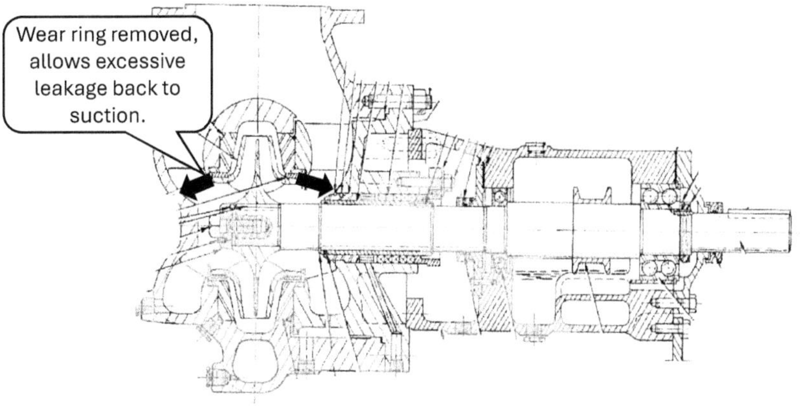

FIGURE 7.12 Enclosed impeller pump cross section showing recirculation flow from missing wear ring.

pump cross-section in Figure 7.12. So, with this internal recirculation flow at this pressure, a considerable pump BHP was added to the motor load.

Hydraulic pump efficiency goes down as a result. While the design condition the BHP was ~410 HP, the actual BHP was ~545 HP. If a flowmeter had been on the system, the flow would not have recorded as high as we thought only with a pressure reading and motor load. The motor HP increased so the overall pump system efficiency decreases. It could be said that the desired flowrate of 1,475 gpm now required 545 HP. The original pump hydraulic efficiency with good wear rings was ~72%. Now the pump system efficiency is estimated ∼54%.

The maintenance technicians solved the initial problem of mechanical friction increasing the drive HP but created a drastic increase in hydraulic HP when one of the wear rings were removed. This also created a hydraulic imbalance on the rotor and significant wasted energy. The root cause of the pump failure was obviously maintenance of the pump. In this case, the pump failure was a result of a pump issue.

Case Study 34 Learning – Check impeller clearances during installation and do not underestimate the negative effect on the pump performance and motor load.

Sometimes troubleshooting the pump and pump system can involve many elements and may even seem to offer contradicting information. There are numerous pump troubleshooting charts from various sources, and these can be a good place to start; however, many times it takes a pump system expert to solve the pumping issue. A quick reference for troubleshooting cause and correction is shown in Figure 7.13. It is a great start and can be used by those with little pump system knowledge, but it does help if the user has a knowledge of pump systems too.

Problem	Possible Cause	Solution
Pump not delivering liquid	Suction Line Or Impeller Clogged	Backflush Pump Or Disassemble to Remove Obstruction
	Wrong Pump Rotation	Change Pump Rotation (Re-wire Motor)
	Pump Not Primed	Reprime Pump, Verify pump is full of liquid
	Air Entrainment - Insufficient Submergence	Increase Tank Level Or Install Vortex Breaker On Tank Or Install Air Vent Line
	Suction Lift Too Much	Increase Tank Level or Shorten Suction Pipe
Pump Not Performing At Rated Flow or Head	Impeller Plugged	Backflush Pump Or Disassemble to Remove Obstruction
	Excessive Impeller Clearances (Internal Leakage & Hydraulic Efficiency Loss)	Adjust Impeller Clearances To Target Clearance
	Worn Impeller	Replace Impeller
	Wrong Sze Or Type Impeller	Replace Impeller With Correct Type & Size
	External Leakage	Repair Leaks
Pump Requiring Excessive Drive Power	Mechanical Friction, Impeller R	Reset impeller Clearances, Verify Bearing Lubrication
	Fluid higher SG than expected	Verify SG and Viscosity
	Stuffing Box Packing Too Tight	Readjust Packing - Loosen
	System Head Lower Than Expe	Throttle Valve or Trim Impeller
Bearings Running Hot	Impeller Clearance Too Big	Reset Impeller Clearances or Replace Worn Impeller or Suction Wear Plate
	Improper Lubrication	Correct Lubrication Issue - level, type
	Lube Cooling System Not Cool	Check Cooling System Cooling Water, Clean Cooler
	Excessive Hot Fluid	Check Fluid Operating Temperature
	Excessive Misalignment	Align Pump and Motor
	High Radial Loads	Verify Pump Operating Point
Pump Noisy or High Vibration	Excessive Misalignment	Align Pump and Motor
	Imbalance	Balance Impeller
	Imbalance - Impeller Clogged	Remove Foreign Object From Impeller
	Bent or Broken Shaft, Impeller	Replace Shaft or Impeller
	Foundation Looseness	Repair Foundation, Tighten Bolts
	Inadequate Piping Support	Correct Pipe Supports & Anchor Points
	Pump Cavitation	Correct Pump System Issue
	Flow Turbulence	Correct Pump System Issue
Excessive Leakage From Stuffing Box/Seal Chamber	Improper Adjustment on Packing Follower	Adjust Packing Follower
	Mechanical Seal Failure	Replace Mechanical Seal, Verify Seal Water Supply
	Stuffing Box Incorrectly Packed	Correct Packing Installation
	Shaft Sleeve Worn (Scored)	Replace Packing Sleeve

FIGURE 7.13 Pump troubleshooting charts.

(*Continued*)

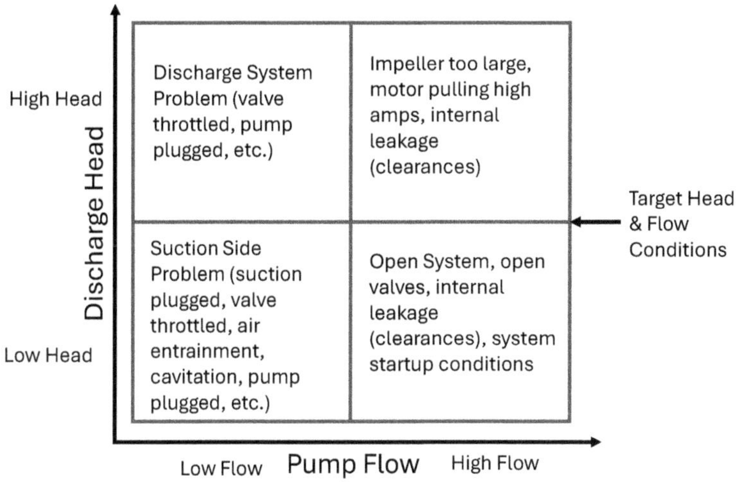

FIGURE 7.13 (Continued)

Another aid to help in troubleshooting and RCFA investigations is getting the right data from the field. It is always good to define the issue or problem upfront before troubleshooting a pump system. I'd suggest making a data collection sheet with the information below. Just want to collect basic facts and data about the pump system and the problem or issue.

- **Pump Information:** description, equipment #, pump manufacturer, pump model, size, serial #, years in service, impeller diameter, # impeller vanes, type sealing, lubricant used, type coupling, design operating flow, and head.
- **Driver Information:** type, HP, speed, fixed speed, ASD, direct drive, or belt drive, if belt drive driver and driven sheave sizes, belt size, # belts, motor nameplate data – rated HP, rated speed, power factor, efficiency, rated full load amps, rated voltage.
- **Motor Conditions:** current (amps), voltage, temperature, vibration, noise, rotation.
- **Pump Conditions:** pump vibration, pump temperature, fluid temperature, noise, suction pressure, discharge pressure, seal water flow and pressure, stuffing box pressure, oil level, baseplate condition, fluid pH, fluid density.
- **Pump System:** stand-alone pump (single), series, parallel, pump system controls, control valve size, model, % open, process flow sheet.

Not everyone may define the issue the same way. Some may better describe what is going on in the system than others. Use caution for end users to go beyond describing what symptoms they are seeing. Describing the symptoms or problem can easily turn into guessing the root cause. Seek several descriptions from several different people if possible. Always verify the initial data gathering before going too deep into system analysis. The next case study is not only a pumping issue resolved but also a case where jumping to conclusions about the root cause before analysis almost cost more money.

CASE STUDY 35 PUMP WON'T PUMP, MAKE PUMP PUMP

A pulper trim pump would not maintain the pulper level under normal trim process conditions. Normal pulper operation has a small trim pump sized to handle the small trim flow and a break pump to handle the full sheet into the pulper for a machine break condition. The trim pump was a small 3×6×14 end suction pump with a design point of 300 gpm at 130 ft. The process diagram is shown in Figure 7.14.

The trim pump ran all the time with the recirculation valve A having a minimum opening of 15%, while the pump out valve B is only opening/controlling when the level was at 75%. The break pump came on when the level reached 80% and did not shut off till pulper level was pumped down to 50%.

The problem was that the break pump was having to come on all the time as the level got to 80% to maintain pulper level even when there were no machine breaks (just handling trim). The trim pump was still running but appeared not to be doing much or at least not keeping up with trim rate into the pulper. Some operators said that it had been an issue for a long time, some even hinted that they were not sure they had ever seen it work correctly (most likely never really noticed what it was doing).

Initial first look at the pump system showed the pump was running the right speed and correct rotation. Maintenance changed the pump, and the pump impeller was verified correct at 11.2″ diameter so it didn't appear to be a major issue with the pump itself. Later, after playing with controls, the process engineers felt like the pump wasn't pumping and that it wasn't large enough for the operation. There were discussions of putting in a project to upgrade the

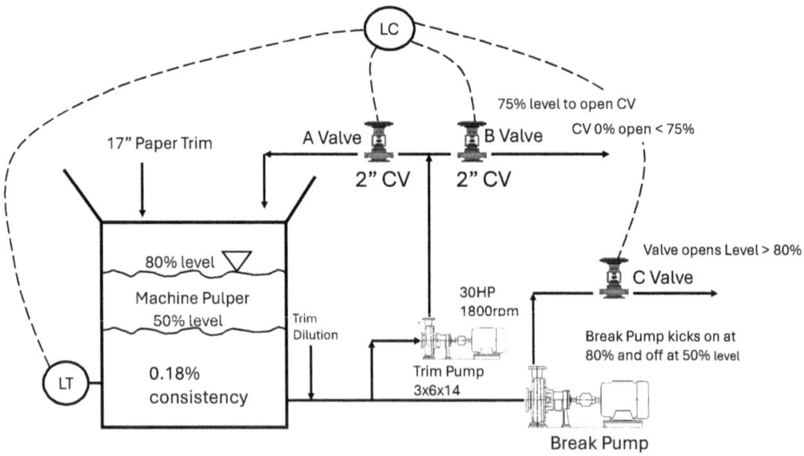

FIGURE 7.14 Process diagram for pulper pump system.

pump – larger impeller and motor or new pump, but before that went further, a deeper look by the pump reliability engineer was initiated. We really needed to understand what we had here to solve the issue, even if that meant a new larger pump. After all, what size pump would it be?

So, the process began verifying the proper size of trim pump as a starting point for the troubleshooting. The machine trim is 17″ total front and back of machine. The heaviest machine basis weight sheet is 33#/3300 sqft with a machine speed of 6,000 fpm. The tons per day rate of the trim into the pulper is 61.2. At 4% consistency coming out of the pulper, the trim pump flowrate to maintain pulper level is 255 gpm. The formulas below were used to calculate the production rate from the pulper.

$$\text{TPD} = \frac{\text{BW} \times \text{TW} \times N}{55{,}000}$$

where:

TPD = tons per day
BW = basis weight of sheet, #/3,300 sqft
TW = trim width, inches
N = machine speed, fpm

$$Q = \left(\frac{\text{TPD}}{C}\right) \times 16.67$$

where:

Q = flowrate, gpm
TPD = tons per day
C = stock consistency in %

So, the design conditions of the pulper system required 255 gpm flow at 4% consistency to handle the trim width of the pulper trim pump. The existing pulper trim pump was a 3×6×14 end suction pump with a 11.2″ diameter impeller and a design operation point of 300 gpm at 130 ft. So initially it did not appear that there was a design issue; however, if the pulper consistency dropped then the flow would have to increase to maintain the level. For example, using the same equation for flow, at 3% consistency the required flowrate would be 340 gpm, which is more than the pump's design capacity. There was a control valve B in the system and recirculation valve A to give some control turndown. With both valves operating, the pump could probably easily get another 150–200 gpm with valves at maximum pump out rate. So even at 2% consistency, the flowrate would be 510 gpm; it still may be possible for the pump to handle the flow.

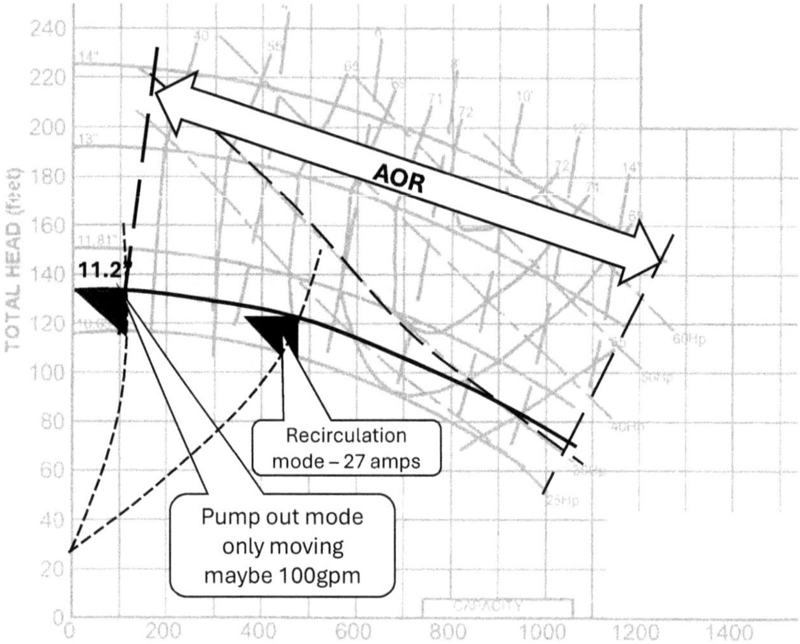

FIGURE 7.15 Pump curve and multiple system curves with operating points.

Data collection of the pump and system details were gathered to begin analysis to see what was going on. There wasn't a good place to get any pressure readings on the pump, so the next best data point was checking the motor load. The motor load was checked with the pump in full pump out mode (valve B 100% open, valve A 15% open) and full recirculation mode (valve A 100% open and valve B 0% open). In recirculation mode, the pump was pulling 23 HP and pump out mode was 10 HP. The pump was driven by a 30 HP, 1,800 rpm motor. The pump design and two operating points measured are shown on the pump and system curves in Figure 7.15.

So, with this data, it appears something is wrong in pump-out mode. The pump is operating near the minimum AOR line or worse and not anywhere near its design point. It appears there is excessive friction head while in pump-out mode. One more comparison to evaluate is what does the design system curve look like versus the as measured or as found system curve. If it doesn't match (I'm already guessing it will not) then there is a system issue.

The theoretical system curve is shown in Figure 7.16 with an existing 2″ control valve and a theoretical 3″ control valve. The 3″ theoretical valve was modeled as a potential solution if we needed more flow without having to install an entire new pump. As shown, the current pump system curve with the

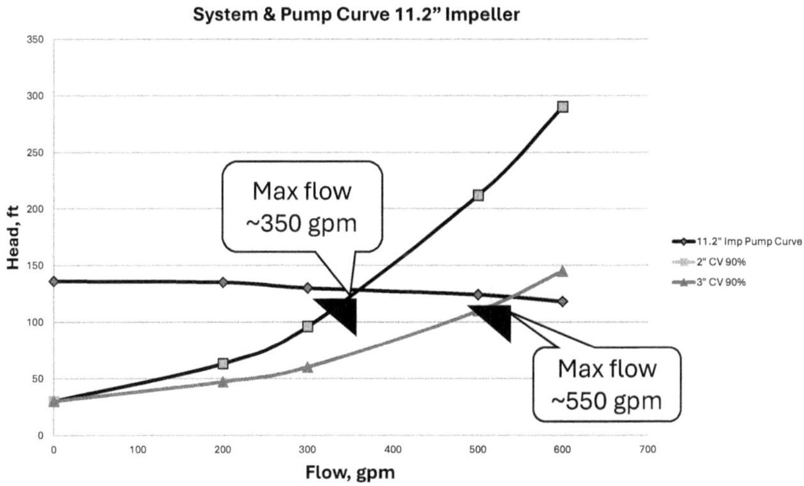

FIGURE 7.16 Pump and system theoretical curves with different control valves.

control valve B 90% open should allow the pump to produce 350 gpm, yet the pump was only producing 100 gpm at best, so there appears to be a system issue, but where and what?

Is there an issue with control valve A or B where a failure is not opening valve B properly? This would be on the next shutdown to verify the control valve function. Before shutdown and inspection of the valves and system, an infrared gun was taken and the system was scanned. With the valves 100% open, the IR photo in Figure 7.17 instantly showed an issue. The line was cold right after the valve (~86°F), while the pump side of the valve B was ~120°F (stock temperature). Either the control valve was broken and not operating the ball or the line was plugged.

The pump system was shut down and the control valve was rolled out of the line for inspection. The valve was verified to be operating but the obvious root cause was revealed. The entire pipeline was plugged with paper stock. No flow was leaving the pulper except when the break pump came on. The trim pump was only recirculating all the time. The curve was flat near dead head, making it difficult to know if any flow was occurring, but the pump was producing zero flow. So, the line was cleaned and normal design operation later resumed. It appeared that the pulper system had been operating this way for a long time, potentially for years even, but the exact duration is not known.

So, what led to this condition and pump system failure became the million-dollar question? There were several factors but the main place to start is the control setup for range of control on valves A and B. B valve was operating 0% open for much of the time. It wasn't starting to open until 75% pulper level. At 80% level, the break pump kicks on and pumps the tank down to

FIGURE 7.17 Pulper trim pump system control valve IR picture.

50%. Once level drops down below 75% again, valve B shuts off again. This continued shut operation can allow dewatering to the stock in the line as there is static head backpressure on the valve.

Manual operation of the system could also have created some conditions to plug the trim line. With three control valves A, B, and C in a complicated split range control, manual operation is not a great operating option. One solution here is to put a minimum open percentage on valve B to maybe 15%. Also open up the operating window for valve B. Original only gave a 5% level for valve B to get trim out of the pulper, maybe lower than 60% level instead of 75%.

Consistency control is also important on the pulper system. Higher consistency with lower required flowrates would make it easier for the valve to plug during operation. Low consistency would require much higher flowrates and then the pump may not be able to keep up.

While not the root cause of the line plugging, there was no cleanout of the control valve B, so adding a cleanout and maybe a flush line would be a good proactive action. Another option to consider is to maybe increase the valve B size to a 3″. This should improve the capacity slightly and make a little more difficult to plug and easier to unplug.

This pump has a design issue with the system controls, which allowed one of the control valves to operate 100% closed much of the time. Valve size and cleanout are also design elements to prevent plugging. The pump issue was obviously a pump system problem and not a pump problem. A costly pump upgrade project was avoided by using fundamental pump troubleshooting to determine the root cause.

Case Study 35 Learning – Before doing any major upgrades on an existing pump system, perform a full pump system analysis to understand the capabilities in the application.

Having a clearly defined timeline of when the issue begins can be highly beneficial for troubleshooting. It can also be misleading, like the previous case study, if the problem start time is unknown, or worse, some make a wrong assumption about when it began. This can make troubleshooting for the pump professional all the more challenging. As the saying goes, "Trust God, all others bring data". If the problem has come up suddenly and the pump system has a history or performing well, then that can be a great benefit to troubleshooting. The next case study discusses that very situation. The analysis detail still must be done but at least there is some success history.

CASE STUDY 36 PUMP PERFORMANCE SUDDEN DROP

A scrubber recirculation pump system was not functioning properly with the pump not producing the flow it was expected to produce. The pump was a small end suction pump sized 528 gpm at 49 ft (125 m^3/hr at 15 m). The pump was driven by a 15 HP, 1,800 rpm fixed speed motor. The change in pump system performance occurred somewhat suddenly. There was no flowmeter on the system, but there was pressure indication which showed there was low discharge pressure.

One thing to consider: is this problem being caused by the pump or the system? The answer to that question will focus the investigation in two different directions – pump direction or system direction. Still some basic operating data would be required such as pump pressures and pump motor load. Early analysis results will direct the focus to the appropriate area.

First consider the pump as the issue. There had been no pump maintenance anytime in the previous 12 months where the pump was opened up. This being the case, the same impeller and pump parts are the same as at least the past 12 months. If the pump had recent maintenance, it might be possible to have installed the wrong impeller on the pump or set the pump impeller clearance up incorrectly.

While impeller wear would gradually occur over time, the pump issue had come up suddenly. This didn't seem to match with a pump wear-related problem involving the impeller or other hydraulic components. Nonetheless, these things had to be verified to be correct. The pump impeller clearance was verified easily and found to be within range. The impeller clearance range was 0.020/0.028″ and it was found set on the low end at 0.021″, which would be more optimal and have lowest impact on the performance. It was decided to verify the impeller diameter on the next downtime opportunity.

Next, looking at the pump operation within the system, some basic operating data was required. The pump operating pressure, TDH, was 10 psi or 7 m. The pump motor was operating with a motor load of 8.8 HP. Now these two operating datapoints by itself may not reveal a complete picture, but put together with the pump performance curve some observations can be made.

The pump operating pressure said the pump was producing a lot of flow (off the end of the curve as a matter of fact). The pump motor load did not line up with that data at all. The pump motor load indicated it was operating well to

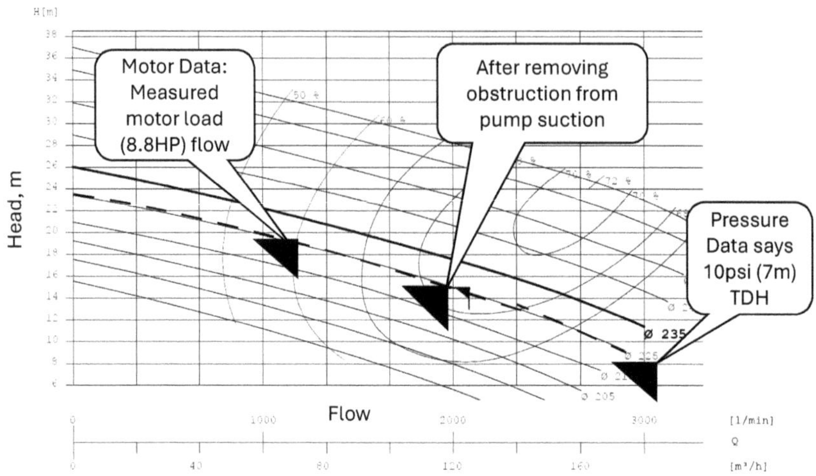

FIGURE 7.18 Pump curve and measured data points.

the left of BEP as shown in Figure 7.18. When things are right on the pump and pump system, all the operating data collected should reveal one operating point (or close to it) for the pump. In this case, there was a stark contrast.

Looking back at Figure 7.13, when the pressure is lower than it should be for a given pump flow (via motor load in this case), one of the possible causes is a plugged pump suction. There was not a place to install a suction pressure gauge, so a physical inspection was scheduled to inspect the pump suction. When the pump was opened up and the suction line inspected, a bump cap (part of plant personal protective equipment, PPE) was found plugging the suction line. After the suction line obstruction was removed, the pump returned to good operation near BEP and original design.

There was a temptation by operations to have the pump replaced, but replacing a pump would have not corrected this pump system issue. Determining if this was a pump problem or a system problem is very important. If that had been attempted, hopefully the maintenance techs would have looked and found the obstruction in the suction line but without instructions to look in the line itself this easily would have been missed. If not, there would have been at least one more maintenance job to correct. This issue was more of an operational issue allowing a contaminant into the pump system.

Case Study 36 Learning – General: When pump operating data contradicts each other, evaluate the troubleshooting guides and basic pump hydraulics to test for what the data is describing as the problem. Specific: Low discharge pressure and pump low flow output can be internal recirculation or plugged suction.

PUMP SYSTEM RCFA

The previous topic mostly involved pump system troubleshooting with some RCFA. Pump system RCFA is in some ways the same and in some ways different. The previous troubleshooting charts in Figure 7.13 are an example of a troubleshooting exercise where there is a problem and looking for the single solution to resolve, but it isn't really a root cause failure analysis. The trim pulper pump case study had troubleshooting and RCFA where the root cause of the line plugging was evaluated. The plugged suction line didn't really complete the RCFA on what caused the obstruction to get into the pump but only completed troubleshooting and correcting the plugged suction.

In another example, a lift pump may not be delivery liquid. Say the reason is that the pump has lost its prime. So the action is to reprime the pump. But what is the root cause of why the pump lost its prime? Maybe that the check valve failed and allowed the suction line to drain back down? And what caused the check valve to fail? These last questions would be part of the actual failure analysis and not troubleshooting. In troubleshooting, the problem may come back again. RCFA seeks to determine the root cause of the trouble and take some action to prevent re-occurrence.

Troubleshooting has a few differences in actual failure analysis. The first being the end goal of the activity. Troubleshooting goal is to find whatever is causing an issue, then to correct that (change the part, etc.), and get back to operating. RCFA's goal is to determine why the part failed. What is the root cause of failure and how to prevent it? Another difference is failure analysis always takes things down to the component level of failure to focus analysis there. Troubleshooting stops at a much higher level. Troubleshooting involves just doing enough to get the process or pump system back in operation, while failure analysis seeks to put things in place to prevent future failure events from occurring.

For a pump system, troubleshooting is very important so that process can get back in operation. However, failure analysis is also vital for pump reliability to make sure failures do not repeat and become chronic. If they are chronic then failure analysis becomes the most important task to eliminate the reoccurring failure events. Below are a few simple steps to complete a pump system failure analysis.

Steps for pump root cause failure analysis (RCFA)

1. Define the failure event. Some questions to answer may include.
   - When did the event occur?
   - What were the operating conditions at the time of failure?
   - What is the timeline before, during, and after the failure event?
   - What is the financial impact of pump failures in downtime cost and repair cost?

   Take pictures of the failure scene, controls conditions, process trends, etc. around the time of the failure. Only have one moment in time to capture this data.

   One thing to remember about pump systems is there can be a functional failure without a mechanical or electrical failure. A pump that fails to deliver the correct flow, gpm, or pressure, psi, in its system as designed then it has experienced a functional failure even though it continues to operate. A mechanical failure also prevents a pump from delivering flow and head, but the RCFA takes on a different nature than purely a pump system functional failure.

2. Determine the RCFA team, budget, tools, etc. The key here is the RCFA leader – there needs to be one or there will not be a structured approach to the RCFA process. The leader will drive other resources to execute the failure analysis and determine the root cause. Identify other key resources such as subject matter experts (pump specialist, process engineering, maintenance, purchasing) needed to complete RCFA tasks. Budget is also important in that it may determine the level of RCFA and confidence in the conclusions. More budget means more analysis data to determine the root cause.
3. Save the failed equipment, parts, components. Without the right failed components, the correct outcome will not result from the RCFA efforts. Not only save the failed parts but preserve them from being contaminated for analysis. Don't clean up the parts until proper inspection had been done. Even the smallest clues from the failure may be cleaned up, covered up, or removed accidently after the failure event. Always try to witness the disassembly of the pump after a failure so "as found" condition is clearly documented and understood.
4. Define the failure mode of the failed parts. Describe how the components failed in as much detail as possible to define the failure mode. For example, the impeller was pitted on the suction side of the inlet vanes a few inches from inlet tip evenly on all four vanes. Pitting covered a 4 square inch area and was 0.187″ deep after only 18 months of operation. The more completely the failure mode is described, the easier and more likely it is to get to the root cause. Some common basic failure modes may include but not limited to the following:

   - Wear – 2 body or 3 body abrasion, erosion, and smearing.
   - Fatigue – bending (rotating, reversed, unidirectional), torsion.
   - Overload – bending, torsion, shear, brittle, and ductile.
   - Fractured – generic adjective for broken.
   - Corrosion – crevice, pitting, galvanic, and fretting.
   - Plugged, fouled, and scale buildup.
   - Contamination and discoloration
   - Cracked – early stage of fracturing.
   - Flaking, scoring, peeling, denting, and pitting.
   - Brinelling and denting.

   It may also be important to identify any symptoms around the failure. This might be that the pump was vibrating or running high temperature or pulling high motor load up to the time of failure. These are all clues to use in the investigation to determine the root cause.
5. Determine the primary and secondary failure modes. If a pump experiences a catastrophic failure, then there will be many failed parts, but most of them will be secondary failures. Secondary failures occur as a result of the primary failure which began the chain reaction of failure. These should be

identified and described, but the RCFA will only concentrate on the primary failure modes going forward as this is what started the entire failure and where the root cause resides. Big mistakes can be made here when secondary failures are wrongly identified.

6. Identify the possible causes for the primary failure mode. Use physics of failure to assist in list of possible root causes. How can this failure mode occur or what can cause this failure mode? Physics of failure is the known physics of how a machine works or in the pump case how the pump operates. Specifically, what physics can cause that specific failure mode? For example, if the failure mode involves cavitation, there are key physics formulas for evaluating NPSHA and pump information to determine NPSHR. If the failure is a broken shaft, the physics of failure helps determine how the shaft fractured. The physics of failure will guide you in the RCFA process. You'll need an SME to help with physics of failure analysis.
7. Perform analysis on the possible causes proving each one to determine the root cause(s) of failure. Each possible cause must have supporting data to confirm or deny it as a possible root cause of failure. This is where the real work happens. It is always good to start with the original design of the pump system, such as a process diagram, with all critical control components, such as process equipment (valves, screens/filters, flowmeters, instruments, and tanks). Also collect design and current pump details such as impeller size, type, and motor specs. Fluid specifications (SG, consistency, viscosity, temperature) should also be verified compared to original design.

   To complete analysis of the pump system, operating data for the range of pump system operation should be obtained such as motor load, pump system pressures (to determine pump TDH), control valve positions, tank levels, and pressure drops. Usually, enough information can be gathered to at least get a few points to construct a simplified system curve. Red flags may arise in this stage when the pump system operating data does not align. This may be a clue to the issue or just bad data from one source or an error in analysis. The analysis step is also where subject matter experts are brought in. A pump manufacturer, seal specialist, a metallurgist, or pump system engineer may be required.

   Reliability failure patterns such as those shown in Figure 7.19 can also help to understand the failure. Failure patterns can point the investigation to certain areas. If the failure is infant mortality (short life after installation), then the rebuild or replacement should be looked at closely especially if the pump has run reliably for years prior to last intervention.
8. Identify the physical, human, and latent root causes. The end of the analysis will be when the physical root cause has been found. Maybe the physical root cause is that the impeller clearance was incorrect. The human root cause may be that the maintenance technician made an installation error or didn't know the correct procedure or clearance magnitude. The latent root cause may be that there was no correct procedure or impeller clearance target in the work order for technician to follow or maybe the technician had not been trained on that particular pump impeller setup?

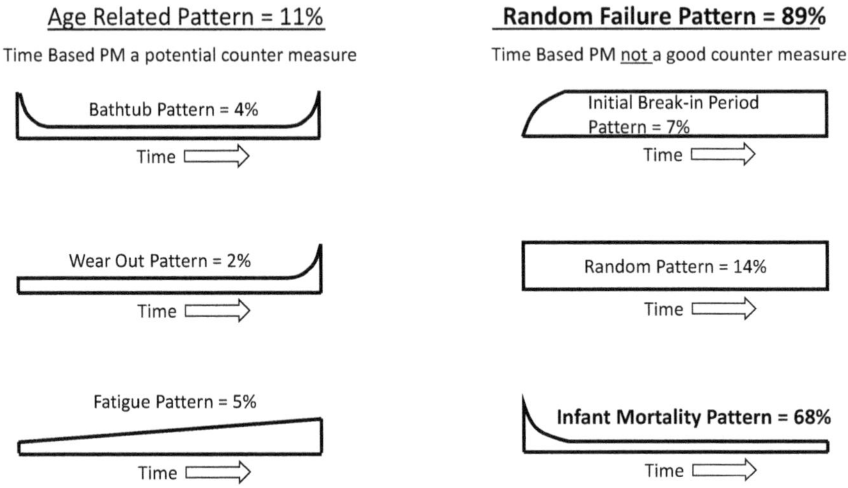

FIGURE 7.19 Failure patterns.

The human root cause can be tricky when the human action involves negligence or insubordination. It is generally not desired to utilize the RCFA process or investigation for discipline, so use caution at the level. But for every physical root cause, there is a human action or inaction that led to the problem.

The latent root cause is the organizational systemic root cause and solution here can branch out and prevent failures in other areas. The systemic root cause is what enabled or allowed the human root cause to occur.

The analysis is also stopped when the last level of control is reached by the organization. It doesn't do any good to say an error happened at the pump factory if the end user can't do anything about it. It is not practical for the plant to disassemble every pump and check pump quality or assembly for an error made at the pump manufacturer. RCFA is not an absolute solution for every event. Use common sense here.

9. Identify the corrective actions to solve each level of root causes. That means action items for the physical root cause, human root cause, and latent root cause. The physical root causes are usually the easiest to execute. If all the RCFA process has been completed but the action items are not accurate and completed, then the RCFA effort yields nothing. Identifying the person responsible for the action and a due date are also important to establish clear accountability and follow up. This is a critical step to prevent reoccurrence not only on a specific problem but also sharing learnings across similar applications.
10. Compile RCFA report. It is always good to document RCFA findings to share with various teams. Not only will some of the learnings benefit other areas but reliability must continually sell and sustain itself with each success in order to earn the right to continue. Many in the organization have not been won over yet to RCFA, and reporting on successful RCFA will help show the benefit to others.

When looking at root cause failure, the findings can also be categorized. At a high level these can be broken down into three categories – design, operations, and maintenance. A pump system is designed, then it is operated, and lastly it is maintained at that original design condition for the life of the pump system. For a failure to occur, it will fall into one of these three categories. The only reason to categorize failures is not to blame but to understand better where to invest to resolve in the future. I see too many not put any investment into design for reliability and then accept a poor reliability product which only leads to premature failure later. If maintenance or operations training is needed to correct the failure, then this must be an action item to prevent reoccurrence, or it will be doomed to repeat itself.

Design sets the stage for what is to come. I throw in original installation into the design category as well. Design is normally not an easy thing to change. Get the design wrong or not designed for reliability and a plant may be stuck with it for a long time. Design cements its DNA into its own reliability. To change a design, usually capital investment will be involved, although sometimes small changes to the pump or system can resolve design issues. Chapter 10 has some surprising results on pump system reliability as it was broken down in a study of pump system failures.

Operations play a critical role in how pump systems are started up and where they are operated within the process the pump system supports. Operations have little control over the grades or rates they operate, but they can make good choices about the way the pumps operate within the process limits. Bad habits or bad strategies can only be a choice by the operator in the seat. If an operating parameter changes away from an original design point, the operator must still run the process even if it comes at the detriment to the pump system. Sometimes, how the system is operated is directly responsible for pump system failure. An example would be operating a tank level well below minimum submergence which will contribute to pump issues.

To keep the pump system in good condition, regular maintenance and, when necessary, breakdown maintenance will be performed. Restoring the pump system back into good condition is the job of maintenance. Maintenance can't operate the pump better and it can't redesign the entire pump or pump system. Maintenance maintains what is already there. Not putting the pump system back in the correct condition to operate reliably would be a mistake by maintenance execution and not one of design or operation.

Most of all the case studies show the categorical root cause areas as well as the specific root causes. Design issues can have a lack of competency or experience by pump system designers. It can also have a lack of proper budgeting to engineer the pump system properly or support design for reliability upon installation where shortcuts were taken. Operations and maintenance usually fall into a lack of training and/or experience to execute proper pump system functions that result in a reliable pump system. These things are not easy to correct, but this is the power and influence of pump system RCFA and is the key activity for continuous improvement in pump system reliability.

Note

1 1991, Paper Process Engineering, Rich Alger, Champion International.

8 Pump Maintenance

Pump maintenance can greatly affect pump performance and reliability. Anytime the pump is worked on, the chance for infant mortality (premature failures) is possible. Maintenance has the assignment to restore the pump and system to its designed condition for optimum performance. A functioning pump is now being disturbed with maintenance, and there is always a chance for something not to be just right after maintenance is completed. The problem or failure may not show up right after startup, but the issue may originate from poor maintenance practices. This chapter will highlight some common areas where maintenance must focus and execute with precision if the pump system is to return to full performance and reliability. It all starts with pump maintenance safety.

PUMP MAINTENANCE SAFETY

As with any maintenance work, always ensure the pump has all energy sources isolated and potential energy removed with proper equipment lockout. Wear proper PPE (safety glasses, gloves, and steel-toe shoes) when working on a pump. Gloves are a must when handling a pump impeller as there are very sharp edges. Before disassembly, ensure that the pump has been drained and if a hazardous chemical is in pump, the pump has been flushed and drained properly. Verify with the process area where the pump was removed to confirm pump contents and then get the material safety data sheet (MSDS) or safety data sheet (SDS) for specific hazards and protection before working on the pump. In waste handling pumps, be cautious that there can be all kinds of foreign objects (needles, etc.) that can be dangerous when disassembling the pump.

Also, special gloves or other PPE may be required at pump disassembly. If the pump is hot from the process, either allow to cool to a safe disassembly temperature or wear proper high-temperature PPE when working on the pump. On installation of hot coupling hubs or bearings, wear proper insulated gloves.

Pump rigging is also very important. Use appropriately rated slings, hooks, shackles, etc. when lifting pump or pump parts. Some components may even need special lifting on larger pumps. Some pump power ends (bearing housing assemblies) may weigh as much as 700 lbs on large end suction pumps. The pump weight in pump manuals for these does not include the pump coupling or forces used to pull out of casing, so be sure to have a safety factor in selecting rigging materials.

Also, for changing some pump components, use proper rigging. Large pump impellers and suction wear plates can be very heavy beyond what can be ergonomically handled by a mechanic. Large suction wear plates may weigh as much as 120–140 lbs and impellers 200–300 lbs. Special procedures are typically recommended for changing these heavy components on large pumps; therefore, consult pump OEM installation and maintenance manuals.

DOI: 10.1201/9781003729525-8

A general rule is to never rig bolts through a pipe flange hole to lift pumps. Acceptable rigging points are under the shaft and suction nozzle or under the main body of the pump. Consult pump manufacturer for special rigging instructions. Guard and protect slings from sharp corners when rigging. In general, do not lift a pump and baseplate together by rigging to the pump. Rig to the baseplate at appropriate rigging locations if moving an entire pump and baseplate assembly.

When changing a pump, drain the used oil from the removed pump before taking it to repair area or preparing for shipping. Pumps tend to leak from seals and will create unsafe conditions. Use special storage and shipping containers for pumps if available, as typically these have the pump secured in a safe shipping position and anchored to a skid or wood crate. Inspect pump shipping container for damage as any damaged pump containers (skids, boxes, or pallets) can allow the pump to shift during shipment that could cause personal injury or damage to shipping vehicles.

Ensure that pump shaft guards are bolted down at all fastened locations. Also, ensure that guard protects any body parts from contacting the pump shafts at a safe gap and distance. The OSHA ramped gap to distance "gotcha sticks" are a good tool to ensure this is done correctly. The farther away from the shaft, the larger the gap can be.

CASE STUDY 37 RISKS IN PUMP MAINTENANCE

Pump maintenance can have considerable risks with many not being obvious. This case study is just several different safety-related issues around pump maintenance where significant injury or fatality occurred. Pump maintenance seems like a safe and easy task, and it is most of the time but even the smallest of mistakes can have serious consequences. These examples are just reminders of how easy a simple task can become life-threatening.

PUMP CHANGE EXPLOSION

A routine pump change was scheduled, and the job was progressing along good. The pump was removed from the casing with no issues. The rotating assembly was in the shop to swap over the impeller to the new rotating assembly. It was a screwed-on 316SS impeller and a fairly large one. Attempts to get the impeller unscrewed from the old rotating assembly were not successful.

Like any mechanical part removal that might be frozen to the shaft, a little heat and a hammer will usually break it free. So, the maintenance technician proceeded to use a torch to apply heat on the impeller. After a few minutes of heating, the impeller exploded with metal shrapnel going in every direction. One of the maintenance technicians was hit in the shoulder with one of the metal pieces. He had cut a major artery and was in serious trouble. This may have resulted in a fatality had it not been for one of the first responders arriving soon enough to put his hand inside his shoulder on the severed artery and stop/slow the bleeding until the air flight helicopter arrived. Fortunately, great response by fellow employees and medical responders saved the man's life.

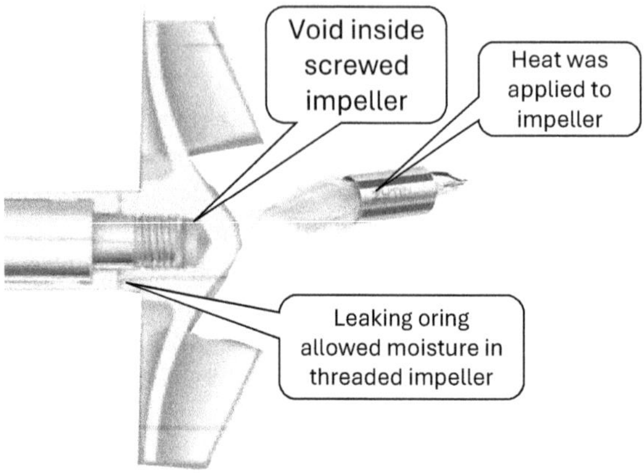

FIGURE 8.1 Screwed impeller cross-section.

So, what caused this failure? This probably was not the first time that some maintenance technicians had used heat on a pump impeller, but it may have been for this technician. After the investigation, this was the most likely theory developed for what had happened. The impeller oring has an important function to seal off against the shaft shoulder. If the impeller oring is missing, damaged, or reused, it may fail and not seal off from the pumped fluid. Fluid then may find its way into the impeller and shaft-threaded void area. Many times, this will cause some corrosion, and the screwed impeller will be very difficult to remove. There was likely trapped water inside the impeller thread area, as shown in Figure 8.1. When heat was applied, that water turned into steam. With the drastic volume change of water to steam, the pressure climbed rapidly until it exceeded the impeller's strength and exploded.

PUMP CHANGE FATALITY

During another routine pump change at a mill, maintenance technicians were changing a condensate pump. Another job that had been executed numerous times. There wasn't anything particularly special about the job. Once the pump was removed, it did have to be lowered to the floor below with a beam hoist crane through an opening in the floor. The same ceiling hoist was also used to lift the pump out of the casing as it was being removed.

In preparation for the pump change, the floor plate was removed. Therefore, as the pump was removed from the casing, it was ready to move down with the trolley and lowered down to the floor below. During the removal of the pump rotating assembly from the casing, a sudden gush of hot condensate was released onto the maintenance employees as the assembly cleared the casing. One of the maintenance technicians lost his balance and fell through the opening in the floor to lower the pump (it was very near the pump location). The employee suffered a fatal fall in this accident.

This tragic accident probably had many elements where it could have been prevented. I never heard what occurred with the lockout that allowed the pump to dump hot condensate when the pump was removed but there were some obvious misses on the lockout to ensure all the condensate was drained from the pump. Some potential items below that could have prevented this fatal accident.

- Was the lockout executed correctly? Was a valve not shut or leaking?
- Was there a bleed valve on the pump to ensure it was drained and depressurized?
- Why was the employee not tied off while working at a dangerous height? Was all PPE applied when doing a line break for hot condensate?
- The floor plate should be removed after the pump rotating assembly has been removed.

PUMP CHANGE HAND INJURY

Maintenance was being performed on an end suction pump as the pump rotating assembly was being changed. The injured employee had a hand on the pump coupling hub as it was being removed from the pump casing. When the pump suddenly released from the pump casing, it swung into the motor shaft/coupling, and the hand on the pump coupling/shaft suffered a contusion.

Couple of learnings from this accident. Pump rotating assemblies can be difficult to remove from the pump housing. Even with casing removal bolts, it may require additional force to remove. Many times, the mobile crane will pull slightly so when it does release, it pulls the pump assembly uncontrollably. It is a good idea to have some secondary rigging to control this release potential energy upon removal to limit the motion of the pump rotating assembly.

While the use of hand personal protective equipment (PPE) helps, it does not prevent hand contusion. However, getting the hand out of the line of fire does prevent that. One option is to tie a rope to the pump or similar item to control the sudden motion while keeping a safe distance from the moving parts.

All these safety events are maintenance execution issues. There could also be a design-related element that if designed better, then some of these accidents may not have happened. Two of the three real examples here involved fatal and near-fatal incidents, therefore, you can never take pump maintenance safety for granted.

Case Study 37 Learning – Never heat an impeller when removing from a pump shaft, always replace impeller oring with maintenance, wear proper PPE, and stay away from line of fire safety hazards.

PUMP STORAGE AND SPARES

Another pump reliability-related topic on the front end of pump specifications and life cycle cost is preparing a purchased or spare pump for long-term storage. Pumps are typically not stored with lubricants; however, if the storage exceeds one year or less, bearing parts should be protected with a vapor-phase corrosion inhibitor or lubricant. The pump should be sealed up as much as possible to hold the vapor-phase inhibitor.

There are several factors attacking reliability of stored pump equipment such as oxidation, corrosion, gravity, vibration, and low material turnover just to name a few. Relative humidity above 50% have large increase in corrosion rates on steel. Shaft rotation every three to four months is a best practice on rotating equipment so that fretting corrosion and false brinelling do not occur on bearings. This also recoats them with lubricants so it also protects against general corrosion.

An uncoated shaft, as shown in Figure 8.2, can have significant corrosion rates in the right storage conditions. For example, the corrosion rate in high humidity environments can be as high as 0.001″ per year. If a shaft bearing target interference fit is a k5 fit for a 3.5″ diameter shaft, which is a target fit of 0.0001/0.0015″, then in a year of corrosion the bearing fit can be lost. Also consider a 3.5″ shaft where it takes a coupling hub fit of 0.0015″/0.003″ interference fit; this proper shaft fit could be compromised in two years of unprotected storage.

The vapor-phase corrosion inhibitor protects the bearings and shafts inside the pump but many times there is little done to protect the outside of the pump. Corrosion of shafts can greatly affect the resulting shaft coupling hub interference fits. A bare shaft could easily lose 0.001–0.003″ of shaft diameter over many years of storage in a high corrosive environment. This would erase the coupling fits on many applications. Some shaft protection may help prevent corrosion and others like Veril

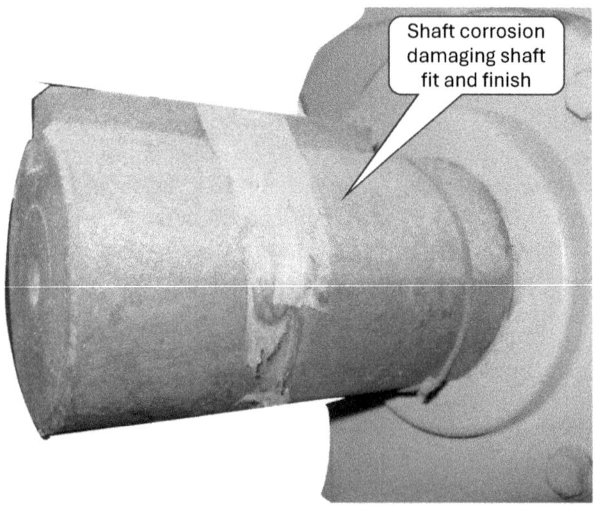

FIGURE 8.2 Shaft corrosion on shaft in storage.

Pak is protective packaging material that combines a highly formulated adhesion wax and a high strength fabric, which acts as a total barrier against high moisture, acidity, bacterial attack, or galvanic attack.

Do not store spare pumps in areas where vibration at the pump skid may exceed 0.05 in/s. False brinelling or fretting corrosion could occur if the period in storage is long enough in high vibration conditions. If a storage area does have or prone to vibration, rubber padding can be used under the storage container or pump to dampen vibration transfer to the spare pump.

Spare pump rotating assemblies should be secured in proper storage containers where the pump is bolted down and can't move around in shipping and handling. This prevents shipping and handling damages – brinelled bearings, bent shaft, broken mounting, etc. The containers can be made of wood or even steel. Having the pump in proper-level storage position also helps keep lubricant in the pump without leakage. Having a container also provides a safe method to pick and handle moving spare pump rotating elements. If the spare pump is complete with casing, never rig to the pump flange to lift and move a pump.

Typically, only part of a pump assembly is kept for spares in a plant at the site storeroom. The spares may consist of a complete power end (rotating assembly – bearing housing, shaft, adapter). This is usually sufficient for spare rotating assembly that spares many different pumps. If a spare pump only spares one pump location, then it may include a stuffing box, adapter, and impeller. Do not have a spare built complete through the impeller if it spares more than one location. It is easy then to install the wrong rotating assembly and wrong impeller to the wrong application and pump system. If spare impellers are needed in the plant, generally a full-sized impeller should be set up so it can be trimmed to fit whatever application it is being installed in. I've seen this failure occur many times over the years and these types of failures can be very costly. Case Study 40 is an example of this very thing happening.

Other components that will need spares will include shaft sleeve, impeller, suction wear plate, stuffing box, mechanical seal, packing, orings, gaskets, special fasteners, and coupling elements. All the components needed to replace a pump power end during routine maintenance. Pump casings generally should not be needed as a spare as they should not be high wear items unless high wear rates from abrasive fluids are being pumped where replacement is necessary on a frequent basis. If the pump power ends are rebuilt at the plant site shop, then bearings, shafts, shaft oil seals, etc. will also be needed.

IMPELLER MAINTENANCE

One of the main components of a centrifugal pump is the impeller. The impeller is the main part that interacts with the fluid to produce flow and head to operate the process. Therefore, the impeller must be right or the pump will not be right. Meaning, it will not match the performance curve for which it was designed and function (head and flow) it was selected to produce. Maintenance has a huge role in getting and keeping the right impeller in the pump to produce target performance.

If we start with a new pump, there will come a time when the rotating assembly or seal or some invasive work will be executed. The pump could also have a work request to inspect the impeller due to pump performance issues. When the pump is opened at the first opportunity, the impeller, wear plate, and casing all should be inspected. First question is what areas should we inspect? What is too much wear? When do we need to replace? How much wear can the pump handle on this application?

For the casing, inspect the volute area, suction nozzle, discharge nozzle, wear plate area (especially if it is made into the casing), machined sealing areas, and cutwater. It is tough to put a hard line on general wear on a pump especially if the pump is still providing full process function. It depends on the fluid and criticality of what is being pumped too. If spillage of the fluid is critical then monitoring casing wear will take a more aggressive approach. There is a lot of managerial judgment on many pump applications. Consult your resident pump subject matter expert or engineer to help determine the severity of wear and potential action. In an abrasive application, using 1/16″ wear in any area may result in replacing the pump casing every six months, whereas in reality it will probably continue to run for 18–24 months before reaching a critical point. I'll try to give some general comments and a few specifics.

If a leak from the casing is not urgent then a less aggressive approach may be satisfactory. If a machined area is worn or pitted enough that oring or gasket will not seal, then replacement should be scheduled. In general wear, more than 1/16″ on wear parts (impeller, wear plate, casing critical areas in Figure 8.3) might be a candidate to be replaced. Impeller wear that reduces the thickness of the vane is a direct link to loss of pump capacity. Some pump manufacturers may supply what a full-sized impeller vane thickness is. If you have this and a full-sized impeller, you can easily determine how much is the wear by measuring the thickness. If you can document this on a new trimmed impeller or request the manufacturer to provide you the details, impeller wear can be precisely monitored.

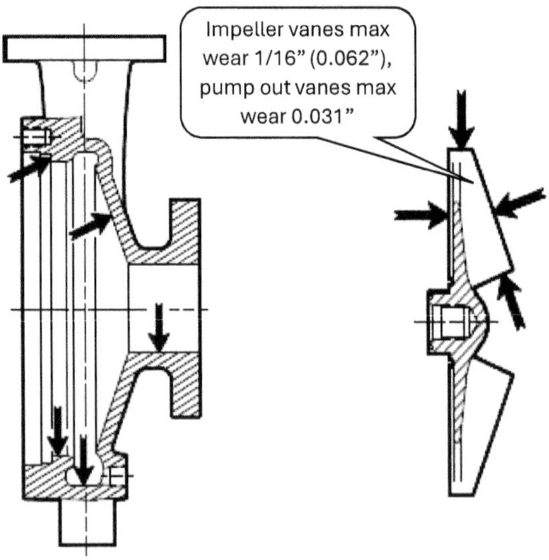

FIGURE 8.3 Pump inspection areas.

Impeller wear can cause many issues with one being hydraulic inefficiency. This not only leads to wasted energy but will also at some point lead to a functional failure where the pump will not meet system demand for flow and head. The next case study is an example showing the impact on hydraulic efficiency for a pump.

CASE STUDY 38 PUMP WEAR AND HYDRAULIC EFFICIENCY IMPACT

One example of the impact of these critical wear areas on the pump impeller or suction wear plate area can be seen in Figure 8.4, where the suction wear plate had significant wear. The wear was ~0.187″ at the impeller diameter. The impeller clearance had been previously reset so the actual running clearance was less than 0.187″ but there were significant wear and loss of hydraulic efficiency on this pump system. With the significant wear, as maintenance was performed on this pump for the last time, the wear was greater than the total travel of the pump shaft and bearing assembly as it never touched off against the suction wear plate.

The pump was an 8 × 8 × 14 end suction pump with a design point of 1,700 gpm at 48 ft. The pump was driven by a 30 HP, 1,200 rpm VSD motor. Evaluating the pump performance led to what initially was conflicting performance results. The motor load, pump system flow, and pressure readings were taken and plotted on the pump performance curve, as shown in Figure 8.5. The system had a process flowmeter on the pump discharge which was reading ~1100 gpm. The pump pressure gauge readings were putting the pump TDH ~35 ft.

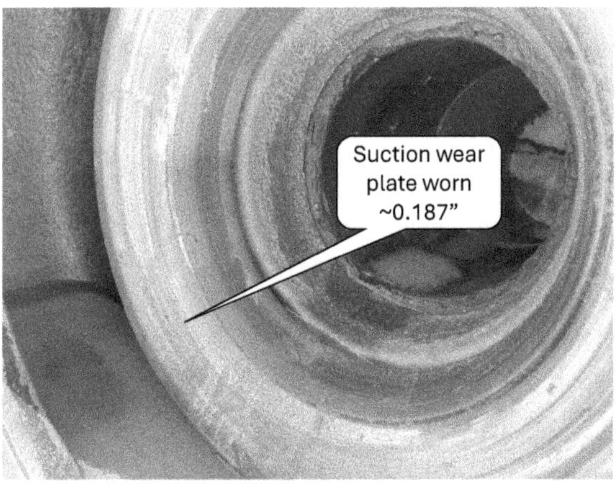

FIGURE 8.4 Suction wear plate wear.

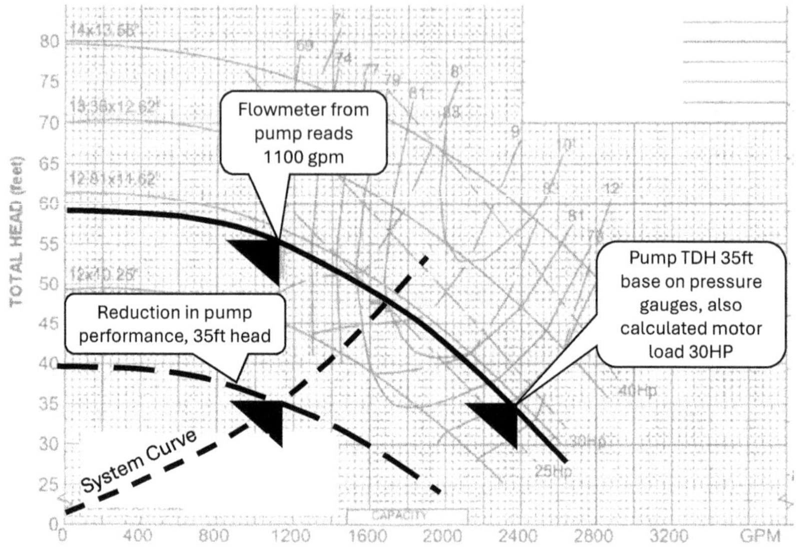

FIGURE 8.5 Pump and system curves with operating measurements.

The motor load was measured at 39 amps at 469 volts. The motor was a 30 HP, 1,200 rpm motor with a 38.2 FLA. It was operating on a VSD at 100% speed. With a pump efficiency of 92.4% and power factor of 0.79, the calculated motor BHP was 30.3. The three pump performance measurements gave a conflicting picture at first glance but made sense considering the pump wear previously noted. The pump TDH pressure is a legit system pressure as the pump performance has been reduced due to hydraulic inefficiency. The motor load is also legitimate as the pump internal recirculation increases the motor HP. The discharge process flowmeter is correct in the actual pump flow produced within the system.

Estimating the pump efficiency off this data can be done by calculating the WHP of the pump and BHP of the motor. The WHP is calculated at the system head (35 ft) and flow (1100 gpm) it is operating. The WHP was ~9.7 and the motor BHP ~30.3. The actual pump hydraulic efficiency was 32%, while the pump curve efficiency was ~69%. The loss of hydraulic efficiency caused ~16 HP energy loss at a cost of $6,000 annually. If the pump is brought back to curve efficiency, the BHP would be ~14 HP. Due to the pump inefficiency, the VSD was not being utilized as the VSD was operating at 100% to compensate for the pump efficiency loss.

In another example, a 16 × 16 × 19 end suction pump was operating at 8,400 gpm (process flowmeter on system) at 84 ft and 83% efficiency. The pump was operated with a 250 HP, 1,200 rpm fixed speed motor with FLA of 33.5. The pump system had random events of tripping out. When the multilin was observed during normal operation, it was pulling between 97 and 103% of FLA.

Calculating pump hydraulic efficiency revealed the issue with high motor load. WHP was estimated at 176 HP. The motor had an efficiency of 95.4 and power factor of 0.81. At 33 amps, the motor BHP was bouncing around 250 HP as the multilin power feed indicated. The calculated pump hydraulic efficiency was 75% instead of the 83% it should have been. The 9% efficiency loss pushed the motor near overload. If the pump was at curve efficiency, the pump BHP would be near 212 HP. The corrective actions at the next opportunity were to reset the impeller clearance to the target to restore hydraulic efficiency.

Case Study 38 Learning – Any significant pump wear will decrease pump hydraulic efficiency and ultimately wasted energy and can also affect pump system reliability.

Cutwater wear is generally somewhat straight or U shape. Wear in cutwater over 0.125″ will begin to affect pump performance and should be monitored for potential corrective action. Again, if pump system design margins allow operation, continued operation with some cutwater wear is acceptable. With excessive cutwater wear, shown in Figure 8.6, exceeding 1″, pump performance becomes increasingly uncertain.

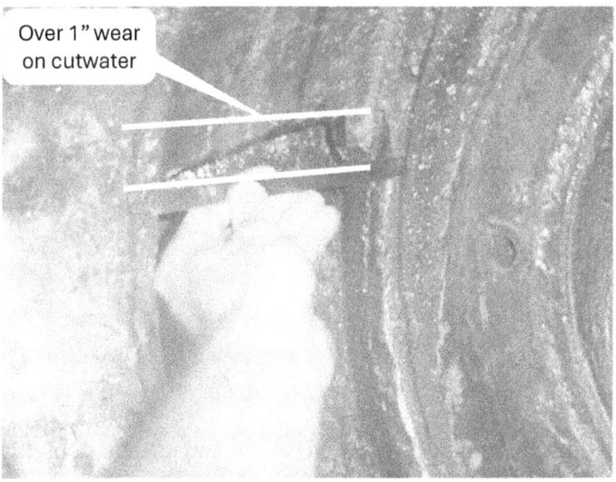

FIGURE 8.6 Pump casing cutwater wear.

In general, pump wear is proportional to the cube of the speed, and the reason why slurry pumps are big and slow. If wear is occurring in a pump, consider the wear rate (and application speed) when determining when to change components.

For an enclosed impeller, the critical wear area from a performance standpoint is the impeller and casing wear rings. This is similar to the impeller and suction wear plate clearance on an end suction pump as far as performance impact. Unlike an end suction pump, the wear and increased clearance on an enclosed impeller cannot be adjusted to restore pump performance. Only replacement of worn wear rings (on impeller and casing) will correct a severe wear issue.

To estimate the hydraulic efficiency loss for a double suction pump, there are several calculations to complete. Basically, the analysis is looking at flow across an orifice which is the area between the wear rings driven by the differential pressure across that clearance area. Below are some steps to estimate the flow from the wear ring area.

1. Solve for diameter of eye from:

$$\text{Impeller Eye Area} = \pi/4\left(D\text{eye}^2 - D\text{shaft}^2\right)$$

2. Solve diameter of ring:

$$D\text{ring} = \text{ring thickness} + D\text{eye}$$

3. Pressure (head) at Dring:

$$H\text{ring} = \left(\frac{D\text{ring}}{D\text{impeller}}\right) H\text{operating}$$

4. Ring area:

$$A\text{ring} = \pi D\text{ring}\left(\text{ring}\frac{\text{clearance}}{2}\right)$$

5. Volume flowing across ring:

$$V\text{ring} = 0.61(2gH)^{0.5}$$

6. Ring flow:

$$Q\text{ring} = V\text{ring} \times A\text{ring}$$

In general, let the process and pump system design and current operation determine the criticality of the wear. If the wear rate and degradation of performance will not lead to functional pump failure before the next scheduled maintenance, then rebuild can be delayed. If the process control valve or VFD is still in a good control range as I monitor the pump wear, then I probably continue to operate the slightly worn impeller.

Impeller back pump-out vanes are on some impeller designs to reduce the pressure behind the impeller. Impeller balance holes also provide a similar function to balance the pressure on each side of the impeller. Balance holes may reduce the axial load as much as 3X. Both back pump-out vanes and balance holes reduce hydraulic efficiency as energy is consumed.

There is one area where impeller wear is worth a different look to go ahead and change even though the process function is still good. That is the back pump-out vane clearance with stuffing box or seal plate. As the back pump-out vane wear or clearance increases, so does the thrust load on the pump shaft and bearing assembly. There are several factors that can increase the clearance on the back pump-out vanes. Wear on the suction wear plate, impeller wear, back pump-out vane wear, and stuffing box wear. All these influence pump clearances and reaction forces. The cumulative amount of impeller wear, suction wear plate, and back pump-out vanes can combine to create a high axial thrust condition on the pump shaft and bearings. Having the wrong parts for pump assembly may also result in wrong clearances on the pump. The next case study will show an extreme case of this and what the results can be from high pump axial thrusts forces.

CASE STUDY 39 PUMP THRUST SPLITS BALL BEARINGS

An end suction pump experienced a strange bearing failure early into its life (infant mortality). When the pump was disassembled, the drive end thrust bearings experienced catastrophic failure where the balls were split in half, as shown in Figure 8.7. This was an unusual type of pump bearing failure so a closer look at the physics of failure would assist in finding the root cause.

The inner race had a rolled edge where the thrust ball bearings heavily rolled on edge. This is likely what led to the split in the balls. The thrust bearing was a Fafnir 5313 K C3 with a load rating of 37,800 lbs. The markings on the inner race confirmed the position on the shaft as being the drive end. The balls showed to have a high contact angle on the inner race as the inner race edge was rolled over indicating a very high thrust load. It was concluded that the high contact angle and high thrust load is what eventually fractured the balls.

The pump was a $12 \times 10 \times 17$ end suction pump with a 16.375″ semi-open impeller. Checking the pump operating point, it appeared to be operating just to the left of BEP with an operating point at 5,300 gpm at 235 ft. Operating near BEP, the radial loads will not be excessive on the pump.

A high thrust load would occur if the pressure balance on the impeller was significantly off design. The inlet pressure was very low as the suction head from the feed tank was only ~12 ft and the discharge head on the pump was ~235 ft (102 psi). The resultant thrust load on the thrust bearing would push the contact angle on the edge as found on the failed bearing and as shown in Figure 8.8. All this matched the pitting on the bearing inner race as well.

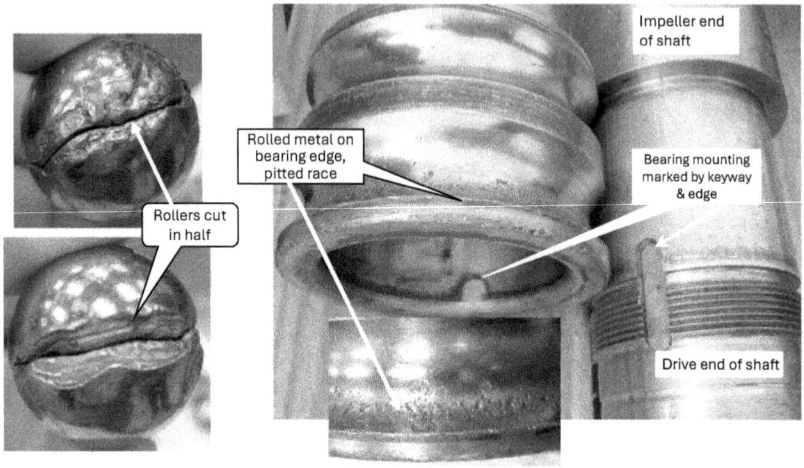

FIGURE 8.7 Failed drive end thrust bearings.

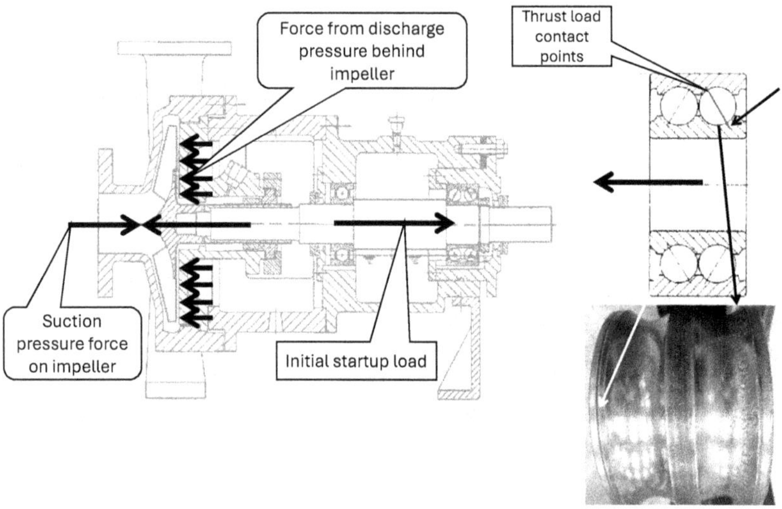

FIGURE 8.8 Pump bearing resultant bearing load.

So, the question then became what made the thrust load higher than normal operation on this pump? Looking at the pump upon disassembly, a large gap was observed between the back of the impeller and the stuffing box, as seen in Figure 8.9. This gap was ~0.375″ where the target gap should have been

FIGURE 8.9 Pump impeller and stuffing box gap.

~0.020″. Under normal circumstances, the proper impeller gap with impeller back pump-out vanes would reduce the pressure behind the impeller and stuffing box. With this large gap, the pressure was very high (probably very near discharge pressure 102 psig).

The impeller and suction wear plate were in good condition with basically no wear. Upon installation, the impeller clearance was installed correctly which should have left the gap between impeller and stuffing box at the target 0.020″, but it was very far off. Further investigation revealed that the stuffing box and adaptor design was for a different assembly, resulting in excessive impeller and stuff box gap.

This pump had previously run for many years with numerous changeouts without this type of bearing failure, before suddenly experiencing a catastrophic failure. Another infant mortality failure which can point to an assembly error, installation error, or improper maintenance as the root cause of the failure. The failure was obviously a pump failure not a pump system failure.

Case Study 39 Learning – Never operate with excessive impeller clearance between impeller and stuffing box. Verify all pump parts are correct for the installation.

Let's assume that now it is time to change the impeller. There are several details that must be executed correctly to get the pump back to proper pump design performance. Each pump performance curve is a specific pump design. The same pump size and casing many times may have several different pump impellers and performance curves. Therefore, the first thing to verify is that the correct impeller is supplied for that pump. On the rear of most impellers is a cast number that is the impeller pattern number. This number must match the impeller pattern number on the pump performance curve if you want to match the performance of the pump. Sometimes the differences may be the number of vanes on the impeller but other times it may not be obvious. Many of these same impellers will fit onto the pump but will not have the correct hydraulic performance.

CASE STUDY 40 NEW IMPELLER, DIFFERENT PERFORMANCE

A pump was rebuilt including a new three-vane 15″ diameter impeller. Upon startup, the pump was not performing as expected and was as good as the old pump. The system in which the pump operated, pressure was more of a factor than flow. As the pump was removed and inspected, it was observed that the impeller clearance was very large and also the fit of the impeller with the suction wear plate did not seem correct. The impeller diameter was verified to be correct at 15″. The impeller was also a three-vane impeller as expected. It wasn't until the original pump performance curve was pulled up that necessary detail revealed the issue.

This particular pump had multiple impellers which could fit the size and assembly. Now this is not unusual for some sizes but many times the different curves with different impellers may be an impeller with a different number of vanes. This specific pump had two different three-vane impellers. The impeller pattern numbers on the pump curves and cast on the impeller confirmed that the wrong impeller was installed. The wrong impeller design was installed on the pump and the lower-performing impeller was installed by mistake. The difference in the pump performance for the two impellers is shown in Figure 8.10. At higher heads, the difference in performance was ~500 gpm.

In some systems, this slight change may have gone unnoticed, but this system did not hide the difference. Not only was the pump performance from a head and flow standpoint but other performance parameters were also different such as NPSHR, BEP, inlet vane area, and max sphere size. Impeller 1 had a BEP of 73% while impeller 2 had a BEP of 66%. This means a potential motor load issue if the design did not have full-curve runout at the connected driver HP. With impeller 1 having an inlet area of 28 sq in versus 50 sq in for impeller 2, the suction NPSH characteristics and impact could have potentially been an issue at some point.

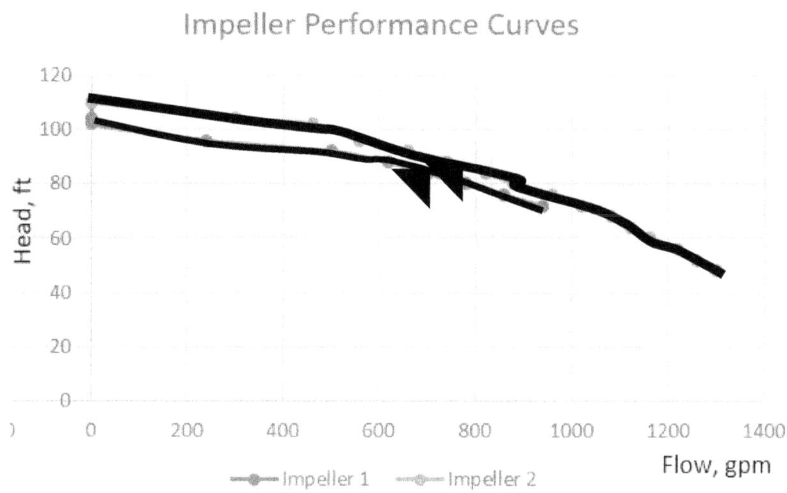

FIGURE 8.10 Different impeller pump performance curves with different pattern.

This pump performance issue was due to improper maintenance by installing the wrong impeller for the application. Even though it physically fit, had the correct number of vanes, and the correct diameter, the different impeller produced a different performance. This was obviously a pump issue and not a system issue.

Case Study 40 Learning – Verify not only impeller diameter and number of vanes but verify impeller pattern number whenever installing a new impeller.

Once the correct impeller pattern is verified, the impeller diameter must be checked to match the pump system design. There are several ways to measure the impeller diameter in the field outside the machine shop setting. If the impeller is a semi-open or closed impeller (back shroud) then a pie tape on the back shroud can easily be done for a good measurement down to 0.010″. If the impeller is an open impeller with even number of vanes, then measuring diameter from vane tip to tip may work. If it has an odd number of vanes, then measure the radius to center or impeller hub OD. Then add impeller hub OD as shown in Figure 8.11. Venier calipers can also be used to measure diameter.

$$D = A + 2B$$

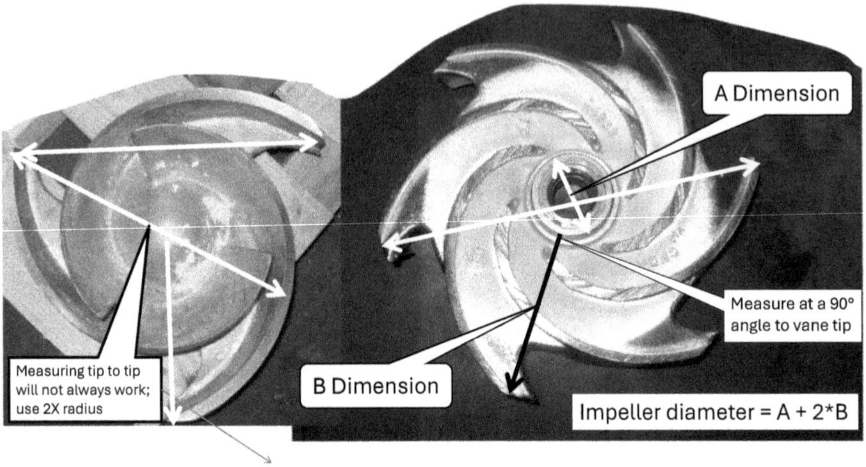

FIGURE 8.11 Impeller diameter measurement.

where:

D = impeller diameter, inches
A = impeller hub OD, inches
B = impeller radius to hub OD, inches

Some impellers have a major and minor diameter. The curves will sometimes show the average of these two diameters, or they may show both diameters. The major diameter is the larger diameter, and it is on the suction side of the pump (side of impeller eye). The minor diameter is on the back side of the impeller. It is important that the impeller trim match the correct angle from the manufacturer, so the performance curve matches what is installed in the pump. It is always good to measure both sides of the impeller in the field as some tapered cuts are small and difficult to notice.

CASE STUDY 41 PUMP MOTOR RUNNING HOT

A medium consistency paper stock pump system suddenly began having issues pulling high amps and occasionally tripping. Further investigation also revealed that the motor was operating very hot, and the pump vibration appeared worse than normal. The pump was a 4 × 8 × 18 medium consistency end suction pump equipped with an inducer. The pump design called for a 16.25″ × 14.80″ diameter impeller with an original design point of 606 gpm at 235 ft. The pump pulled from a suction standpipe with a level control valve on the discharge of the pump. The pump had a 100 HP, 1,800 rpm fixed speed motor as the driver.

The pump had been pulling 95–110% of full load amps pending on the process flow rate through the system. The pump was running between 400 and 500 gpm most of the time. The motor skin temperature was up to 262°F, so the core temperature had to be pushing 300°F. With the standard F class motor insulation, the max winding temperature is 155°C (311°F) so the motor temperature was very near the insulation rating.

The pump discharge-level control valve was operating in a very closed position most of the time. The pump was demanding over 100 HP much of the time and tripping the motor. The motor load appeared to be real, so the pump was shut down to check the pump internals and impeller to determine the cause of the high load. The findings were that a maximum-sized impeller (18″ × 17.5″) had been installed in the pump. The pump curve with two different sized impellers is shown in Figure 8.12. At 600 gpm, the maximum impeller would require ~118 HP to operate the pump.

The specified impeller for the application 16.25″ × 14.80″ was installed in the pump and normal operation resumed. The specified impeller now pulled only about 75% of motor full load amps as normal. The motor skin temperature dropped down to 126°F. The vibration dropped over 50% when compared to the previous oversized impeller. The original sized smaller impeller did not give full curve runout for the motor, but the process seldom got up to 600 gpm, so it never overloaded the motor.

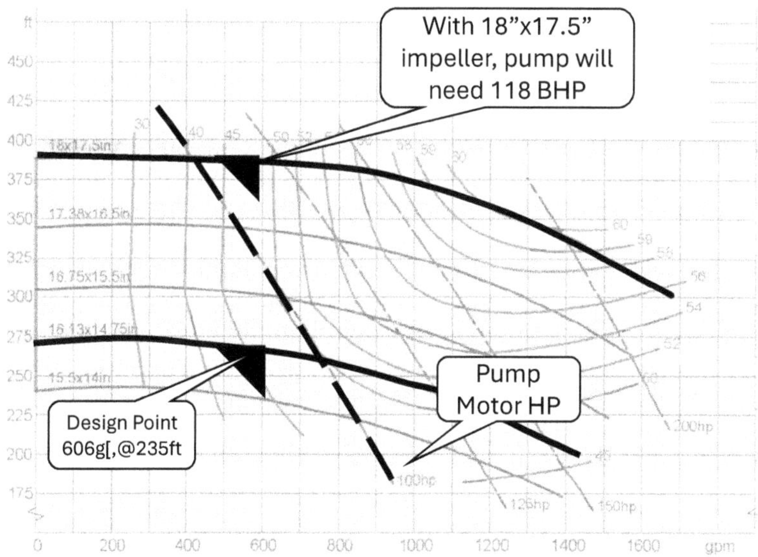

FIGURE 8.12 Pump curve showing alternate operating point and impeller.

So, what caused this failure? The investigation revealed that the spare pump, along with its impeller, was used for two different applications, but each having different impellers, operating points and, motor sizes. So, there were three pumps in rotation with two of them with large impellers and one with the small impeller. Looking back at history, there had always been issues at times on both pump systems and likely due to having the wrong size impellers on the spare pumps.

One correction to this issue was to setup separate spares with the correct size impellers for each. Several years later, another mistake happened with the two spare pumps. Both were in the shop at the same time and the impellers accidently got installed on the wrong spares. The wrong size pump got installed again on this same pump. The entire motor overload event repeated itself. Again, the process area was taken down to correct the impeller to resume normal operation.

While one design mistake on the front end was having the wrong spare pump setup, there was another major learning from this series of failures – verify the impeller size every time a pump is changed. Two separate times, the maintenance technicians could have discovered and prevented the failure event, but it was missed. The pump removed from the application was grossly a different-sized impeller from the one being installed but was never noticed or measured.

While the spare pump design from the original setup was incorrect for long-term reliability, maintenance execution is a root cause of the pump performance failure. Also, another obvious pump issue and not a system issue.

Case Study 41 Learning – Verify the correct impeller diameter every time a new pump is installed.

Once the correct impeller has been verified, the impeller can be assembled on the shaft. Impellers may either be screwed or secured with a shaft key and impeller nut design. A screwed impeller has opposite threads where it tightens on the shaft during operation. Wrong shaft rotation and the impeller may unscrew or loosen from the shaft. In the impeller with the nut, the impeller is keyed to the shaft and held with an impeller nut. The impeller nut usually has an oring to seal off against moisture and fretting corrosion between shaft and impeller. Figure 8.13 shows a shaft where that occurred after reusing an oring previously. The impeller was difficult to remove from the shaft after the fretting corrosion. It is important to install a new impeller/sleeve/nut oring during every pump assembly.

When assembling the impeller nut, be sure to have the correct torque wrench to properly preload the nut. Larger impeller nut torque may be as high as 1,500 ft-lbs. Some designs may also recommend using a retaining compound like Loctite 243 use up to 360°F or Loctite 272 for high temperatures. Failure to follow the installation guidelines of the pump manufacturer can lead to pump component failure if the impeller assembly gets loose.

FIGURE 8.13 Corrosion under impeller from moisture after reusing oring.

CASE STUDY 42 TANK LEVEL OVERFLOWING WITH PUMP RUNNING

A machine pulper tank started overflowing during operation and it appeared that the pump system would not maintain the level. The pulper has one pump with a recirculation line and control valve and a pump out line and control valve, as shown in Figure 8.14. The two control valves work together to maintain the level in the tank during normal operation. The pump on this system is a 6 × 10 × 16 end suction stock pump with a design point of 1,900 gpm at 56 ft. With the full sheet into the pulper, 275 bone dry tons per day (bdtpd) is the stock production rate into the pulper. With 277 bdtpd pumping out of the pulper at 4% consistency equates to 1,154 gpm. This leaves 746 gpm for recirculation. The original system design appears to have plenty of margin to manage the pulper level. Also, the system has never had any level control issues since original machine startup.

One of the first things checked was the operation of the control valves to make sure they were actually opening as they should. The valves were confirmed to be opening as they should to pump out the pulper. The recirculation valve did not appear to be bleeding stock off the system. The pulper tank was also drained early on when the high level started to become an issue. Nothing was found inside the tank. The agitators are side entry agitators, and some wear was noted on the extraction plates, but everything looked good. The level control continued to be an issue and the tank continued to overflow.

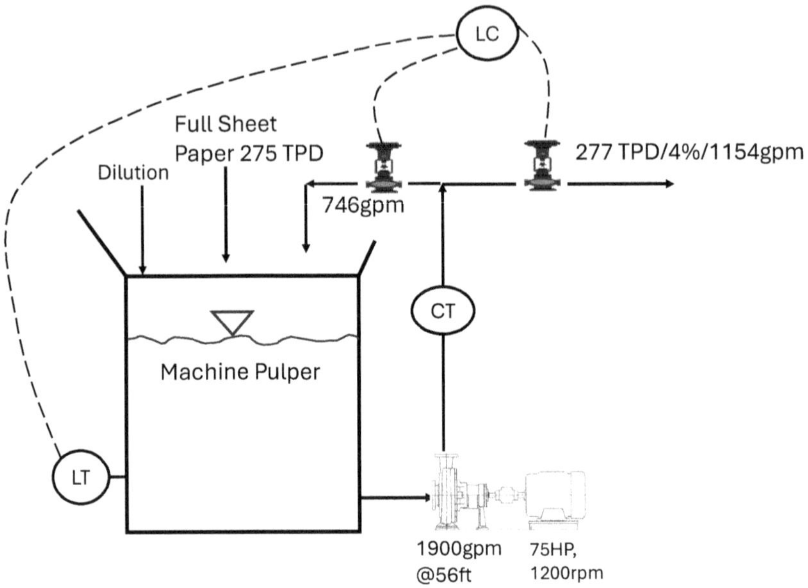

FIGURE 8.14 Process diagram of machine pulper pump system.

The vibration analyst was called in to check the pump vibration. The pump vibration was very low with an overall 0.0179 in/s and 0.6 gs which was significantly less than normal. Looking at normal vibration, it typically ran 0.0557 in/s and 2.6 gs. Normal vibration was still pretty low, but the pump vibration with the high tank level was ~3 times lower than normal pump system operation, as shown in Figure 8.15. This was a red flag and another clue or symptom of what was wrong.

There wasn't a good location to get a pump pressure reading on the pump so the only real operating data available was the pump motor load. The pump curve and operating points is shown in Figure 8.16. The pump has a 14.875″ diameter impeller and at the design point would be requiring ~53 HP. The pump had a 75 HP, 1,200 rpm motor with FLA of 89. When the motor load was checked, it was only drawing 40 amps, so the motor was very lightly loaded with an estimated 29 HP. Looking at the curve, the pump could not be pumping more than 100–200 gpm at that low HP load with that impeller.

At this point, it was obvious that the pump appeared to be the problem but just not sure what was wrong with the pump. The pump was taken down and pulled to investigate the root cause. The impeller nut had come off and worn the impeller against the suction wear plate. Around 75% of the impeller vanes

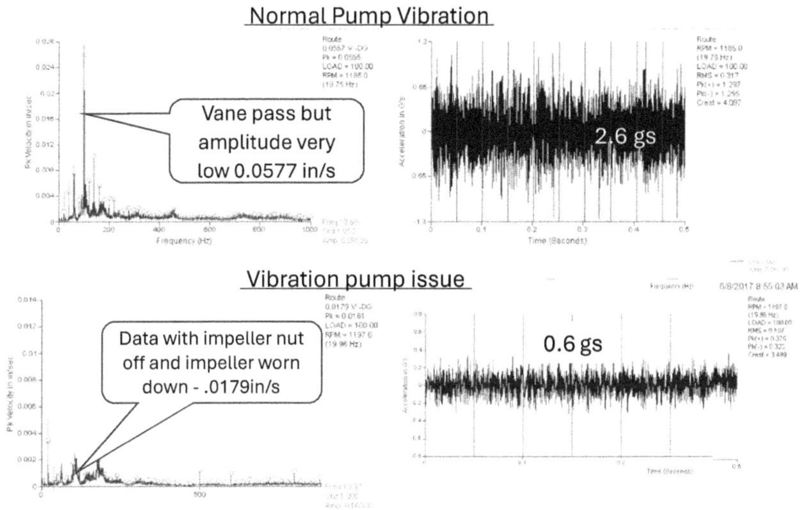

FIGURE 8.15 Normal and failed pump vibration data.

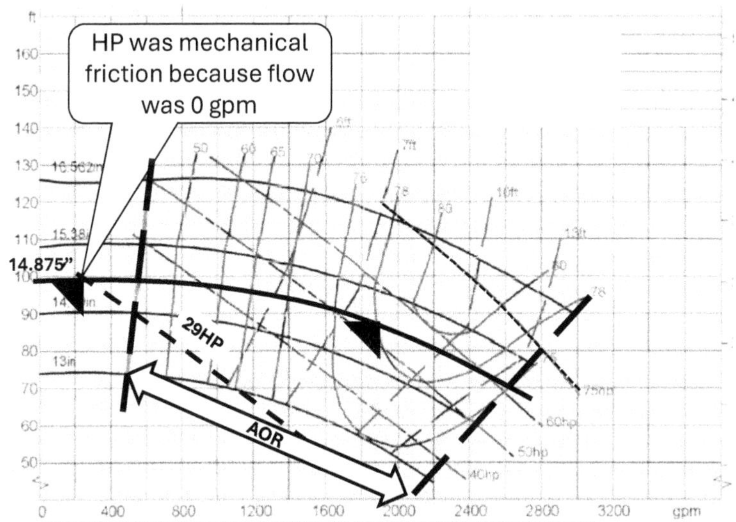

FIGURE 8.16 Pump curve with design and current operating points.

was worn down. Likely most of the HP load was probably mechanical drag from the impeller on the wear plate. A worn-down impeller could not break the system static head to produce any flow, which is why the pulper level went high.

Later an operator mentioned finding a large cap that washed out of the system on a previous tank draining. Sure enough the cap was the impeller nut from the pump. When the impeller nut came off, the pump pressure behind the impeller instead of thrusting the shaft now pushed the impeller off the end of the shaft into the suction wear plate. As the impeller moved and the impeller to stuffing box/seal chamber clearance increased, the pressure behind the impeller increased significantly. The excessive pressure wore down 75% of the impeller vane thickness which dropped the performance head of the pump. Eventually, the pump was running dead head and not producing any pump flow. From this point, it didn't take long for the tank to begin overflowing. Since the tank had trim going in and no flow going out, eventually the consistency got very high in the pulper tank.

This is not the only time I've witnessed a pump impeller nut coming off on this type of pump design. Each time not only was the pump performance erased but the pump damage was significant. Impeller nut torque on installation is a critical step of the impeller as is having a good impeller nut oring installed. On this particular pump, the impeller nut was an M42 size nut and required 600 ft-lbs of torque. This would require a large torque wrench not a standard toolbox sized torque wrench that most maintenance technicians would have in their toolbox. The specification also called for a thread locking compound, which is also not typically used by maintenance most of the time.

The root cause of this pump impeller failure is incorrect pump maintenance during a pump change. A latent root cause is inadequate training for proper pump maintenance installing the impeller and nut. This was obviously a pump problem and not a system problem.

Case Study 42 Learning – Use torque specifications and thread locking compound if specified for impeller nut installation.

The pump impeller typically has the clearance set between the back of impeller and stuffing box prior to installation. For typical end suction pumps, once the end suction pump assembly is installed into the casing, the impeller clearance is set with respect to the suction wear plate. This clearance is mostly based on operating temperature of pumped fluid to allow for thermal growth of the shaft assembly from the held bearing into the suction wear plate. For small end suction pumps, this impeller clearance may be as small as 0.005″ and up to 0.020″ for larger end suction pumps. This clearance is typically good for pump fluid temperatures in the 125–200°F range. As temperatures increase, the impeller clearance also increases.

There are some exceptions to setting the impeller clearance off the suction wear plate. The recessed impeller pump (vortex pump) typically sets the clearance off the stuffing box. There is a large clearance gap between the impeller and suction side of casing (there is no wear plate). In a typical medium consistency pump, the impeller is set in the middle of the stuffing box and suction plate which typically is in the 0.045–0.070″ range. A vertical turbine pump basically sets the same impeller clearance, but it is vertical and accomplished with the coupling gap since the motor bearing is supporting the vertical shaft assembly. It is always best to refer to the specific pump manufacturers installation manual for proper impeller clearance setup.

While the reason for impeller clearance is thermal growth, the importance and effect of proper clearance lie in the hydraulic efficiency. The impeller clearance creates a recirculation path from discharge head pressure to the suction head pressure between the impeller and suction wear plate, as shown in Figure 8.17. Clear water-like fluids may see a 1% hydraulic capacity loss with every 0.002″ excess impeller clearance. For high solids or higher viscosity fluids, the losses will be reduced as flow across the impeller clearance is much lower.

Since impeller clearance is determined by process fluid temperature, pump manufacturers typically will provide a chart for clearances for various fluid temperatures. Usually, these start with a clearance good up to around 200°F. Above 200°F, there may be increased clearance every 50°F or so. If impeller clearance is too tight, there is a chance that the impeller could rub on the suction wear plate and create premature wear.

Before moving the impeller to set impeller clearance, check that the seal or packing is in proper position for adjustment. For packing, loosen packing gland follower to relieve pressure on the shaft sleeve. For mechanical seals, install the seal clips to secure the rotating and stationary faces, then loosen setscrews that are fixed to the shaft sleeve or shaft. Make the impeller adjustments and reverse the process to reset the seal to the shaft/shaft sleeve.

There are three methods to measure and adjust impeller clearance. Depending on what tools might be available as to which one is chosen to use. Most craftspeople will use the dial indicator to setup and set the clearance.

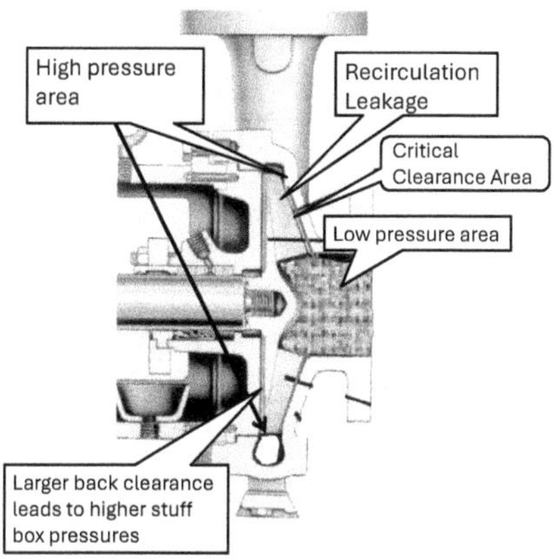

FIGURE 8.17 Impeller clearance hydraulic efficiency loss.

1. Feeler gauge method (Figure 8.18): Use feeler gauge under the head of adjustment screw on bearing end cover to measure clearance. Once the impeller has touched the suction wear plate and shaft does not turn easily by hand, back off the screw by the target impeller clearance. Use a feeler gauge between the bearing end cover and under the head of the screw to set the correct clearance. Do this for each end cover screw around the end cover. Then adjust the jackbolts to move the impeller off the suction wear plate until the end cover contacts the locking end cover screw. Tighten the jam nut on the jackscrew and torque the end cover screws to target amount.
2. Dial indicator method (Figure 8.19): Once the impeller has been run in to touch the suction wear plate, set up a dial indicator on the end of the shaft and set the indicator to zero. Adjust end cover until target axial movement (impeller clearance) has been obtained observing the dial indicator.

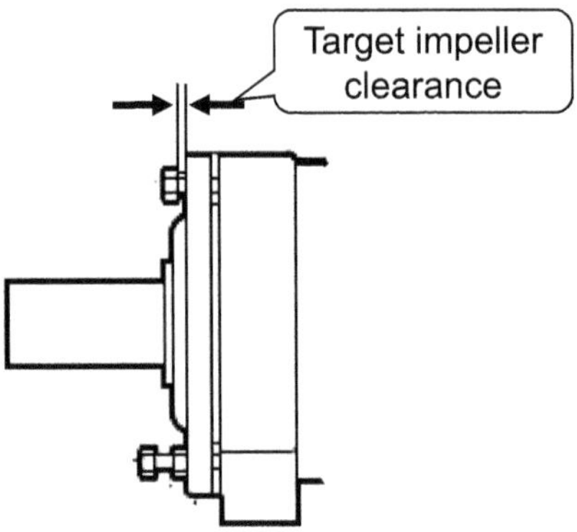

FIGURE 8.18 Feeler gauge method.

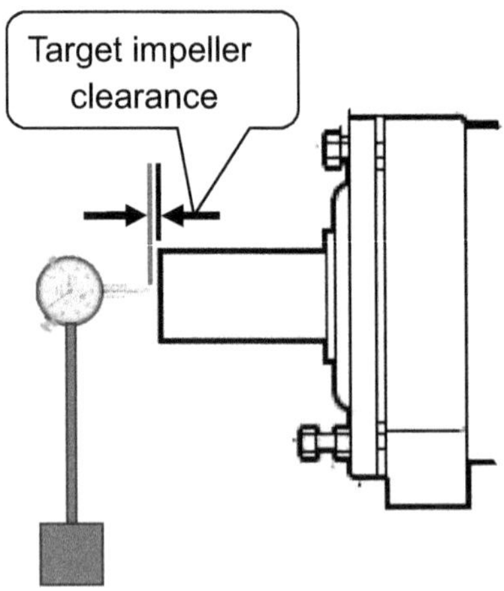

FIGURE 8.19 Dial indicator method.

Nominal Dia. (in.)	threads per inch
1/4	20
5/16	18
3/8	16
7/16	14
1/2	13
5/8	11
3/4	10
7/8	9
1	8

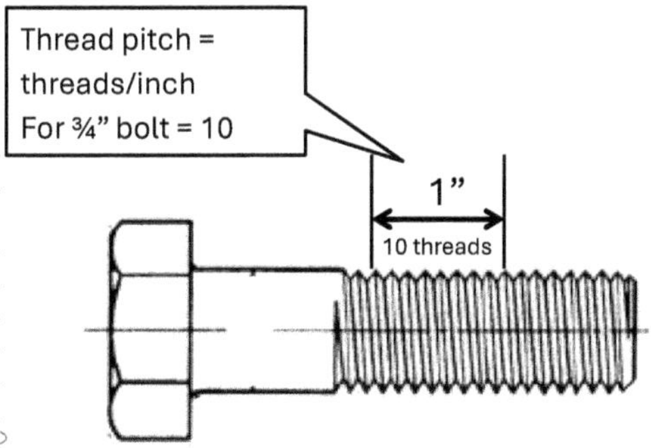

FIGURE 8.20 Screw pitch method for impeller clearance.

Some will indicate one side of the bearing end cover. This will work but extra precautions should be taken as the end cover can get misaligned (cocked) if the indicator is only on one side. Prefer to put indicator on end of shaft if there is room. Verify end cover is square around bolt circle when tightening.

3. Turn of nut method (Figure 8.20): A less common method is turn of nut method. Knowing the thread pitch, the axial distance is known for each revolution of the screw. This method is probably not as accurate and more difficult to execute in the field. Nonetheless, it is a viable option especially if other precision tools are not available for some reason. Because of thread clearance, remove any end play in the threads before making adjustments. As an example, for a ½″ screw, thread pitch 1″/13 threads = 0.076923″ per thread or one revolution of screw head. Each flat is worth an amount of axial movement or 0.076923/6 = 0.0128″ per flat. If impeller clearance target is 0.008″, then move screw 0.008″/0.0128″ per flat or 0.62 flats.

One caution on moving bearing end cover bolts is to ensure you are tightening and loosening the right bolts. The end cover screws move the shaft into the pump suction wear plate. The jackbolts move the shaft away from the suction wear plate. The jackbolts have jam nuts with them. Whichever screws you are moving, the other set must be loose and allow movement. If they are both tight, then the result will induce bending stress on the bearing end cover to the point where brittle fracture of the cast bearing end cover can occur, as shown in Figure 8.21.

While the general default is to set the impeller clearances to the manufacturer target to obtain best hydraulic efficiency and pump performance, there are times where increased impeller clearances are desired and targeted. Increased impeller clearances are needed to help address pumping fluid with entrained air. The extra recirculation across the wear plate into the impeller eye breaks up the air core at the impeller eye to help pass the air. The other condition in which a larger impeller clearance might be targeted is when handling trash or large solids in the fluid that can plug a pump. Larger clearances can pass some solids that might otherwise plug the impeller.

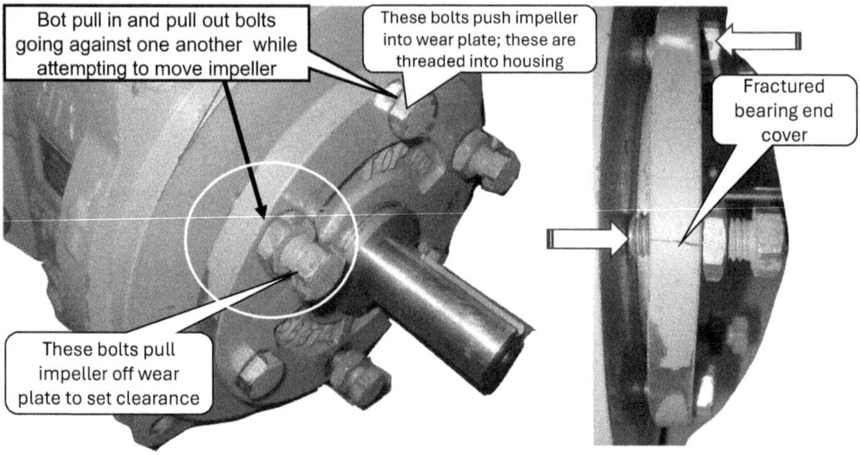

FIGURE 8.21 Cracked bearing end cover.

When increasing the impeller clearance, ensure the pump and application have the hydraulic and motor capacity to handle the decreased hydraulic efficiency. Having a pump with full curve runout is especially desired when clearances are increased. Head will go down, flow from pump goes up (although flow to the process may go down in pending-type system controls), and motor power goes up.

CASE STUDY 43 REBUILT PUMP BUT NEW ISSUES

A stock transfer pump system had been in operation for many years without a major pump replacement (including main wear parts such as casing, wear plate, impeller). A pump rebuild overhaul was scheduled to restore the pump to like-new condition. The resulting performance was very unacceptable and unexpected. The troubleshooting and correction led to a custom solution that was also very different than typical.

The pump system was a simple stock transfer setup, pumping fluid from a collection tank to a mix tank, as shown in Figure 8.22. The pump flow varied but a typical flow was ~1,000 gpm with the control valve ~40% open. As the flow varied, the control valve typically operated in a range of 25–40% open. The pump was an $8 \times 8 \times 14$ end suction stock pump with a $12.4'' \times 10.9''$ diameter, five-vane impeller. The original design point was 2,100 gpm at 50 ft so the pump operated to the left of BEP but within the AOR range for the pump. The pump had a 40 HP, 1,200 rpm direct drive motor as the driver.

After the pump rebuild and startup where a new impeller, casing, and suction wear plate was installed, the pump and system operated pretty good for about 24–36 hours and then the performance began to drop off. The pump could only produce about 450 gpm with the control valve 100% open. When the pump was opened, it was found plugged with stock and debris (plastic and wires). This was a recycle stock system, so heavy trash was a common variable but nothing on the supply end had changed when the pump was rebuilt. The pump impeller clearance was maximized after several plugged pump and cleanout events to help improve plugging of the pump. The impeller clearance ended up ~0.090″. The pump continued to plug, as shown in Figure 8.23, photos of the impeller during one of the pump cleanout events.

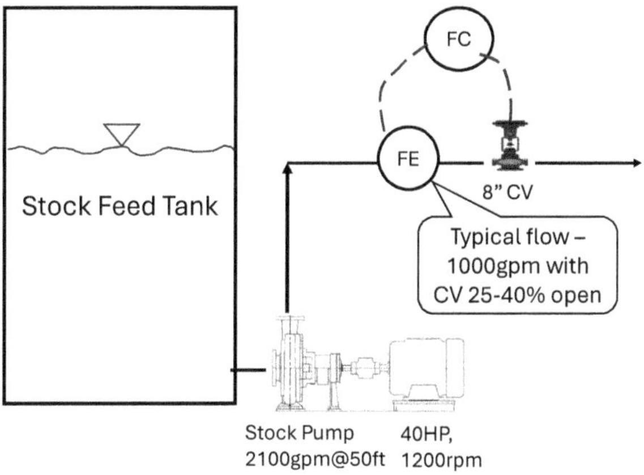

FIGURE 8.22 Process diagram for stock transfer pump system.

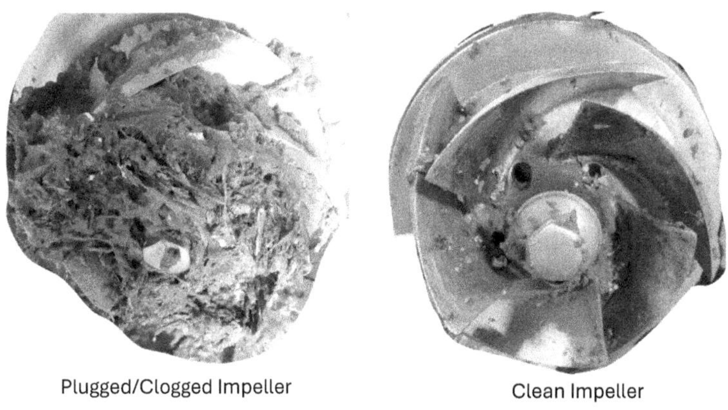

FIGURE 8.23 Plugged impeller versus clean impeller.

The process area was losing a lot of tons for shutting down one to two times per shift to clean out the pump and could not maintain production with these issues. At this point, the old extremely worn impeller was removed from the dumpster (fortunate it was still there) and reinstalled on the pump. The pump returned to normal operation with the old, worn impeller and stopped plugging.

The issue with the new impeller was the size and shape of the new impeller when compared to the old, worn impeller. The old impeller had lost much vane thickness so there was a very large running clearance. The impeller inlet vane was rounded off and had the main inlet corner of the impeller eye worn off. The edge of the vanes was also worn down and rounded on the leading edge. All the larger, sharp edge corners of the new impeller became grab points for plastic and wires mixed in with stock to plug the new impeller.

It is important to note that the pump was not rebuilt because of poor performance, but due to overall wear, including areas of the casing worn through and leaking stock. The original design point of the pump had the pump oversized (2,100 gpm versus 1,000 gpm) for what the process actually needed at current production rates. This difference allowed for a severely worn pump to continue to operate without any process issues; it just operated inefficiently as the control valve was throttled.

The best option would be to install a different pump, resized for the application and the correct pump type to handle the high trash load (a recessed impeller pump). However, there were two pump systems that would have to be replaced, and the capital cost was too high to solve immediately.

There was only one real option without changing the entire pump design and that was to make a custom impeller to operate similar to a worn impeller and handle the high trash condition. The pump needed a large operating impeller clearance and some rounded edges. A 3D map of the worn impeller compared to a new impeller was modeled and showed how much trimming had to be done on the new impeller to match the worn impeller pump performance for high trash solids. This contrast model is shown in Figure 8.24. The outer shape is a new impeller, and the inside reduced contrast is the worn impeller.

The impeller had the following notes on modifications to match the worn impeller which did not plug.

- The impeller had an additional 0.250″ vane thickness removed from the special air handling impeller (which had 0.125″ thickness removed).
- A large triangular shape (2″ × 2.5″) was cut off the impeller inlet vanes.
- Round edges of impeller vane on discharge side edge (no sharp edge like a new impeller).

FIGURE 8.24 3D impeller model of worn and new impeller.

This design change was all made possible by the pump being oversized for the application when compared to the original design. Had the pump needed that 2,100 gpm at a higher head, this impeller reduction for wear would likely not be possible. This is the same process when oversizing an impeller to handle air entrainment, but this case study involved high trash load plugging the pump. This new but customized impeller was later installed, and the pump functioned back to normal condition without plugging just as the worn impeller was performing.

This pump performance issue was due to pump impeller design. Initially, a new impeller was installed correctly in this pump, but process conditions caused it to chronically plug. This was not something that original system designers would have known. A recessed impeller pump would have likely been chosen had this contamination level been in the original design. The process contamination at this point was much worse than originally designed or expected. Even the standard modified impeller from the pump manufacturer would not handle the amount of trash in the process stream so a special designed impeller was the action to resolve the pump performance issue.

Case Study 43 Learning – While increased impeller clearances reduce pump performance, it can improve pump performance when trash load or air entrainment is high.

SEAL MAINTENANCE

All packing must leak some to properly function and have a full life. Leakage is controlled to dissipate heat from sealing surfaces and enhance surface lubrication. Leakage is controlled by a small amount of lubricant (oil, silicone, etc.) forced out of the packing by compression. A lube film is formed along the friction surface. Over time, the packing set will experience a loss of lubricant. This also results in a loss of packing volume and more packing tightening to prevent excessive leakage. Eventually, the packing lubricant will be depleted and dried out and will need to be replaced. Figure 8.25 shows a worn-out packing ring where lubricant is depleted, and packing has dried out. Also noted on this is the effect of the shaft radial forces causing shaft deflection into one side of the packing ring.

The packing sleeve will need to be replaced at some point as it is also a wear item before damage to the core shaft is done. Verify the correct shaft sleeve has been used for packing. Shaft sleeves for packing generally are hardened sleeves or have hard surface hardening to minimize wear. Some pump designs will have slightly different sleeves for seal or packing arrangements. This can greatly affect sealing, packing lubrication, and overall reliability. Packing fit in stuffing box should be approximately 0.005″T. Most pump packing arrangements are the standard 2L3–2 rings packing, latern ring, three rings of packing and packing follower as shown in Figure 8.26.

The packing radial load is highest where the packing gland follower is pressing on packing rings. These rings will also see the most wear and will need replacing first. Many times, the inside row of packing becomes washed out from lubricants as the seal flush water from latern ring goes into the product. As a result, the packing becomes hard and brittle. Some latern rings can be difficult to remove as well. Always good to verify that the latern ring is still lining up with seal water port when repacking the stuffing box, to ensure seal water can get to the packing.

Packing should run on shafts or shaft sleeves with 16–32 Ra surface finish. Any significant scaring or burrs will damage the packing. Packing tightness should be adjusted to where there are 8–10 drops per min per inch shaft diameter leakage or

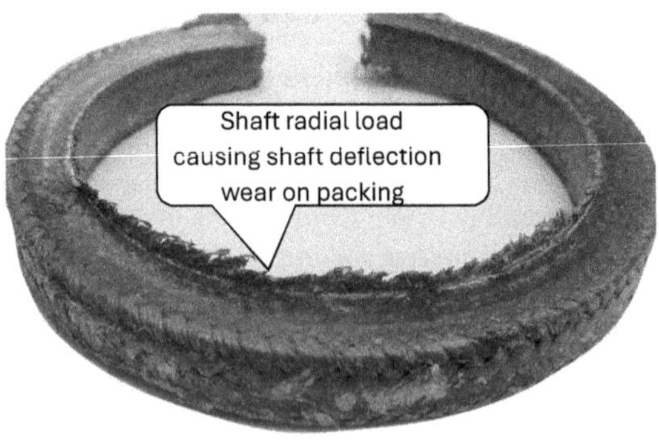

FIGURE 8.25 Worn-out packing ring.

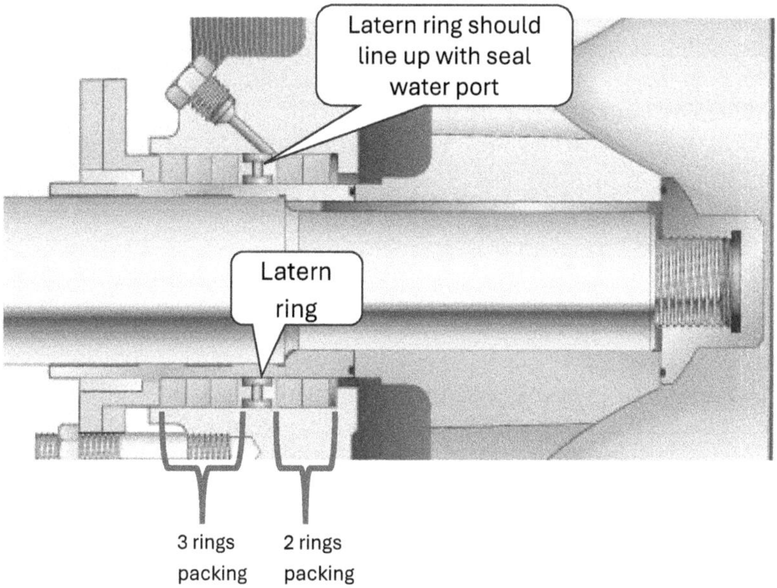

FIGURE 8.26 Typical 2L3 packing arrangement.

1–2 drops per second regardless of shaft size as a general rule of thumb. Packing tightening one flat every 15–30 minutes. Retighten packing after break-in period, which is typically 1–2 hours. Excessive tightening prior to startup may burn the packing, glaze the surface, and permanently damage packing surface and function.

Some useful quick reference equations for packing installation such as packing size, number of packing rings, and packing ring length are shown below.

$$\mathrm{PS} = \frac{\mathrm{SBB} - D}{2}$$

where:

SBB = stuffing box bore, inches
D = shaft or shaft sleeve diameter, inches
PS = packing size, cross-section, inches

$$\#\text{Packing Rings} = \frac{\text{SB Depth} - \text{LR Thickness}}{\text{PS (cross section)}}$$

where:

SB Depth = depth of stuffing box, inches
LR Thickness = thickness of latern ring, inches
PS = packing size, thickness of square packing, inches

There should be an excess/fraction number of rings from the calculation. For example, if the calculation had 5.2 rings of packing, only five rings would be used. The excess in the calculation is to allow for the packing gland follower to fit inside the stuffing box from ⅛″ to ¼″. If the gland packing follower does not stick inside the stuffing box, the packing will tend to extrude outside the stuffing box as the packing follower is tightened and the outer ring of packing will be damaged.

$$\mathrm{PRL} = (D + \mathrm{PS})\pi$$

where:

PRL = packing ring length, inches
D = shaft or shaft sleeve diameter, inches
PS = packing size, thickness of square packing, inches

Besides the formula above, a good way to get the correct packing length is to wrap the shaft with some packing and cut. Other packing rings can be cut after that the fit but be very careful or the length may change (usually shortens) in the cutting process. Packing should be tamped into stuffing box to seat it properly on installation or packing will not load evenly or properly upon startup. Packing joints should be staggered every 90° for each ring during installation, so that no joints align, preventing leak paths. Angled cuts are preferred but straight cuts are acceptable.

The stuffing box pressure on a pump is determined by the pump discharge and suction pressure as well as the impeller design features, such as a semi-open impeller with back pump-out vanes or open impeller with balance holes. The following two equations can estimate the stuffing box pressure for each condition. Keep in mind if pump-out vanes, impeller, or the stuffing box is worn then the stuff box pressures will differ from those calculated.

$$\mathrm{Psbv} = \frac{\mathrm{Pd} - \mathrm{Ps}}{4} + \mathrm{Ps}$$

where:

Psbv = pump stuffing box pressure w/back vanes, psig
Pd = pump discharge pressure, psig
Ps = pump suction pressure, psig

$$\mathrm{Psbv} = \frac{\mathrm{Pd} - \mathrm{Ps}}{10} + \mathrm{Ps}$$

where:

Psbv = pump stuffing box pressure w/holes, psig
Pd = pump discharge pressure, psig
Ps = pump suction pressure, psig

The seal water pressure typically needs to be 10–15 psig above the stuffing box pressure to ensure adequate flow to packing or seal. There are several key pieces to include for seal water to the packing or seal for product flush or quench and drain hookups. There are seal water flow and pressure modular units on the open market, or a standard hookup of components can easily be done, as shown in Figures 8.27 and 8.28. A throttle valve, pressure gauge, and check valve are a minimum for seal

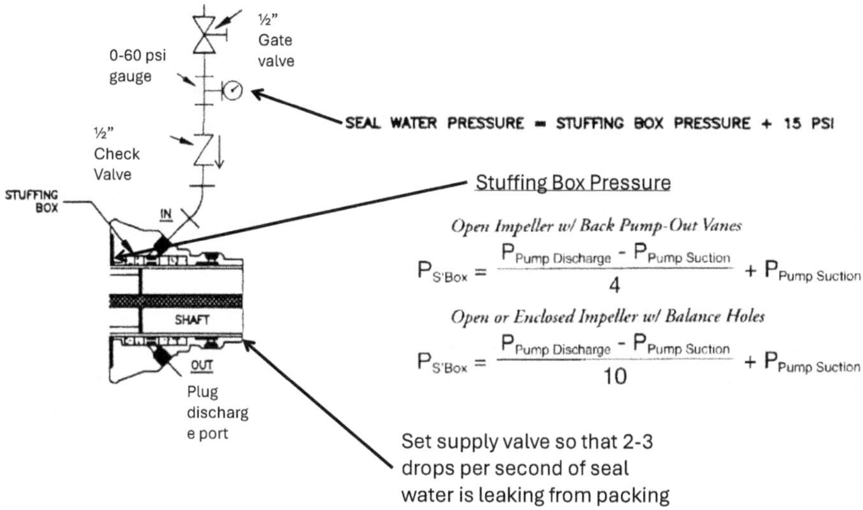

FIGURE 8.27 Seal water supply to packing/seal product flush.

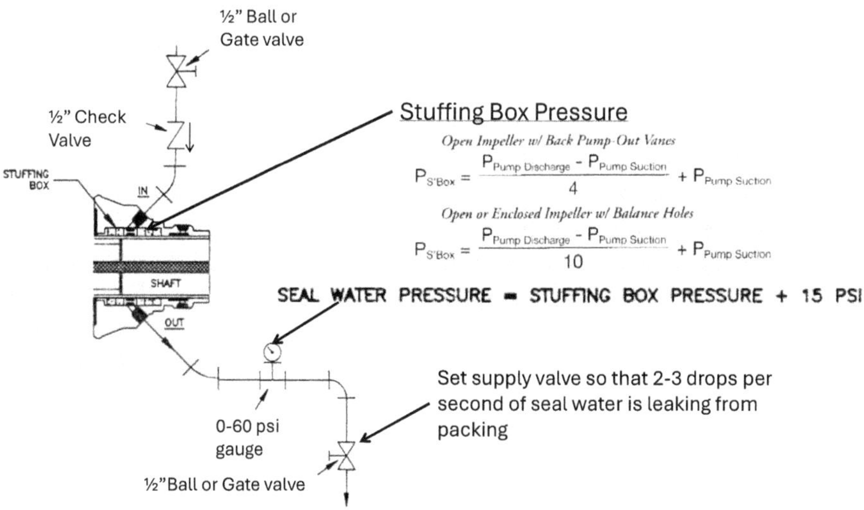

FIGURE 8.28 Seal water supply to packing/seal quench and drain.

water supply. The seal water supply should also be filtered for best reliability, especially for a seal application. Filtration media for seal water supply system should have a 30-micron filter for soft or carbon face seal materials and 50 microns for hard or tungsten/silicon carbide-type seal materials for optimal mechanical seal life.

A common mistake on seal water hookups that happens on a large number of installations for a quench and drain is not setting the backpressure in the stuffing box correctly. Many times, the quench and drain are just an open line to the sewer. In this case, there is little to no stuff box pressure from the seal water. When this happens, stuff box pressure overrides the seal water, and the seal water just goes in and straight out of the latern ring or seal chamber without lubricating or flushing. If the product is abrasive and not clean, then the packing will burn up or the seal will fail in a short time. The discharge throttle valve must be throttled until the target stuff box pressure (back pressure) is achieved to maximize seal or packing life, as seen in Figure 8.28.

Another common error during packing is tightening too much before or upon startup and burning packing surface from excessive gland pressure. Tightening and adjustments should be made slowly and gradually. Similarly, loss of packing lubricant from high seal water flush flow can lead to short packing life. Loss of seal water flow from a plugged seal water line can also lead to similar failure. Seal water quality from contamination creating abrasive flush water or chemical attack also leads to short packing life. Packing extrusion due to poor fit inside the stuffing box or protruding from the stuffing box will also cause short packing life.

If a throat bushing is used instead of a latern ring or standard 2L3 packing arrangement, note that some of these components are directional in assembly. It may function with the rotation of the shaft and fluid to remove contaminants from the stuffing box (sealing area). Look closely at the rotation arrows on the throat bushing and follow the manufacturer's instructions.

Some common errors on mechanical seal installations are forgetting to tighten setscrews before removing seal clips, which does not properly load seal faces. Not tightening gland bolts/nuts evenly can misalign the seal in the seal chamber. Damaged orings from shaft burrs or not lubricating orings during seal assembly may also cause premature seal failure.

One final item that I have seen done from time to time is using the seal water supply system as a chemical addition point where the chemical is added into the seal water supply line before the pump stuffing box or seal chamber. While this may be acceptable to some degree, caution should be taken to ensure that the concentration of the chemical has full compatibility with the pump components (stuffing box, shaft sleeve, shaft, impeller), mechanical seal materials, or packing material. A seal reliability issue may occur if this is done without proper compatibility with the chemical addition.

CASE STUDY 44 PUMP MECHANICAL SEAL FAILURES

A series of paper stock pumps in a lineup was experiencing mechanical seal failures and short life. All of the pumps were the same with the same design points. They were 14×16×22 end suction pumps with single mechanical seals utilizing seal water product flush. The original design point was 6,600 gpm at 67 ft. When the operating pressures and motor loads were checked, the actual operating points were identified as shown on the pump curve in Figure 8.29. All the pumps were operating within the AOR, and all except #2 pump were operating within the POR, indicating minimal hydraulic effects on seal reliability.

Further inspection of the pump system revealed that the seal water lines were plugged feeding the mechanical seal flush lines. The arrangement was a basic product flush where seal water went into the pumped fluid. The lines were full of mud and sludge. The seal water system is equipped with a dedicated seal water pump and filter system, but the filters are only 150 micron and not sufficient to remove enough contaminants. There were also times in which the filter system was bypassed due to filter plugging and valve issues on backflush system. Over time, sludge/mud had built up in the headers and plugged the seal water lines.

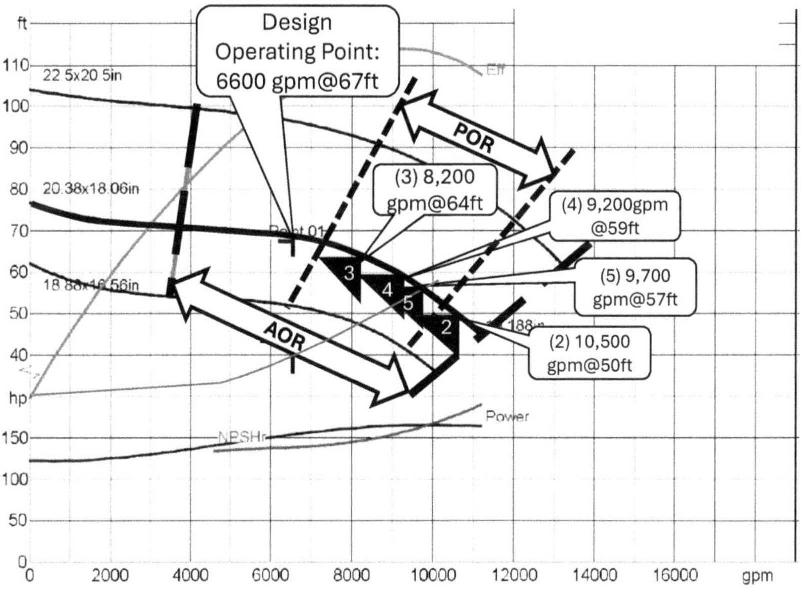

FIGURE 8.29 Pump curve with design and current operating points.

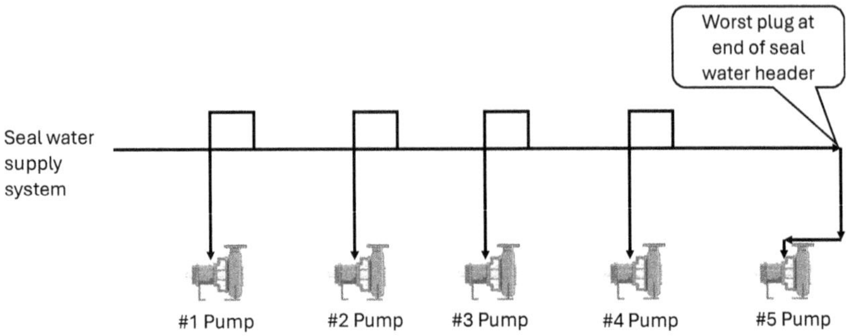

FIGURE 8.30 Seal water distribution system on pump lineup.

The worst was #5 pump which was at the end of the seal water header of the pump lineup as shown in Figure 8.30. Due to the piping arrangement, the #5 pump tended to have most of the sediment settle to the end of the header into the low point which was at the seal water piping and hose in the pump seal gland. The #5 pump had about 3 feet of mud/sediment in the ½″ seal water piping to the mechanical seal.

The lines were flushed out and put back together but a slight modification was made to the seal water piping – a mud leg was added, as shown in Figure 8.31. This allowed the seal water lines to be flushed of contaminants from time to time. This was not only done on the other pumps in the lineup but also a modification on any future seal water systems where they were found plugged with sediment. The old saying about when repairing equipment "don't just fix it, improve it" applies to every pump reliability problem. There are solutions and insurance design modifications to improve the probability of good performance.

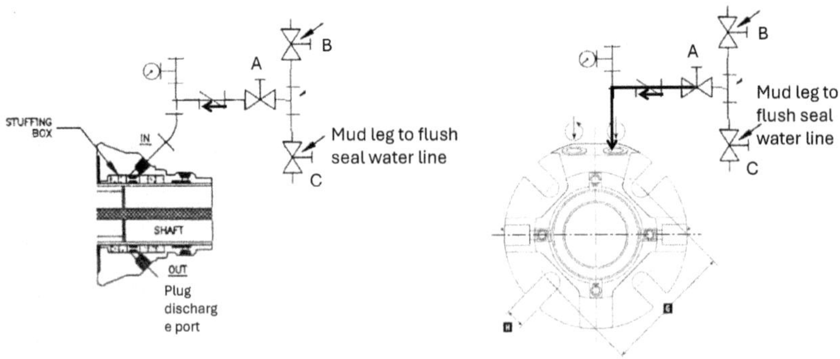

FIGURE 8.31 Seal water supply system with mud leg to flush header.

The seal water flush plan in Figure 8.31 had a set procedure for executing on the run as well. When flushing, we wanted to keep seal water in the supply line and not let product flush backward through mud leg drain. If the system had a check valve functioning properly, this would be all that was needed but if it only had shut-off valves then they had to be operated in order. For example, valve A should be shut first, then open valve C. If valve A is not shut or if a check valve is not installed, product could flow out through the mud leg, contaminating the seal chamber and seal.

The seal water supply system plugging was partly a design issue in how it was piped up and also from a seal water cleanliness issue. There was a seal water filtration system, so it was also a possibility that in the past, it was not working properly. This could have been either a maintenance issue or operator issue in cleaning the filter bank. This was more of a system issue than a pump issue as the pump was operating in a very good condition.

Case Study 44 Learning – Ensure seal water has ample pressure, flow, and cleanliness for optimum life and reliability.

Packing is fairly straightforward but still has a number of details to execute correctly for good reliability and longevity. The next case study involves a few of these things to get right on packing.

CASE STUDY 45 PACKED RIGHT?

Packing is a relatively easy and straightforward pump and shaft sealing design arrangement; however, there are still things that can go wrong and lead to poor sealing reliability. A pulper tank pump had packing as a sealing method. After changing the pump from a bearing failure, the pump began experiencing stock leakage. Initially, the running adjustments to the packing follower was not done within a day or two of startup which caused the initial leakage, but after that adjustment was made the leak returned in a few weeks.

After a few adjustments over two to three months, the packing follower appeared to be bottomed out in the stuffing box and could not be adjusted further to stop the leak. The pump was taken down to repack the pump. During the repack job, several things were learned as to why the packing did not last as long and would not seal properly.

The pump stuffing box packing design was a 2L3 packing arrangement which is typical. What was found was a 3L2 arrangement, as shown in Figure 8.32. The packing size was 3/8″ square packing. The problem when this happens is the latern ring is now out of position to function properly which

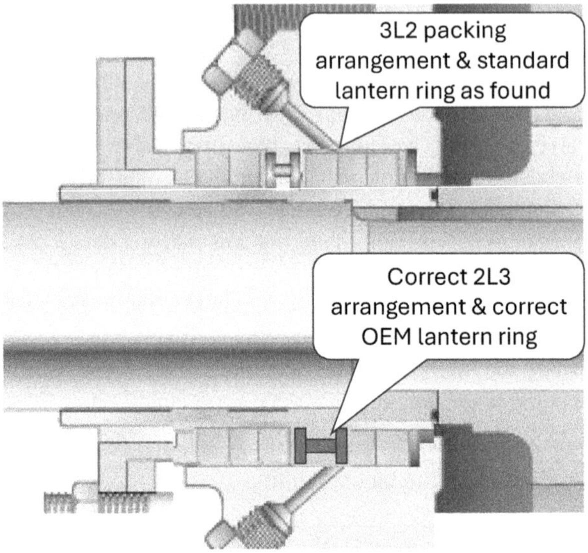

FIGURE 8.32 Stuffing box packing arrangement – top incorrect 3L2 and wrong latern ring, bottom correct 2L3 and wider latern ring.

is to let seal water into the stuffing box to lubricate, flush, and cool the packing. With a 3L2 arrangement, the latern ring is pushed away from the seal water port. With this the packing was not being lubricated with seal water properly and the packing was burned up.

The other issue we discovered was there was a difference in the latern ring size from the pump original manufacturer versus a standard latern ring stock for 3/8″ packing. The standard latern ring thickness along the shaft is ½″. The pump manufacturer had a custom latern ring which was 1″ thick. Therefore, using standard latern ring material for 3/8″ packing would offset the seal water into the latern ring by ½″. Not only is the seal water port blocked with the packing errors found but now another ½″ of packing adjustment is lost for the packing follower. This allowed the packing follower to bottom out in the stuff box bore in a short amount of time.

The packing issues in this case had a root cause category of maintenance. A special latern ring width added to the error but the lineup of the latern ring with the seal water port was not verified on the previous installation.

Case Study 45 Learning – Always verify latern ring lineup with seal water flush port and the correct packing arrangement is installed.

Another key item to verify when using packing is that the correct shaft sleeve has been used. A shaft sleeve for a mechanical seal is not the same as the one for packing. This not only involves the metallurgy and hardness but many times it also includes the physical dimensions. An example of this can be seen in the information below on a large sized pump shaft sleeve outer diameter.

Packing Shaft Sleeve OD – 3.738″/3.740″
Mechanical Seal Shaft Sleeve OD – 3.748″/3.750″
Stuffing Box Bore – 4.724″/4.728″

The difference in the two shaft sleeves OD is 0.008″/0.012″ with the mechanical seal sleeve being larger than the packing shaft sleeve. For packing, the pump would use ½″ packing. When evaluating the packing fit for each sleeve on the pump, the resulting fit should be verified.

Packing Fit (using packing shaft sleeve) = 0.005″/0.008″ tight
Packing Fit (using mechanical seal shaft sleeve) = 0.012″/0.013″ tight

The resultant packing fit in the stuffing box is potentially two to three times tighter if a mechanical seal shaft sleeve is used. Most of the time a mechanical seal shaft sleeve is a soft material so if this is combined with the tighter fit, very short packing and sleeve life would result. It is a good practice to measure shaft sleeves to verify the correct fit, with a typical target of 0.005″ tight.

Another watch out on packing is staying with the correct dimension of packing, such as metric or standard, which can also affect the resultant fit in the stuffing box. A 25 mm metric packing is similar to a 1″ packing (25.4 mm), but if 25 mm packing was used in a 1″ packing stuffing box, then the loss of fit would be 0.0157″, which may create clearance in the stuffing box.

Mechanical seals can be very sensitive to pump operating conditions. Determining the root cause of mechanical seal failure will involve the pump itself and the pump system. The next case study shows several aspects to consider in getting total reliability from mechanical seals as various aspects of the pump design affect seal reliability.

CASE STUDY 46 CHRONIC MECHANICAL SEAL FAILURE

A fine screen feed pump system began to have chronic mechanical seal failures. While the pump had suffered mechanical seal failures in its lifetime, the average life was typically three years. Starting with a pump changeout through the next eight months, there were six seal failures which involved changing the mechanical seal and sometimes the pump. A new pump rotating assembly was installed on three of the six seal changes, while the remaining three involved reusing the existing rotating assembly. The time for seals leaking from installation ranged from a few days to three to four weeks. Some seal failures occurred after a few months of operation, while others failed within a few weeks. It was obvious that there was a significant issue with the seal life over the past nine months of operation.

What was the mechanical seal failure mode? The seal had a tungsten carbide hard face and carbon soft face seal material. The hard face had very little wear and the carbon face had some wear from stock but mostly there was stock past the seal faces in the spring area of the stationary seal face. Two things would have likely had for this to happen – seal faces open up and lack of seal water flushing through the seal. Figure 8.33 shows the seal detail.

The mill had been rebuilding some mechanical seal assemblies where a new seal cassette would be installed in a reused seal gland. The seal assembly was a new seal but reused the seal gland housing in the cartridge seal assembly. Some thought this seal gland reuse may have created some issues causing the seal failures. Investigation of the six seal failures showed that out of the six failed seals, five of them were completely new seal assemblies and only one rebuilt seal assembly (new seal cassette and reused seal gland). Therefore, using rebuilt seals seemed to support that it had nothing to do with the failures.

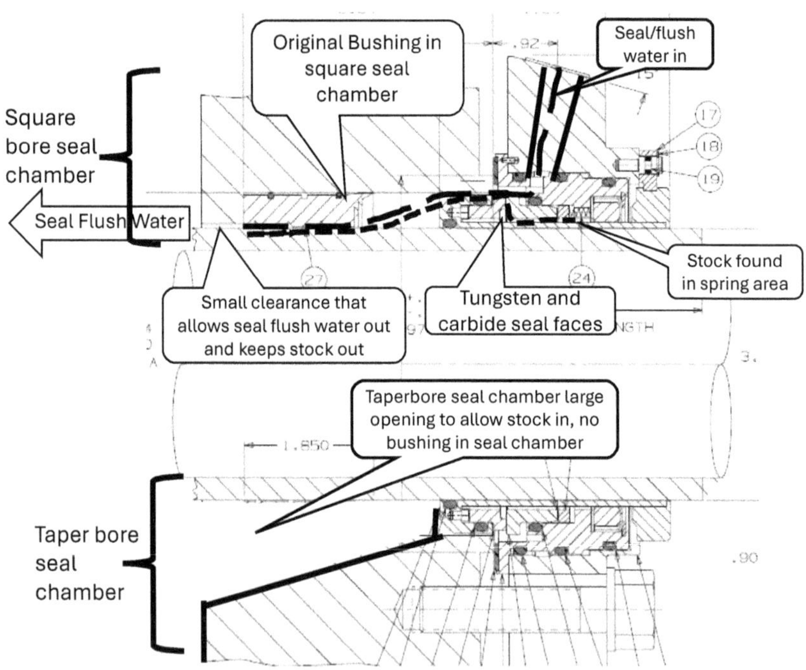

FIGURE 8.33 Seal failure contamination and seal water path – top: square bore versus bottom: tapered bore seal chamber.

In the middle of the failures, several findings were noted. The seal water was hooked up to the wrong port on the seal gland. This wasn't a major issue as the opening is still in the same general area but having seal water in the port with the shaft rotation is a best practice in seal water hookup. The more significant finding was after installing a pressure gauge on the seal water supply, low stuffing box pressure was discovered. The previous seal water supply line didn't have a pressure gauge so one had to be installed. The impeller with balance holes and system pressures resulted in a stuffing box pressure of 15 psi (35 ft). Adding 15 psi to the stuff box pressure and under normal conditions, the pump needed ~30 psig seal water supply pressure. The pump only had about 18 psi with the current valve settings once the pressure gauge was installed. This was all corrected, but the pump still had two more seal failures, so this was not the main root cause but only a contributing cause to the poor seal reliability.

During the pump system investigation, other operating points were also observed when taking readings on the system, as shown in Figure 8.34. It is a large end suction pump – $16 \times 16 \times 19$ with an $18.3'' \times 15.6''$ diameter impeller and driven by a 250 HP, 1,200 rpm motor. The pump normal TDH ranged from 88 to 107 ft. Most of the system head was friction head, as the total static head was only ~4 ft. The flow control valves normally operated in a mostly closed position.

The pump curve is shown in Figure 8.35 where several operating points were outside of the AOR, which would result in high shaft forces and potentially cause shaft deflection and seal issues. The operating conditions

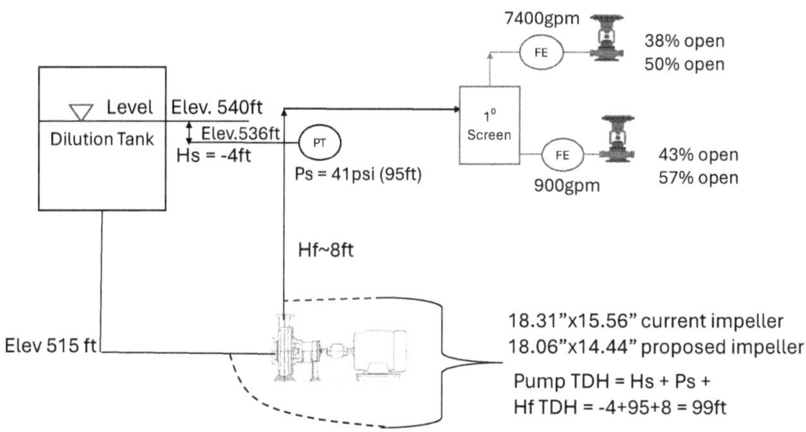

FIGURE 8.34 Process diagram for stock fine screening pump system.

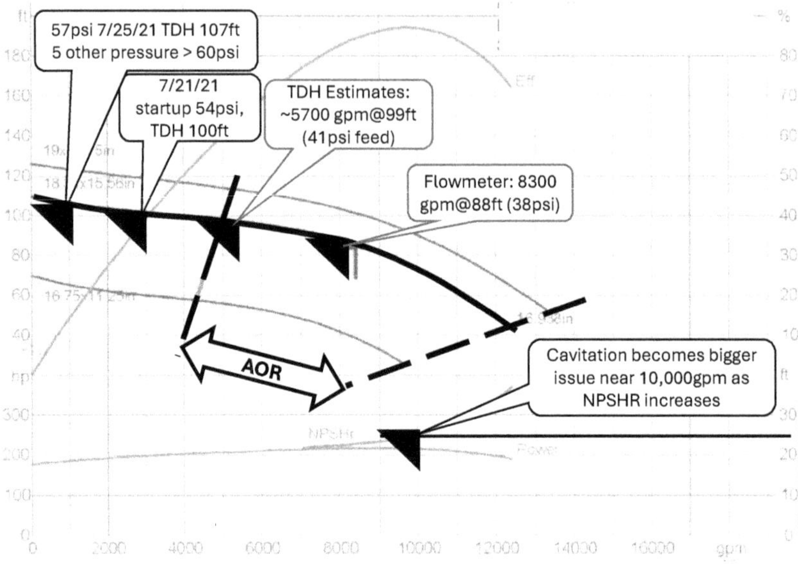

FIGURE 8.35 Pump curve with various operating points.

that were outside AOR were all during startups for short periods of time. While none of these operating conditions was corrected, this type of operation had been going for at least 15 years with no chronic seal failures. There was some small cavitation damage on the pump inlet vanes. Under normal operating conditions, the pump should not be seeing cavitation issues, but at flows over 10,000 gpm, cavitation may become destructive.

Further into the investigation to determine what had changed nine months earlier, it became apparent that the pump had a new seal chamber installed on the pump rebuild. Looking closer, the new seal chamber was a tapered bore seal chamber. The original seal chamber was a straight bore seal chamber. The original straight bore seal chamber included a throat bushing to further support the shaft, while the new tapered bore seal chamber did not include this support. The tapered bore seal chamber (stuffing box) had a large opening to the seal area which greatly increased the pressure and possibility of stock being around the seal. This made seal flush water even more important. Once this was realized, a new straight bore seal chamber was installed with the throat bushing. After the installation, the seal failures completely disappeared. Both style seal chambers are shown in Figure 8.36 and Figure 8.33, displaying the impeller or process side of the seal chamber.

FIGURE 8.36 Seal chamber (stuffing box) design comparisons.

Since the pump system operated with control valves heavily throttled, there was an opportunity to trim the impeller and save some energy and improve system reliability. An $18.06 \times 14.44''$ impeller would allow the control valves to open in the 50–55% range. This would also lower the system pressure and save 30 BHP or $11,000 annually. The smaller impeller would also further reduce shaft radial loads to support the seal reliability.

This case study identified the root cause of pump failure as a change in the seal chamber design at both ends. It was proven when the problem began and when the problem ended with the same action causing both, which was the seal chamber design change. There were also some operating conditions which were intermittently operating the pump outside the AOR. This failure was due to the pump and not the pump system, as many of the case studies in the maintenance-related area typically are.

Case Study 46 Learning – Pay attention to seal water pressure supply and stuffing box designs for optimum mechanical seal reliability.

BEARINGS AND LUBRICATION

Typical pump bearings for end suction pumps are angular contact ball bearings. Larger end suction pumps may utilize a roller bearing on the radial end due to the size of radial loads. The thrust bearings are typically 40° angular contact in a back-to-back arrangement.

The back-to-back bearing arrangement work together for two pump operating conditions – startup and normal operating condition. During startup, the resultant thrust on the shaft assembly is toward the drive as the suction pressure on the impeller is larger than the pressure behind the impeller (pump system pressure is not established yet on startup). Once the pump is up, the speed and system pressure are fully developed, the pressure behind the impeller is greater, and the axial force is toward the pump as shown in Figure 8.37. Bearing thrust is toward the casing due to high pressure behind the impeller.

In each condition, a different bearing is loaded on the thrust end assembly. Each condition has one bearing running unloaded. Some pump specifications may specify a 40° and 15° contact angle for these bearings due to the startup bearing running unloaded most of the time. The bearing operated unloaded will likely be under the minimum load and can cause bearing skidding/smearing. This combination of bearing contact angles reduces skidding on the lightly loaded bearing (startup bearing).

As mentioned previously, the bearing thrust load is directly proportional to the pressure behind the impeller which is affected by the impeller design, discharge pressure, and impeller gap with stuffing box.

Some pumps may have different thrust conditions. For a medium consistency (MC) pump, the thrust is opposite due to low pressure behind the impeller (vacuum behind the impeller). The MC pump is also typically higher thrust and runs at a higher temperature so bearing housing cooling is typically specified.

Lubrication in pumps is just like other rotating equipment – it is the lifeblood of the equipment. There must be the right lubricant, the right quality, and the right

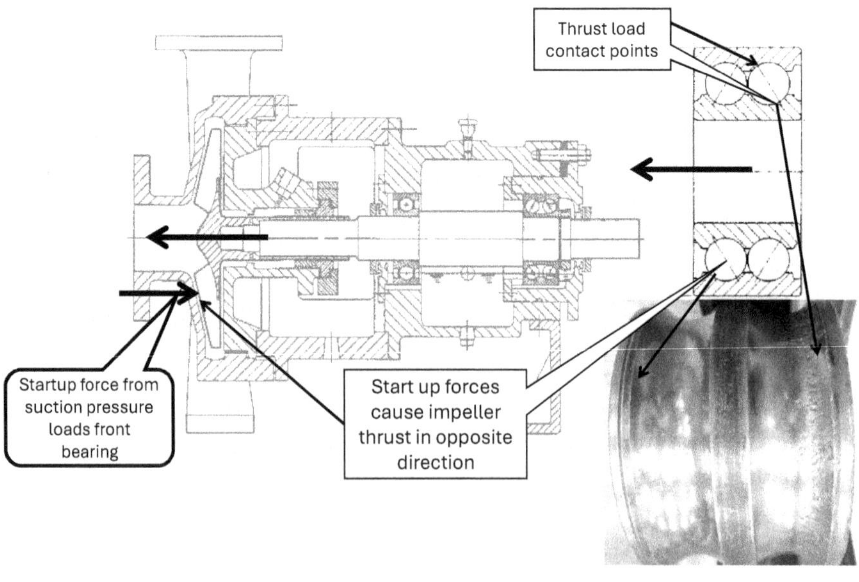

FIGURE 8.37 Pump bearing arrangement and thrust load.

amount. If any of these lubrication basics are done incorrectly, the pump will not achieve its full operational life.

What is the right lubricant? The most important lube property in selecting a lubricant for any application is viscosity. The viscosity is what creates the film thickness to prevent excessive wear on bearings and other wear parts. The minimum lube viscosity for ball bearings is 15 cst at operating temperature with most pumps in a range where a 68 cst lube oil viscosity is the best median choice. Some general guidelines on lube selection based on oil temperature may be as follows:

Oil temp < 150°F use 46 cst.
Oil temperature 150–180°F use 68 cst.
Oil temperature 180–200°F use 100 cst.

In general, pump lubricants should have good additives for oxidation, corrosion, and demulsifiers to separate and shed water contamination. A rust and oxidation (R&O) lubricant is a good choice. Pumps may perform well with either mineral or synthetic oil. Some pumps run with greased bearings. There are pros and cons of using synthetic or mineral oil-based lubricants. In moderate temperatures, a mineral lubricant can perform well. In very cold or hot temperatures, a synthetic lube oil is a better choice. Synthetic lubricants have a higher viscosity index and are less sensitive to temperature changes which makes them good for wide temperature extremes.

A general guideline for pump temperatures is a maximum pump oil temperature ~180°F for continuous operation or ~210°F maximum intermittent temperature. Lube oxidation doubles for every 18°F increase above 140°F. A general pump temperature scale for operation is shown in Figure 8.38. This is for general pump application and can vary depending on the process temperatures. Above 210°F, a pump shutdown or trip may be warranted. Ball bearings used in pumps generally have a bearing stabilization temperature of ~257°F, so temperatures below this and the bearing should generally not experience catastrophic failure in a short-term condition.

Oil temperature is affected by pumped fluid temperature and bearing loads. In general, it is best to not use housing cooling unless necessary. Some designs may create housing cooling which reduces bearings clearances as outer race shrinks as inner race grows. This can reduce bearing clearances and increase friction. Some bearing housings have cooling fins to improve natural convection cooling of the bearing housing.

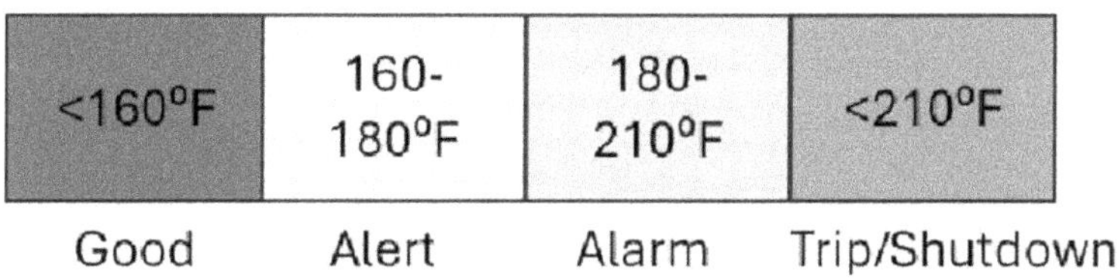

FIGURE 8.38 General pump allowable bearing temperatures.

The choice of mineral or synthetic oil may come down to service intervals or oil changes. Many pump manufacturer manuals may call for an oil change at 2,000 hrs which is once per quarter but in most cases, this is overkill. Most pump oils do not run high enough temperatures to oxidize the oil in 2,000 hrs. If a mineral oil is degrading due to temperature, then a synthetic might be a good choice to extend drain times before an oil change.

Most oil changes I have encountered on centrifugal pumps have been due to water contamination where a leaking seal has contaminated the oil. I have conducted several two-year studies on many pump systems, conducting oil analysis every three months. Excluding a contamination event, mineral oil never had any degradation after a year, and I never had a synthetic oil show any degradation after two years. Whenever water contaminates the oil, a hot flush and drain is required, but not for the oil having an oxidation, viscosity, additive depletion, or any other oil condition issue. All of these pump studies ran less than 160°F oil sump temperature.

The flooded sump lube strategy is typical for many pump designs. Pumps operate at moderate speeds so that this design is simple and provides adequate lubrication. The oil level must be at the center of the bottom of most roller, as shown in Figure 8.39. This is where all oil level indicators are set. Oil levels above the target will begin to increase fluid and rolling friction as the roller moves through the oil, thus increasing the operating temperature. This is seen in the Stribeck curve on lube regimes. If the pump is run more flooded, the oil level will rise to the shaft level, leading to leakage at the shaft seal. There is a fine line in pump bearing lubrication oil levels where there is enough lubricant, no lubricant, and too much lubricant. In many cases, there may only be 0.25″ from a proper lubricated bearing to no lubricant, therefore, close monitoring of oil levels is necessary to ensure good reliability results.

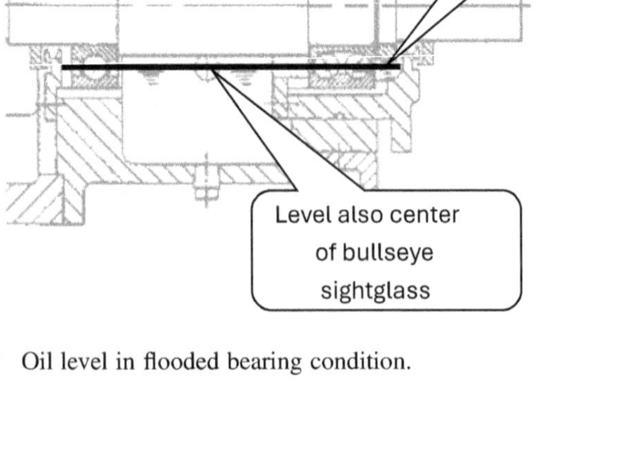

FIGURE 8.39 Oil level in flooded bearing condition.

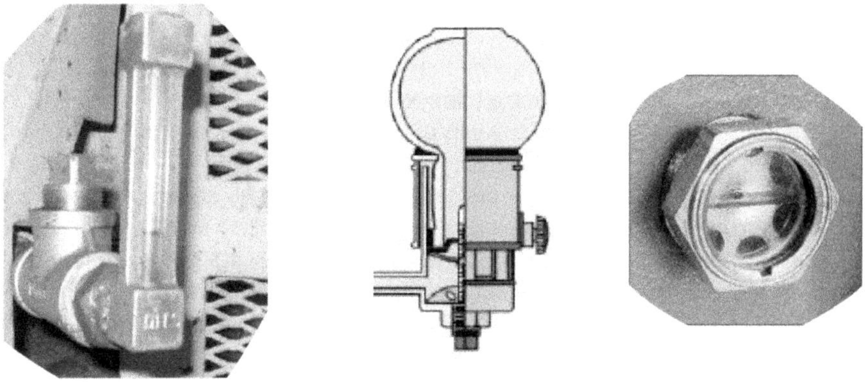

FIGURE 8.40 Oil level indicators: (a) tube sightglass, (b) constant-level oiler, and (c) bullseye sightglass.

There are several different types of oil level indicators used on pumps, as shown in Figure 8.40. Most pumps come equipped with a bullseye sightglass on the pump housing. When a bullseye sightglass is on any equipment, the target is always the center of the bullseye. A bullseye has a crosshair (two intersecting lines) indicating the target level. This sounds simple but I've seen many errors with this over the years.

Another common oil level indicator used is the tube sightglass. It includes a vent at the top to equalize pressure and provide an accurate level reading. If this vent gets plugged, it will not give a correct oil level reading.

Another oil level device often used on pumps is a constant-level oiler. These function to automatically (depending on setting) add some oil if the oil level drops during operation. There should still be oil level indication somewhere else on the pump housing, as the constant-level oiler does not give a visual indication of what the oil level is. It only shows the oil level in the bulb reservoir (this is not the oil level in the housing). A couple of watchouts on constant-level oilers is ensuring the installation is correct in regard to shaft rotation. Correct location will not draw out oil from the bulb reservoir. Every time the reservoir is removed to add oil, a gulp of oil goes into the sump and some overfilling will occur. It is easy to trapped in a catch 22, where adding more oil leads to increased leakage, prompting more oil to be added. It is a dangerous cycle.

Any oil level indicator has some common watchouts. Pump installation being level is one of them. An unlevel pump will not give a true oil level reading. Depending on which side of an unlevel pump the sightglass is on, the level may read low or high. A mechanic would not park his car on a hill to check the oil level, so the same is true for a pump bearing housing. A true oil level must be verified while the equipment is down. While some designs may not show a big difference, some applications will. To check on the run, a best practice is to have a mark for down level and operating level. This is more of an issue with tube sightglasses.

Another common lubrication design on centrifugal pumps is the oil flinger design. This design has a ring or flinger attached to the shaft that pulls oil and flings it through the housing to where the bearings are lubricated. The oil level on these designs is below the rollers. Flinger diameters are typically in the 1.7X shaft diameter range, with flinger immersion into the oil between 0.156 and 0.375″.

There are several watchouts for oil flinger designs where failures can be initiated. Since the bearings do not sit in a sump of oil, long idle periods can lead to static corrosion of bearings. This is particularly a higher risk for pumps where there is an installed spare that sits for weeks or months at a time. If the vibration of the idle pump is high enough, then false brinelling can also occur since little to no oil film is present.

Housing vent locations can also be a problem on flinger designs especially if the flinger is too close to the radial position. Oil can escape the oil vent as it slings out causing oil leaks. The key is not to overfill a flinger design, as high speed and higher operating temperatures can lead to bearing skidding. Oil flinger designs will also have a minimum speed to sling oil enough to lubricate the bearings. One example was on a paper machine fan pump operating on a VFD. The VFD in normal operation was not an issue; it stayed within flinger design speeds. However, there was a condition during a machine boil out where the speed was drastically reduced below the minimum speed for flinger operation. The solution for this condition was to add oil to pump housings so bearings ran flooded at the low speeds during boil out conditions.

Oil cleanliness can be a key factor for centrifugal pumps, especially from water contamination. Ideally, pump lube oil should have less than 150 ppm of water contamination and have an ISO cleanliness of at least 17/15/12. Rolling element bearings have clearances of 0.1–3 microns so oil cleanliness is important to prevent two body and three body abrasion. Since pumps have such a low oil sump volume, oil analysis is difficult to justify as it is easy and cost-effective to just change the oil. Pumps may be fitted with a bottom sediment bowl for contamination visual inspection and ability to drain off contaminants, especially water.

Lube oils should be filtered as it is received before adding to pump bearing housings. Lube oil should be properly stored in sealed containers (non-metallic). There are many types of oil containers such as those from Oil Safe where the lube oil can be properly sealed until adding to the pump bearing housing. Some may even be equipped with a pump and wand to eliminate the use of funnels. The use of funnels is generally not a good practice as these are not good to control contamination ingression.

Pump reliability can't be achieved unless proper lubrication is executed. So, let's look at a few case studies around pump lubrication and failures that can occur.

CASE STUDY 47 PUMP LUBRICATION FAILURE – RIGHT AMOUNT

There are many ways in which a pump can have a lubrication-related failure. This case study will quickly highlight three instances where lubrication failures occurred in a centrifugal pump system due to the wrong amount.

A pump had an unexpected bearing failure during normal operation. The pump had a sight level gauge for oil level indication. The pump oil delivery design was with a flinger ring, so the bearings were running outside the oil level. During the previous days and weeks leading up to the pump failure, the

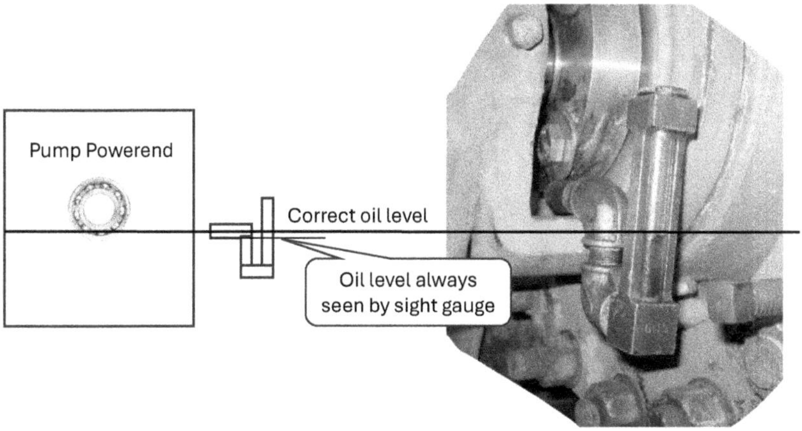

FIGURE 8.41 Pump sight level gauge piping error.

pump oil level was inspected on maintenance rounds and found to have the correct oil level by the gauge.

Upon inspection of the pump once it was changed, the shop was surprised to see that the pump had very little oil inside. However, the sightglass tube showed a good oil level. The problem was the oil level in the sightglass tube was an illusion. Since the port used was near the oil level, a change in elevation was necessary to put the sightglass tube in a readable range. However, this elevation change resulted in a fluid trap very near the target oil level, as shown in Figure 8.41. Most pumps are equipped with a bullseye sightglass. The maintenance tech attempted to enhance his oil level monitoring but created an oil level error when the oil level went low. The pump had been experiencing some oil leak issues, so the lube technician wanted a larger range of oil level monitoring. Maintenance was identified as the failure category for this lube failure.

Another pump experienced a similar rapid catastrophic bearing failure. Upon inspection of the pump when maintenance arrived to remove the pump, it was observed that the pump showed a good oil level in the sightglass tube on the side of the pump. Unlike the previous example, it was piped out without a pipe trap, so the oil level was thought to be correct. Upon inspection of the pump, it was revealed that contamination had plugged the bottom port, causing the oil level to give a false reading. As a result, the pump was operating with no oil in it.

I never got to see the type of contamination so not really sure of the root cause, but one lesson is sightglasses can give false readings. There are many possibilities of the contaminants such as dirty castings if it is a new pump,

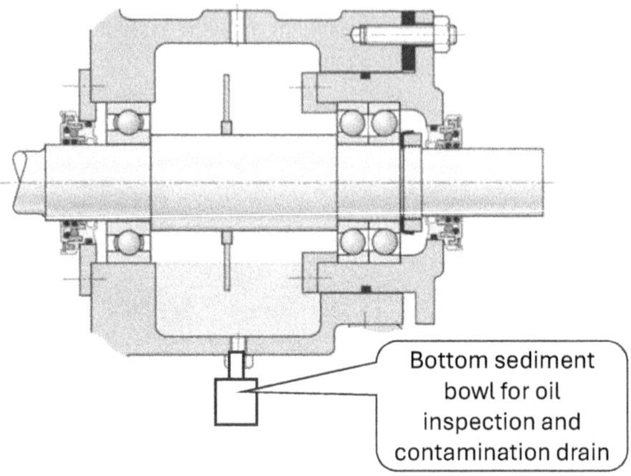

FIGURE 8.42 Bottom sediment bowl for pump power end drain.

paint flakes, or just dirt. One risk on this application was setting up the oil level sightglass coming off the bottom of the bearing housing. That made it a high-risk reliability situation for contamination issues on monitoring the oil level. The bottom drain port on the pump power end is a good place for a bottom sediment bowl to see and drain contaminants out of the pump, as shown in Figure 8.42.

Pump oil leaks can be an issue from time to time; however, I've found that if things are installed and maintained correctly, leaks need not be a chronic issue. Since pumps have a low volume of oil and a small volume of oil flooding the bearings, it is critical to keep a close check on any oil leaks. A small drip can lead to an oil starvation situation pretty quick. For example, one drop every 10 seconds may lose almost 3 quarts of oil in a week, as shown in Figure 8.43. So, for many pumps with an oil reservoir capacity of 4 quarts, this may be enough to cause a catastrophic failure in less than a week.

Pump leaks are fairly simple as there are only a few leak points on a pump power end – shaft seals, vents, and flanged connections. I did experience some pump oil leaks that was problematic until we understood why. The pump was slowly and chronically leaking from the air breather vent. It was occurring on most similar models to varying degrees. Changing breather elements did not solve the problem.

Looking closely at the pump design revealed an interesting fact that was causing the oil leaks. The pump utilized an oil flinger for lubrication, which isn't a problem in most designs. There were many models and sizes of this particular pump model that didn't have oil leak issues, but it was the oil flinger

Oil Lost From Leakage

Leakage	Loss in one day (Gallons)	Loss in one month (Gallons)	Loss in one year (Gallons)
One drop in 10 seconds	0.11	3.38	40.0
One drop in 5 seconds	0.23	6.75	81.0
One drop per second	1.13	33.75	405.0
Three drops per second	3.38	101.3	1,215.0
Drops break into stream	24.00	720.0	8,640.0

FIGURE 8.43 Oil leakage rate estimate.

design combined with one other design feature that created the chronic issue. It was the size of the power end oil flinger relative to the air breather location. The sizes where the oil flinger was very close to the air breather vent had oil leaks out the vents. The flinger was directing the oil straight up the housing into the air vent port, as shown in Figure 8.44.

The modification that ended up working was to branch the vent out so the oil that entered the air vent port did not have a clear path out of the pump. Another design change that also worked was a splash guard on the inside of the bearing housing over the vent port. It also gave protection from the direction of rotation from the oil flinger, as shown in Figure 8.44. The only disadvantage of this option was the pump had to be completely disassembled

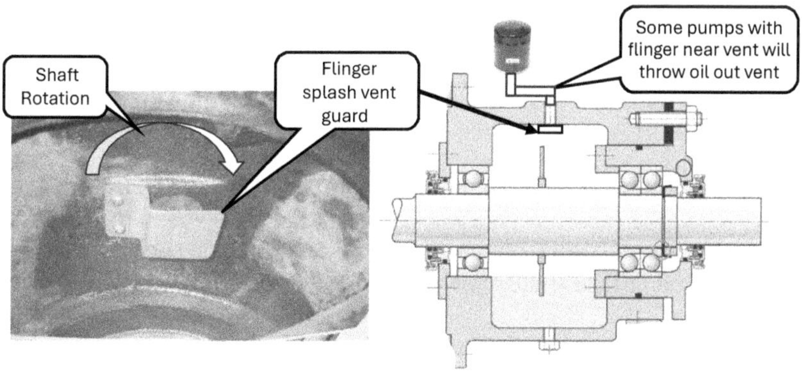

FIGURE 8.44 Pump leak from air vent solutions.

to install where the external vent filter piping could be done on an installed pump. When either of these options was installed, the oil leak issues were resolved.

Another lubrication mistake can occur by targeting the wrong oil level. A double suction pump ran for years with oil leaks. Constant-level oilers were added so they wouldn't run out of oil. Every time a shift maintenance came by and saw oil leaking, they would add some to the oil reservoir to make sure it made through the shift. Repeated addition of oil to the constant-level oiler resulted in excess oil into the bearing sump, fueling the leak; however, part of the issue was the initial target oil level. Maintenance had believed that the oil level should be that of a flooded sump bearing (center of bottom-most roller); however, this pump had oil rings and had a level lower than that.

Another issue is there was no oil level indication setup on the pump bearings. There may have been at one time, but it got replaced by the constant-level oiler. The constant-level oiler does not give a visual of the oil level. Two things were done to correct the issue. First, the correct oil level was established; second, an easy-to-read sight level gauge with a marked target oil level (while running) was installed. After these things were completed, the chronic oil leaks stopped. This did not cause a catastrophic mechanical failure in this scenario; however, the assumption or error the other direction would have. Assuming it was an oil flinger design and targeting a lower oil level outside the bearing rollers would result in catastrophic failure.

The oil leak problem out of the pump housing vent was proven to be a design issue that needed a design change or modification to stop the oil from slinging out. Some of the other pump leak issues were maintenance-related.

Case Study 47 Learning – Never underestimate the correct lube oil levels and small leaks. Look for sightglass errors when monitoring oil level.

One question that comes up in most plants is when should we change the oil in our process pumps? There are three types of oil change management methods – time-based, condition-based, and both time- and condition-based oil changes. Most plants either lack an oil change strategy or if they do, typically, rely on a time-based oil changes.

For time-based oil changes on pumps, most will adopt an annual oil change. Many pump manufacturers will recommend changing the oil every 2,000 hrs which is about once per quarter. The other factor on oil changes is what type of oil is used, particularly mineral or synthetic.

Condition-based oil changes can be a good strategy where oil analysis determines oil change necessity. This is typically done on large volume systems, which most pumps are not. An oil change may be necessary if any of the following is observed from oil analysis results.

- High water content above 0.03% water (300 ppm). This generally means the seal has leaked water into the oil so the seal may need repairing as well.
- Excessive ISO particle count (example 21/20/18) indicating heavy particle contamination. Generally, target 17/15/12 for a pump lube oil. Look for wear debris, such as copper, iron in oil analysis report.
- Viscosity changes $> 15\%$. This could be caused by wrong oil added, oxidation, water, etc.
- Any big changes in additives showing depletion.
- Changes in oxidation testing (total acid number, TAN). If TAN changes by 0.2–0.3 KOH/gm, then oxidation may trigger an oil change.
- Other clues to oil condition might also include odor, color (darkening), sludge, etc.

The best approach I believe is both time and condition monitoring. The pump can be on a time-based oil change unless an oil condition triggers an oil change. Pumps do not hold much oil so oil sampling on process pumps is not a feasible strategy for determining when to change an oil, but visual inspection on the sight glass can be. On a single pump the oil cost is not much either, but when there are hundreds of pumps in a plant, regular oil changes can be expensive and manpower-intensive.

For many plants this might be a challenge getting the downtime opportunity to execute annual oil changes on pumps. Although, I've seen it carefully done on the run too, but this is risky. With the advancement of synthetic lubricants, pump lubrication has seemed to also shift to using synthetic lubricants even though they may not be necessary from an application condition standpoint. Synthetic lubricants typically allow for longer drain intervals, which may change the approach to annual oil changes. If a pump has a chronic leak, it would not be advisable to use synthetic oil from a cost standpoint.

Another factor in oil change intervals is operating temperature. An average temperature of ~75 process pumps in the pulp and paper process was 140°F. For oil oxidation life, the oil life can be estimated by the equation below. For the process pumps average temperature of 140°F in this study, the estimated oil life was ~1,500 days or four years. At 150°F, the life drops to 891 days or 2.4 years. At 170°F, the life drops to 319 days. Therefore, a theoretical estimation for almost any pump would be at one year.

$$\text{Log } L = K1 + \frac{4750}{T + 273}$$

where:

L = Oil life in hours
$K1$ = oil life factor, -8.45 for R&O oil
T = oil temperature, °C

Depending on where you get your information, your strategy for oil changes in process pumps could be from three months to four years. For the time-based piece of pump oil change intervals what is the correct time? Several different times in my career, I've tested pump oil from pumps which did not have any significant contamination (seal leaks) issues after a one and two years of service. Each time, there were never any significant issues with the lubricant as far as viscosity, additives, and oxidation. All the pumps tested were good running pumps with no seal leakage or contamination issues.

With the oil analysis trial results on pump lube oil, the target time-based oil drain interval was set at one year for mineral oil and two years for synthetic oil. Also, a good idea to use a lube record tag on the pumps to keep up with oil changes. Good to attach these tags to a fixture that does not get changed outside of the time-based oil change intervals; pump oil condition monitoring by visual inspection should be routine. When a change was noted on the lube oil visual, an oil change was initiated and scheduled. The next case study is one example of where this was executed very well by area maintenance personnel.

CASE STUDY 48 PUMP BEARING HOUSING CONTAMINATION

A series of pump rotating assemblies (power ends) began to experience severe particle contamination right after being installed. This was first noticed by the area lube technicians as they made visual inspection rounds on the pumps. The power ends that had the issue were random sizes and not every power end that was installed had the issue. There was no real tracking of specific power ends through the repair process, therefore, there was no way to track each power end that had the problem. Of course, the power ends coming out of the storeroom were rebuilt at various times and pulled out of the storeroom at various times as well.

The pump power ends ran either 1,200 rpm or 1,800 rpm on the applications. Speed did not seem to be a factor, as both higher- and lower-speed pumps experienced issues, while some did not have contamination problem. All the pumps ran a synthetic 68 cst viscosity rust and oxidation oil. All the pumps (those with the issue and those without) were running with an oil flinger design. The issue occurred on various sizes of power ends and various size pumps in different areas of the mill.

Looking deeper at a few of these pumps to understand the failure mode, each pump that had the same basic issue or symptom. The pump oil would turn black within ~24 hours of installation and operation. The pumps with the issue operated in good conditions with the pump power end temperatures in the 125–140°F range. Figure 8.45 shows a comparison sample bottle from a pump after 24 hrs and how dark the oil got. The other sample bottle is another pump without the issue that had run for three months, and the oil still looked like new oil.

FIGURE 8.45 Oil sample bottle visual comparison.

The oil was so contaminated that it sometimes would not even test to show the particle count. It would max out the particle tester to 24/24/24, where a typical pump oil from new but dirty oil would test around 19/17/15. The particle tester itself would have to be cleaned from all the ferrous contamination. Figure 8.45 also shows one of the samples where after a few days the ferrous particles would finally settle in the bottom of the sample bottle.

After the issue appeared to become chronic on subsequent pump rebuilds, oil analysis was started to understand the issue further. All the pumps saw big increase in iron particle count. Figure 8.46 shows a trend of one pump after new installation and three successive oil changes. The Fe contamination was high on this pump starting at 455 ppm, then eventually dropped to 60 ppm after two oil changes and flushing the power end.

A few of the power end oil analysis samples also saw an increase in silicon. One saw a silicon contamination level of 83 on the first oil analysis sample. The contamination was so severe that on some of the pumps where magnetic bearing seals where used, these pump seals got plugged with ferrous metal. The plugging caused some of the pumps to begin to leak oil.

Oil was typically changed several times to flush the pumps out and the issue seemed to disappear. It appeared to be a one-time contamination issue that was resolved after flushing out of the system. This led to more focus on rebuilding and storage practices of the spare pump power ends. The pumps were rebuilt in the plant's central shop and never left the plant site.

With the high silicon showing up on some of the contaminated pumps, pump housing surface became a query point. The new pump housings were not coated with any internal paint coating. Some of the rebuilt pumps did have an internal coating. It was not exactly known as the issues came up which

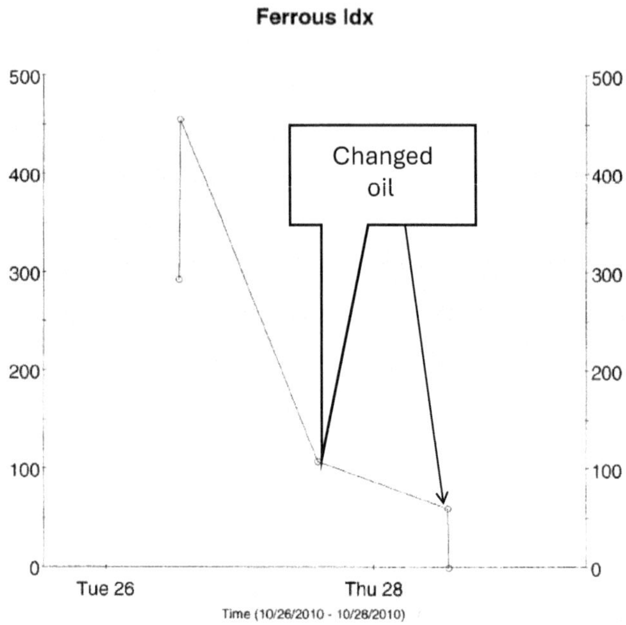

FIGURE 8.46 Ferrous particle trend from oil analysis.

pump housings were coated, and which were not. With high silicon it was suspected that some may have had some casting issues which were causing or contributing to the contamination issues since these were cast housings. New pump casings from some pump suppliers have been known to come in dirty at times. However, since all the pumps with the issue were rebuilt pumps, new pump housings were dismissed as the root cause of contamination.

Pump housings that did not have contamination issues were also inspected on future pump rebuilds and the internal housings looked very good. Some were coated and some were not. The final inspection point for pump rebuild assembly was the parts washer where the pump housings were cleaned prior to reuse and reassembly. There was not a consistent history of cleaning the parts washer.

On initial cleaning and inspection, a large amount of iron deposits was found in the bottom of the parts washer sump. The parts washer was initially only equipped with a coarse screen on the shower cleaning loop. While the parts washer seemed to do a good job cleaning parts, the iron contamination built up in the system with no filtration or tight enough frequency to flush and clean the parts washer. The parts washer cleanliness was the root cause of the pump housing contamination. After the parts washer was flushed and cleaned, the pump contamination issues eventually stopped. A filter upgrade was also purchased for the parts washer to ensure that clean washer fluid was showered on pump housings going forward.

The contaminated pump housings had a root cause category of maintenance as the parts washer was not maintained properly. It also has a root cause in design as the parts washer was not initially equipped with a filtration system.

Case Study 48 Learning – Housing cleanliness is affected at each step in the repair process including at the parts washer.

BEARING MAINTENANCE

Pumps are relatively simple machines which operate with a shaft and bearings in a housing. However simple, get this bearing assembly wrong and the pump is wrong. Pump bearing maintenance is another critical element for overall pump reliability results. Maintenance may be involved in pump rebuilds or repairs from failure and may have to execute precision bearing maintenance to ensure pump reliability so understanding some critical elements around precision bearing installation are vital. Just throwing a bearing on a pump shaft will not deliver optimum pump reliability.

Pump bearings typically are either angular contact or deep groove ball bearings, cylindrical roller bearings, or in some cases spherical bearings. The smaller end suction pumps will utilize ball bearings and larger end suction pumps may have a cylindrical roller bearing due to higher radial loads. Most pump bearings are cylindrical bore, but some may be tapered bore.

CYLINDRICAL BORE BEARINGS

Many common practices for installation of a cylindrical bore bearings involves temperature mounting of the bearing to the shaft. Most of the time, the verification or measurement of the actual bearing fit to the shaft is never checked or measured. As long as it turns easily after it is mounted on the shaft, that is usually considered good enough and many times it is, but sometimes it isn't. A considerable risk is taken without measuring and calculating the bearing fit.

Cylindrical bore bearings rely on the shaft size to set and determine the bearing fit to the shaft. The bearing already has a bore tolerance adhering to ABEC-1 or RBEC-1 standards, so the shaft is the only component the end user has some control to manage bearing fit and assembly quality. Pumps generally have bearings with a rotating inner ring. The rotating element (inner ring or outer ring) on a bearing is generally the component that has the interference fit so the pump would have an interference fit on the inner ring to the shaft and a clearance fit in the bearing housing.

The proper interference fit is required to hold the bearing on the rotating component, so it does not spin. Excessive fit will increase bearing preload, friction, and probability of premature bearing failure. Excessive fit could also potentially lead to a race locking up on shaft if all the radial internal clearance (RIC) has been removed during installation and the bearing locks up in operation. A low bearing fit on the shaft can also lead to the bearing spinning on the shaft and catastrophic failure.

The magnitude of this fit is affected by magnitude of the load (light, normal, heavy) and type of bearing (ball, cylindrical roller, spherical). Light bearing load would be the bearing dynamic load rating (C) to actual bearing load (P) ratio $C/P > 20{:}1$. Normal load would be in the $C/P = 7–17$ and heavy load would be C/P 4–8.

Many pump manufacturers may have their own bearing fit tolerances, but most will not be far from ABEC1 or RBEC1 bearing industry standards. If the pump manufacturer bearing fit class or standards are not available, then default to bearing industry standards like the table shown in Figure 8.47.

For an example of measuring the bearing fit and shaft diameter on the pump rebuild or repair, take a pump bearing and shaft using a 6317-ball bearing. Using the last two digits of the bearing number multiplied by 5 mm to get the bearing nominal bore of 85 mm. An 85 mm ball bearing with normal load would call for a k5 bearing fit tolerance.

Looking at the shaft bearing diameter table for class fits shown in Figure 8.48, there are three pieces of important information here – bearing bore tolerance, shaft diameter tolerance, and shaft bearing fit tolerance. The bearing fit tolerance is a result of the bearing bore tolerance and shaft diameter tolerance. When the high and low tolerances are paired, they determine the range of the assembled fit. It is important to note that these are just target bores, diameters, and fits. This is not the actual fit; however, if the k5 fit is the target fit for this bearing application and the measured shaft diameter is within range then the k5 fit will be satisfied.

Conditions	Shaft diameter, mm Ball bearings[1]	Cylindrical roller bearings, metric taper	Spherical roller bearings	Tolerance Symbol
Rotating inner ring load or direction of loading indeterminate				
Light loads	18≤100	≤40	—	j6
	100≤140	40≤100	—	k6
Normal loads	≤18	—	—	j5
	18≤100	≤40	≤40	k5 (k6)[2]
	100≤140	40≤100	40≤65	m5 (m6)[2]
	140≤200	100≤140	65≤100	m6
	200≤280	140≤200	100≤140	n6
	—	200≤400	140≤280	p6
	—	—	280≤500	r6
	—	—	>500	r7
Heavy loads	—	50≤140	50≤100	n6[3]
	—	140≤200	100≤140	p6[3]
	—	≤200	>140	r6[3]
High demands on running accuracy with light loads	≤18	—	—	h5[4]
	18≤100	≤40	—	j5[4]
	100≤200	40≤140	—	k5[4]
	—	140≤200	—	m5[4]

FIGURE 8.47 ABEC1/RBEC1 bearing fit class specifications for solid steel shafts.

Shaft Bearing-Seat Diameters (Values in inches) (cont.)

Bearing Diameters	Bore (inches)		j6			k5			k6			m5			m6		
			Shaft Dia.		Fit in 0.0000"	Shaft Dia.		Fit in 0.0000"	Shaft Dia.		Fit in 0.0000"	Shaft Dia.		Fit in 0.0000"	Shaft Dia.		Fit in 0.0000"
(mm)	Max.	Min.	Max.	Min.		Max.	Min.		Max.	Min.		Max.	Min.		Max.	Min.	
4	0.1575	0.1572	0.1577	0.1574		0.1577	0.1575		0.1579	0.1575		0.1579	0.1577		0.1580	0.1577	
5	0.1969	0.1966	0.1971	0.1968	1 L	0.1971	0.1969	0 T	0.1973	0.1969	0 T	0.1973	0.1971	2 T	0.1974	0.0001	2 T
6	0.2362	0.2359	0.2364	0.2361	5 T	0.2364	0.2362	5 T	0.2366	0.2362	7 T	0.2366	0.2364	7 T	0.2367	0.2364	8 T
7	0.2756	0.2753	0.2759	0.2755		0.2759	0.2756		0.2760	0.2756		0.2761	0.2758		0.2762	0.2758	
8	0.3150	0.3147	0.3153	0.3149	1 L	0.3153	0.3150	0 T	0.3154	0.3150	0 T	0.3155	0.3152	2 T	0.3156	0.3152	2 T
9	0.3543	0.3540	0.3546	0.3542	6 T	0.3546	0.3543	6 T	0.3547	0.3543	7 T	0.3548	0.3545	8 T	0.3549	0.3545	9 T
10	0.3937	0.3934	0.3940	0.3936		0.3940	0.3937		0.3941	0.3937		0.3942	0.3939		0.3943	0.3939	
12	0.4724	0.4721	0.4727	0.4723		0.4728	0.4724		0.4729	0.4724		0.4730	0.4727		0.4731	0.4727	
15	0.5906	0.5903	0.5909	0.5905	1 L	0.5910	0.5906	0 T	0.5911	0.5906	0 T	0.5912	0.5909	3 T	0.5913	0.5909	3 T
17	0.6693	0.6690	0.6696	0.6692	6 T	0.6697	0.6693	7 T	0.6698	0.6693	8 T	0.6699	0.6696	9 T	0.6700	0.6696	10 T
20	0.7874	0.7870	0.7878	0.7872		0.7878	0.7875		0.7880	0.7875		0.7881	0.7877		0.7882	0.7877	
25	0.9843	0.9839	0.9847	0.9841	2 L	0.9847	0.9844	1 T	0.9849	0.9844	1 T	0.9850	0.9846	3 T	0.9851	0.9846	3 T
30	1.1811	1.1807	1.1815	1.1809	8 T	1.1815	1.1812	8 T	1.1817	1.1812	10 T	1.1818	1.1814	11 T	1.1819	1.1814	12 T
35	1.3780	1.3775	1.3784	1.3778		1.3785	1.3781		1.3787	1.3781		1.3788	1.3784		1.3790	1.3784	
40	1.5748	1.5743	1.5752	1.5746	2 L	1.5753	1.5749	1 T	1.5755	1.5749	1 T	1.5756	1.5752	4 T	1.5758	1.5752	4 T
45	1.7717	1.7712	1.7721	1.7715	9 T	1.7722	1.7718	10 T	1.7724	1.7718	12 T	1.7725	1.7721	13 T	1.7727	1.7721	15 T
50	1.9685	1.9680	1.9689	1.9683		1.9690	1.9686		1.9692	1.9686		1.9693	1.9689		1.9695	1.9689	
55	2.1654	2.1648	2.1658	2.1651		2.1660	2.1656		2.1662	2.1655		2.1663	2.1658		2.1666	2.1658	
60	2.3622	2.3616	2.3626	2.3619		2.3628	2.3623		2.3630	2.3623		2.3631	2.3626		2.3634	2.3626	
65	2.5591	2.5585	2.5595	2.5588	3 L	2.5597	2.5592	1 T	2.5599	2.5592	1 T	2.5600	2.5595	4 T	2.5603	2.5595	4 T
70	2.7559	2.7553	2.7563	2.7556	11 T	2.7565	2.7560	12 T	2.7567	2.7560	14 T	2.7568	2.7563	15 T	2.7571	2.7563	18 T
75	2.9528	2.9522	2.9532	2.9525		2.9534	2.9529		2.9536	2.9529		2.9537	2.9532		2.9540	2.9532	
80	[illegible]	3.1490	3.1500	3.1493		3.1502	3.1497		3.1504	3.1497		3.1505	3.1500		3.1508	3.1500	
85	3.3465	3.3457	3.3470	3.3461		3.3472	3.3466		3.3475	3.3466		3.3476	3.3470		3.3479	3.3470	
90	[illegible]	3.5425	3.5438	3.5429		[illegible]	[illegible]		3.5443	3.5434		3.5444	3.5438		3.5447	3.5438	
95	3.7402	3.7394	3.7407	3.7398		3.7409	3.7403		3.7412	3.7403		3.7413	3.7407		3.7416	3.7407	
100	3.9370	3.9362	3.9375	3.9366	4 L	3.9377	3.9371	1 T	3.9380	3.9371	1 T	3.9381	3.9375	5 T	3.9384	3.9375	5 T
105	4.1339	4.1331	4.1344	4.1335	13 T	4.1346	4.1340	15 T	4.1349	4.1340	18 T	4.1350	4.1344	19 T	4.1353	4.1344	22 T
110	4.3307	4.3299	4.3312	4.3303		4.3314	4.3308		4.3317	4.3308		4.3318	4.3312		4.3321	4.3312	
115	4.5276	4.5268	4.5281	4.5272		4.5283	4.5277		4.5286	4.5277		4.5287	4.5281		4.5290	4.5281	
120	4.7244	4.7236	4.7249	4.7240		4.7251	4.7245		4.7254	4.7245		4.7255	4.7249		4.7258	4.7249	

FIGURE 8.48 Bearing shaft diameter fit size and tolerances.

For this specific bearing example, the bearing bore tolerance is 3.3465″/3.3457″ and the shaft diameter size tolerance is 3.3472″/3.3466″ for a resulting 0.0001″/0.0015″ tight bearing fit.

The actual fit is also important. It can reveal how close a bearing is to its upper or lower fit tolerance. Why is this important? In the example, the range of fit is 0.0001″ tight to 0.0015″ tight. That is a pretty large range of fit with only 0.0001″ tight on one end and 0.0015″ tight on the upper end. There is uncertainty in any precision measurements from person to person even with highly trained and skilled professionals. The best machinist may only be within 0.0003″/0.0005″ accuracy between each machinist. Taking this human measurement tolerance into account, a bearing fit on the low end or high end could easily be out of specification.

A shop audit study of 16 machinist in nine different service centers showed that experienced machinist should have no problem measuring journals within 0.0003″. Same study also found that 25% of measurements on certified master rings was off by 0.001/0.002″. This was due to tool and machinist calibration neglect so calibration of machinist in many shops was not accurately executed.

It is always advised to measure all fits on a pump rebuild where bearings are replaced on shafts. A simple sheet could be filed out like the one shown in Figure 8.49. This will reveal the actual bearing fit as the actual bearing bore and shaft diameter will be measured and known. Measurements should be taken with micrometers down to 0.0001″ significant digits. Do not use calipers for determining bearing shaft or housing fits. The housing fits are going to be loose fits. The same type bearing reference charts can be used for housing fits as done for shaft fits to determine the total fit for the bearings. Consult bearing manufacturer books for this information.

Ball bearings have a radial internal clearance (RIC), but you can't measure it. RIC is the clearance between the rolling elements and the outer race. This also makes

Position	AB	CD
Bearing Manufacturer		
Bearing Number		
Inside Diameter		
Inside Diameter		
Average ID		
Outside Diameter		
Outside Diameter		
Average OD		

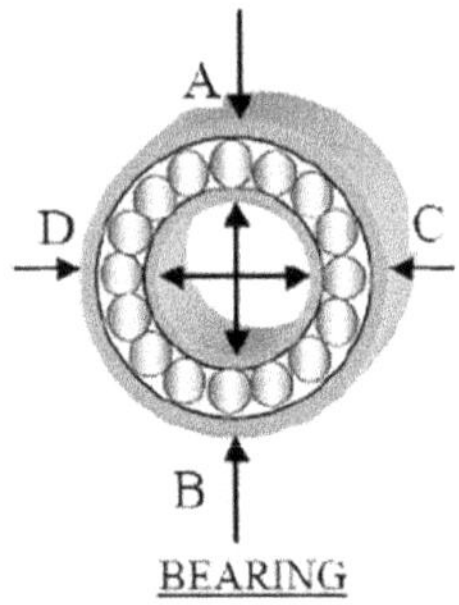

Journal		
Position	AB	CD
H1		
H2		

Average Shaft Diameter	
Bearing Fit	

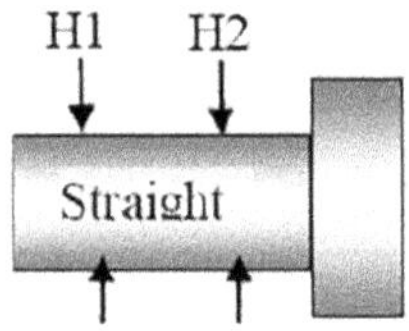

Housing Bore:		
Position	AB	CD
1		
2		

Average Housing Bore	
Bearing Fit	

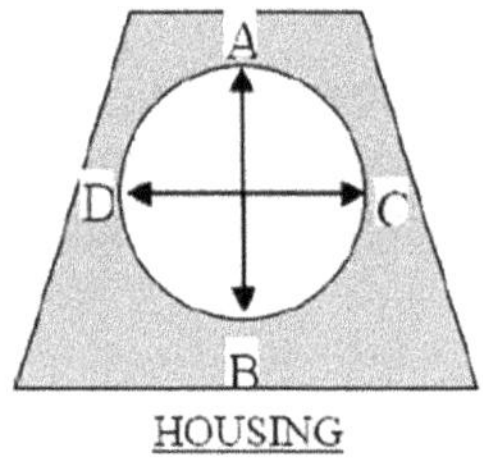

FIGURE 8.49 Sample bearing fit measurement sheet.

measuring the bearing fit all the more important. Most pump roller antifriction bearings are standard clearance (C0) or an occasional C3 clearance. The clearance designation is given in the bearing number suffix except for standard clearance bearings where no designation is provided.

Once the bearing fit has been measured and determined then mounting the bearing can begin. If the bearing fit on the shaft is not within the target fit, then several options may be available. Those options may include the following:

1. Making a new shaft is always a good option. This is a good time to ensure the shaft is within the tolerance size range from the target fit table. The finish for a cylindrical bore shaft should also be ~63 Ra range.
2. If a new shaft isn't a good option, then sleeving the journal may be.

3. If the diameter is only slightly undersized, then a metalizing of the shaft might be a consideration.
4. If the actual bearing bore is not at a minimum diameter (especially if it is at the upper end of the range), then grab a few more bearings off the shelf and measure to see if one has a smaller bore diameter that will get the actual bearing shaft fit to within target.
5. If the decision is made to run with a bearing shaft fit slightly under target (0.0001/0.0002″) then a short-term solution might be to use a bearing retaining compound until a more permanent solution can be executed.

Cylindrical bore bearings can be mounted in several different ways. Cold mounting is popular and effective on small bore cylindrical bearings < 70 mm (4″ OD). Cold mounting involves pressing, pushing, or driving the bearing on the shaft. The key point is to always apply the drive-up force to the race with the fit; for inner ring rotation, such as a pump shaft, this is the inner race. If the opposite was done, then bearing Brinell damage could likely result as the impact force goes across the rolling elements.

There are specially made tools that can be purchased for cold bearing mounting, which are usually better than homemade versions. They are usually non-metal materials. Metal bars, sleeves, and hammers can result in chards of metal getting into bearings. Some bearing manufacturers have estimated that 16% of premature bearing failures are due to improper cold mounting methods.

The other primary method for mounting cylindrical bore bearings is temperature mounting, where the bearing is heated before installation. As the bearing is heated, it expands, increasing the bore size so it becomes larger than the shaft, allowing for easy mounting. There are several accepted methods for temperature mounting a cylindrical bore bearing, such as using an induction heater, oil bath heater, or ovens. Never use a torch to heat a bearing as there is no way to properly control the heat, resulting in localized heat damage. Oil bath and ovens are acceptable but not typically efficient or available. Most pump bearing sizes fit well with the use of induction heaters for temperature mounting.

Bearings have stable temperature that should not be exceeded. In general, the bearing should not be heated over 250°F. Many induction bearing heaters will not allow high temperatures, but if the heater does not have that function, play close attention as bearing heats up. Most induction heaters have a demagnification function but if they don't, reverse the bearing and energize the induction heater to eliminate magnetization of the bearing. Also do not leave any bearings near an induction heater or they can magnetize. Bearings should have no magnetism even though generally < 3 gauss is acceptable.

Once the bearing mounting is complete and cooled, it is always a good practice to apply some lubricant on the bearing (oil or grease) as the pump assembly is finished so some protection for storage is present. Although bearings have a storage slushing compound preservative, a lube film will provide extended protection against vibration and corrosion until installation.

TAPERED BORE BEARINGS

Tapered bore bearings are used many times with spherical-type bearings. Tapered bore bearings may have various mounting arrangements such as direct tapered seating on a tapered shaft, adapter sleeve, or removal sleeve. The two sleeve options have cylindrical shafts, and the sleeve is tapered to match the bearing. Sometimes the tapered sleeve will need to be located on the shaft to set rotor position before final bearing installation and clearance is established.

For discussion purposes, spherical bearings will be used in the example to explain a simple installation for tapered bore bearings. The roller bearing has a bench or unmounted radial internal clearance (RIC) that should be measured first to verify that it falls within the target bench/unmounted clearance (C0 or C3). Hot applications will typically use C3 clearance bearings. For an example take a 22228 C3 spherical bearing for a large double suction pump assembly. A typical bearing RIC table is shown in Figure 8.50, where looking at chart A the unmounted RIC range for a C3 bearing should be 0.0063″/0.0079″.

RIC is measured using feeler gauges. Only stick feeler gauges in one row at a time. Do not roll bearing over feeler gauge as this is not a real clearance. Make sure rollers are centered between races and not cocked. Since you can't measure 0.0001″ with feeler gauges, 0.006/0.008″ is sufficient for a C3 RIC on this bearing. If measuring bench clearance, check bearing RIC at the top of bearing. If the bearing is hanging on the shaft, then measure RIC at the bottom. Bench clearance and

Chart A — Unmounted Radial Internal Clearance of SKF Tapered Bore Spherical Roller Bearings (In Inches)

Bore Dia. d(m)		Normal (in.)		C3 (in.)		C4 (in.)	
Over	Incl.	Min.	Max.	Min.	Max.	Min.	Max.
24	30	0.0012	0.0016	0.0016	0.0022	0.0022	0.0030
30	40	0.0014	0.0020	0.0020	0.0026	0.0026	0.0033
40	50	0.0018	0.0024	0.0024	0.0031	0.0031	0.0039
50	65	0.0022	0.0030	0.0030	0.0037	0.0037	0.0047
65	80	0.0028	0.0037	0.0037	0.0047	0.0047	0.0059
80	100	0.0031	0.0043	0.0043	0.0055	0.0055	0.0071
100	120	0.0039	0.0053	0.0053	0.0067	0.0067	0.0087
120	140	0.0047	0.0063	0.0063	0.0079	0.0079	0.0102
140	160	0.0051	0.0071	0.0071	0.0091	0.0091	0.0118
160	180	0.0055	0.0079	0.0079	0.0102	0.0102	0.0134
180	200	0.0063	0.0087	0.0087	0.0114	0.0114	0.0146
200	225	0.0071	0.0098	0.0098	0.0126	0.0126	0.0161
225	250	0.0079	0.0106	0.0106	0.0138	0.0138	0.0177
250	280	0.0087	0.0118	0.0118	0.0154	0.0154	0.0193
280	315	0.0094	0.0130	0.0130	0.0169	0.0169	0.0213
315	355	0.0106	0.0142	0.0142	0.0185	0.0185	0.0232
355	400	0.0118	0.0157	0.0157	0.0205	0.0205	0.0256
400	450	0.0130	0.0173	0.0173	0.0224	0.0224	0.0283
450	500	0.0146	0.0193	0.0193	0.0248	0.0248	0.0311
500	560	0.0161	0.0213	0.0213	0.0268	0.0268	0.0343
560	630	0.0181	0.0236	0.0236	0.0299	0.0299	0.0386
630	710	0.0201	0.0264	0.0264	0.0335	0.0335	0.0429
710	800	0.0224	0.0295	0.0295	0.0378	0.0378	0.0480
800	900	0.0252	0.0331	0.0331	0.0421	0.0421	0.0539
900	1000	0.0280	0.0366	0.0366	0.0469	0.0469	0.0598
1000	1120	0.0303	0.0406	0.0406	0.0512	0.0512	0.0657
1120	1250	0.0327	0.0441	0.0441	0.0559	0.0559	0.0720

Chart B — Recommended Clearance Reduction Values of SKF Tapered Bore Bearings (In Inches)

Bore Diameter d(mm)		Reduction in Radial Internal Clearance (in.)	
Over	Incl.	Min.	Max.(1)
24	30	0.0006	0.0008
30	40	0.0008	0.0010
40	50	0.0010	0.0012
50	65	0.0012	0.0015
65	80	0.0015	0.0020
80	100	0.0018	0.0025
100	120	0.0020	0.0028
120	140	0.0025	0.0035
140	160	0.0030	0.0040
160	180	0.0030	0.0045
180	200	0.0035	0.0050
200	225	0.0040	0.0055
225	250	0.0045	0.0060
250	280	0.0045	0.0065
280	315	0.0050	0.0075
315	355	0.0060	0.0085
355	400	0.0065	0.0090
400	450	0.0080	0.0105
450	500	0.0085	0.0110
500	560	0.0095	0.0125
560	630	0.0100	0.0135
630	710	0.0120	0.0155
710	800	0.0135	0.0175
800	900	0.0145	0.0195
900	1000	0.0160	0.0215
1000	1120	0.0175	0.0235
1120	1250	0.0190	0.0255

FIGURE 8.50 Spherical tapered bore bearing RIC clearance and RIC reduction tables (Ref. SKF).

unmounted clearance should be almost the same especially for smaller bore bearings but may be slightly different on large bore. Both locations are where the clearance naturally occurs from weight of bearing components.

There are a number of methods to drive the bearing up on the tapered shaft. Cold mounting is pretty much executed with either a spanner wrench and locknut or hydraulic assist. The spanner wrench is only practical on small bore bearings (< 70 mm bore). Hydraulic assist is much easier and involves a hydraulic nut to push the bearing on the shaft. Light oil may be applied on the shaft before installing. Do not use an anti-seize product as these contain soft metals and are contaminants to the roller bearing.

The goal in mounting a tapered bore is to remove bearing RIC to a target amount. This is also shown from bearing manufacturer in the RIC reduction table, as shown in Figure 8.50 chart B. Unmounted RIC and RIC reduction values may be similar for different manufacturers, but it is important to always use the tables from the same manufacturer as there can be variations. It can also change with the vintage (age) of the bearing. These clearance tables are adjusted from time to time due to bearing design changes including metallurgy changes. For the 22228 bearing, the RIC target reduction is 0.0025″/0.0035″ RIC reduction. If 0.007″ unmounted RIC was measured, the bearing would be driven up the taper until a resulting mounted RIC of 0.0045″/0.0035″ was measured.

$$\text{Mounted RIC} = \text{Unmounted RIC} - \text{RIC Reduction}$$

Now the question becomes which mounted/measuring method will be used to achieve the mounted RIC? The basic method is RIC reduction with feeler gauge measurements. Regardless of how the bearing is driven up, there must be a method to determine when to stop. Continuing to measure the mounted RIC with feeler gauges is the most basic way. Care must be taken to not have any measurement errors with this method. For example, if the bearing is mounted in a split bearing housing, do not measure the clearance while mounting after it sets in the bottom half of the housing. This will deform the outer race and egg all the clearance to the top and excessive mounting (RIC reduction) and an "untrue" RIC measurement will result.

On smaller bore tapered bearings, angular drive-up may be used. Here a table is given for the clock angle from a starting position where the bearing is locked into the shaft, but no clearance has been removed. This is only used for smaller bore bearings < 90 mm or so. Know which type of nut is used – standard nut or metric nut. The tables for clock angle will have separate target angles for metric or standard nuts. The angle may be something like 170° for proper mounting. Consult the specific bearing manufacturer for this drive-up information.

Axial drive-up distance is another method to achieve RIC reduction. For a solid shaft and a standard 12:1 taper, the axial drive-up distance is estimated by the formulas below.

$$\text{Solid Shaft} - \text{Axial Drive Up} = 16 \times \text{RIC Reduction}$$

$$\text{Adapter Sleeve} - \text{Axial Drive Up} = 18 \times \text{RIC Reduction}$$

For the 22228-bearing example, the axial drive-up distance would be 0.040″–0.056″. The final RIC should be in range with the drive-up distance. Various methods can be used to measure this distance depending on the specifics of the application.

Hydraulic axial drive-up is another slightly different method where a hydraulic pump and nut are used. The hydraulic nut is taken to a set pressure as the starting point for bearing mounting. This is the line-to-line point for bearing mounting stress on the tapered shaft. Then a dial indicator is installed on the bearing and driven up a target amount as axial distance. The indicator is mounted through the hydraulic nut to measure the distance the bearing is driven up the shaft. Consult the specific bearing manufacturer for this method as not all bearing manufacturers utilize this method and all drive-up data is manufacturer-specific.

These other mounting methods to remove bearing RIC do not require a check with a feeler gauge to verify bearing mounting. The target parameters (axial distance, clock angle, pressure, and distance) are all sufficient for proper bearing mounting. Different applications will render some methods unfeasible to use and some will have advantages. For example, mounting a bearing in a blind housing makes it difficult to use the feeler gauge method.

Always consult your bearing manufacturer for bearing design and mounting guidance as what is shown here is very generic and there are many special designs and applications which may require more detail to achieve precision installations.

PUMP FASTENERS

One last maintenance item to cover to ensure a reliable pump is around all fastened joint assemblies. There are many fastened joints in pump equipment and pumping systems. Fastened joints include impeller fixture to shaft, suction wear plates, seal chamber, sealing, bearing end covers, couplings, pump and motor hold-down bolts, piping flanges, etc. Fastened joints are everywhere in the pump system so to have reliable pump systems, there must be reliable fastened joints. Most of these joints are well designed to handle the external loads they see. When the installation is done properly, there are typically no failures due to fastened joints.

Bolted joint reliability depends on one key thing – establish the proper joint clamp load to withstand external forces during operation so that the joint remains in compression. If the fastened joint separates, then the joint has for all practical purposes functionally failed. When a bolted joint is assembled, the two joints are in compression and the bolt is in tension. The bolt preload is generated by tightening and stretching the bolt while at the same time this same force creates a clamp force that compresses the joint.

While many details must be correct, even well-designed pump system joints ultimately depend on joint assembly. What are the key elements during a fastened joint assembly to ensure reliability? Bolt installation is the main element that maintenance technicians control to achieve a reliable bolted joint. Bolt torque is not bolt preload, but just the mechanism to deliver bolt preload and joint clamp load.

Actually, only about 10% of bolt torque goes into bolt preload in a dry torque assembly. The rest of the bolt torque is consumed by the bolt friction in the assembly. The bolt friction can be divided into thread friction and friction underneath the head of the bolt. A simple expression of the bolt preload using K factor as a representation of the friction condition. For generic dry conditions, the K factor is ~0.2 and lubricated conditions ~0.15. Bolt torque can be estimated by the equation below.

$$T = KDF$$

where:

T = bolt torque, in-lbs
K = K factor for bolt assembly friction conditions
D = bolt diameter, inches
F = bolt preload, lbsf

Maintenance bolted joints typically are preloaded to 75% of proof load (proof strength is slightly less than yield strength) for the bolt material. The bolt is preloaded and operates in its elastic range so when external loads come, the bolt can stretch and not exceed yield. An external load increases the load on the bolt (elongation) and reduces the clamp load (compression) of the joint at the same time. Achieving the target bolt preload in each fastened joint is critical for pump and system reliability.

In assembly, how does the installer achieve target bolt preload? Below are some elements to ensure the fastened joint assembly is in a precise condition. For pump system fastened joints, if these bullet list items are followed, I expect you will see a 99% reliability on most of the pump systems in regard to fastener failures or issues.

1. Have the correct bolt torque for the grade bolt used. First defer to the pump manufacturer operating maintenance manual for the target bolt torque. For many pump manufacturers, they have specific torque tables for certain parts of their pump assembly with target bolt torque values for assembly. These will include casing bolts, bearing end cover bolts, frame bolts, and other specialty assembly areas.

   If that is not available or known, then default to the bolt size and grade material for 75% proof load for that bolt at the target bolt torque. Maintenance joints in general should not be tightened beyond 75% proof load. Maintenance joints are those that are assembled and disassembled. Permanent joints are those typically assembled once for life, such as structural joints and are typically preloaded to 90% proof load. Be sure and note if the bolt torque is dry or lubricated torque and what lubricant is used. If non-pump manufacturer bolts are used, be sure to note if substituted bolts are coated that would change bolt preload at target torque.

   While the best choice to get the target bolt preload for pump and system assemblies is to follow pump and other manufacturer tables and recommendations, there may be times where this is not available. If a general torque table isn't available, then the long way to calculate bolt torque can be done as follows. Take a ¾″-10 SAE grade 5 bolt, what is the dry torque specification for assembly?

a. Calculate the stress area for a ¾″-10 bolt.

$$As = 0.7854\left[D - \left(\frac{0.9743}{n}\right)\right]^2$$

where:
As = bolt stress area, in ^2
D = bolt diameter, inches
n = thread pitch, threads per inch

For this example, with ¾″ bolt and n = 10 the As = 0.335 in ^2 σ

b. Calculate bolt preload.

$$F = 0.75\sigma As$$

where:
F = bolt preload, lbf
As = bolt stress area, in ^2
σ = bolt strength, psi

For this example, a grade 5 bolt has a strength of 85,000 psi. This calculates to be a bolt preload force of 21,356 lbs.

c. Calculate bolt torque from bolt torque formula for dry torque so K factor = 0.2. This comes out to be a bolt torque of 3,203 in-lbs or 267 ft-lbs.

2. If lubricated and dry torque values are given, utilize lubricated torque when possible. Lubricated torque produces more accurate bolt preload and less preload scatter around a bolt circle.
3. For lubricated torque, lubricate not only the threads of the bolt but under the head of the bolt or nut or whichever component is being turned to tighten. Approximately 50% of the bolt friction eating up the torque is friction under the head of the bolt. The remaining 40% is thread friction. Most technicians will lubricate the threads but forget to lubricate under head of the bolt and significant bolt preload error will occur. If joints are improperly tensioned, they may loosen or other failure modes such as fatigue may occur.
4. When tightening the bolt circle, tighten in a star pattern to evenly pull in the parts around the bolt circle. Tightening one side will cock the assembly and create uneven preload or joint clamp around the bolt circle.
5. On larger bolted joints or bolt circles, it can be a best practice to tighten on two to three passes to final torque. The first pass might be 35% of final torque, second pass at 70% of final torque, and then final pass at 100% final torque. This will also evenly clamp the bolted joint surfaces during the tightening process.
6. Torque wrench operation is also important. Torque wrench should be moved in smooth motion for the full range of motion. Any sudden acceleration or jerking motions will cause click error torque spikes.

7. Inspect the thread condition, especially on screw applications where you can't change the nut. Most people generally interchange bolt and screw in describing a fastened joint (I do as well); however, there is a difference in a bolt and a screw. A screw goes into a blind tapped hole in the part. A bolt goes through the joint and has a nut on the other side. Therefore, screws that go into a blind hole need more attention and care since the thread condition can't easily be changed. Poor thread condition from corrosion to damaged threads can change friction and subsequently the joint clamp load after torque process.
8. Make sure fastened joint surfaces are flat. Any bolted joint surfaces that are not flat will eat up bolt torque to bend and flatten the surface. The result will be little clamp load and joint compression which directly puts external forces on the bolt. Related to this if the bolts are used to pull in parts that also eats up torque that can deliver clamp load to the joint.
9. Most pump fastened joints can be torqued with standard 250–300 ft-lb click torque wrenches. However, on large pumps with large, fastened joints, the magnitude for target bolt torque will exceed the standard torque wrenches so other torque methods must be employed. This may either be larger torque wrenches, hydraulic torque wrenches, or other bolt preload methods. One example of this might be large impeller nuts that may have target bolt torques up to 1,200 ft-lbs.
10. In general, don't reuse fasteners more than one to two times. If damaged at all do not reuse. Studies have shown bolt preload can reduce by 40–50% after reusing five times. When larger bolts are used, for example, on some large pump casing, these bolts may not be as readily available, and this can be a problem area for bolts to loosen.
11. If a bolt is not perpendicular to the surface during assembly, significant bending stress may result. This bending stress drastically reduces the fatigue strength of the bolt. As little as 1 degree head angularity may reduce fatigue strength by 50%.

For the 1% of other pump system fastened joint failures, exceptions may occur due to high external load or marginal original design. In these cases, there are other things to consider to provide some reliability margin in the assembly. In general, pump bolted joints have significant design margins. Bolt conditions such as corrosion, high friction, and poor bolt preload installation execution are mostly to blame on fastener failures on pump systems. Low bolt preload results in low joint clamp load.

Many times, preload loss may be due to machine vibration that loosens fasteners over time. It could also be due to low bolt preload or low clamp load initially. Embedment relaxation can cause a 10–15% loss of initial preload over time under the right conditions. If there are external loads or vibration causing a loss of initial bolt preload, there are several installation changes to provide some insurance for the bolted joint. Lock washers can be very effective. Split ring washers are often used but the load is not even around the bolt head. The Nordlock lock washer is a very good solution for vibration-challenged joints to prevent loosening. Loctite also has some good products to help prevent joint loosening. If there is room on the bolted assembly, a double nut can also be a cheap and effective way to prevent loosening.

Another fastener tip for pumps and any other fastened joint is ensuring the bolt materials are appropriate for the applications. Check material compatibility with fluids they operate around. One common error observed in industrial applications involves galvanic corrosion. Typically, whenever there is a carbon steel bolt that has corrosion, maintenance technicians will tend to always upgrade to a stainless bolt. If the bolt or screw threads into a blind hold that is carbon, then there will be some galvanic corrosion of the active metal which is the carbon steel and very poor joint assembly will result. First, when the threads corrode, thread friction greatly increases, and lower bolt preload occurs. At some point loss of threads will occur from galvanic corrosion using the stainless bolt. The bolt can be replaced easily but the threaded part usually cannot, and the replacement cost is also much higher.

9 Predictive Maintenance

Pumps are just like any other piece of equipment when it comes to predictive maintenance, in that centrifugal pumps exhibit specific vibration characteristics and signals that indicate how the pump is functioning. Predictive maintenance is detecting machine defects using non-invasive methods to collect and analyze data, in order to predict corrective actions to manage the remaining useful life of the asset. There are many technologies available in predictive maintenance to utilize, evaluate, and determine pump condition, each having various returns on investment (ROI) as shown in Figure 9.1. Some of those include vibration analysis, ultrasound, infrared, and oil analysis.

While this chapter is not meant to teach vibration analysis or other predictive maintenance technologies, I do want to share some insights as to how pumps exhibit issues in some of these technologies. Most of this chapter will focus on vibration analysis as it is probably the most commonly used and has a very good ROI. In my experience, vibration analysis can be 90% effective in diagnosing the issues with world-class execution. Another important element of diagnosing vibration issues and faults is to understand the physics of failure (POF) in which the vibration originates. The POF can help determine when a vibration defect is a symptom of a condition or the root cause.

Vibration analysis is probably the most common technology utilized. Lower vibration has many benefits to the pump equipment such as increased component life, lower maintenance cost, unscheduled downtime avoidance, and lower energy consumption. Many studies over the years in different types of equipment have shown that lower vibration leads to higher reliability. I have also conducted similar studies around pump equipment and vibration. One of these studies shows the curve, in Figure 9.2, demonstrating the correlation between pump vibration and mean time between failure (MTBF). A regression of the data gave an equation for MTBF as a function of pump vibration by the equation shown below. Essentially, if the pump vibration can be reduced by half, then the MTBF can be doubled. Knowing this, it is important to understand the causes of pump vibration and resolve this, if possible, to increase pump reliability.

$$MTBF = 0.5652 \times V^{-1.042}$$

Having a good vibration program can also be a valuable input into maintenance repair decisions. For example, if a pump is running rough, maintenance must decide appropriate course of action. Sometimes just changing the pump may not resolve the issue, but vibration analysis can help diagnose the issue and it may not be a maintenance or pump-related problem.

Pumps not only will exhibit typical vibration faults but also some specific faults as a reflection of how they are operating. Pump vibration is a response to an excitation

DOI: 10.1201/9781003729525-9

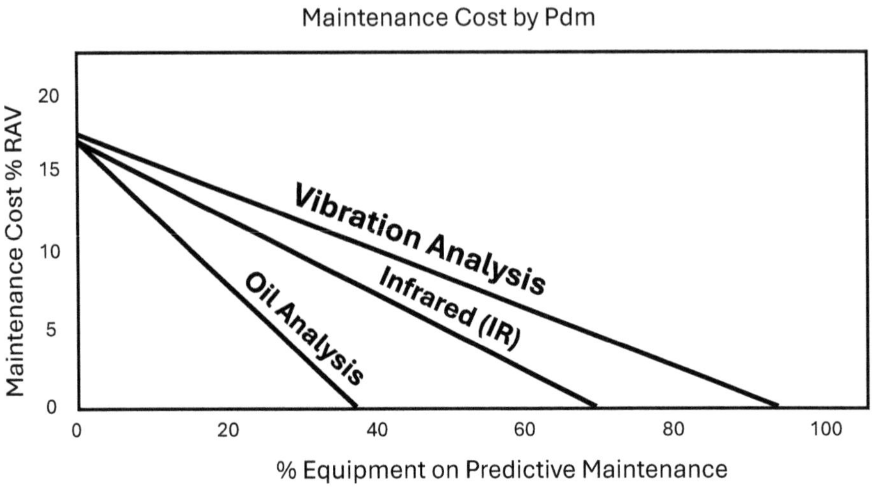

FIGURE 9.1 Maintenance cost by predictive maintenance type.

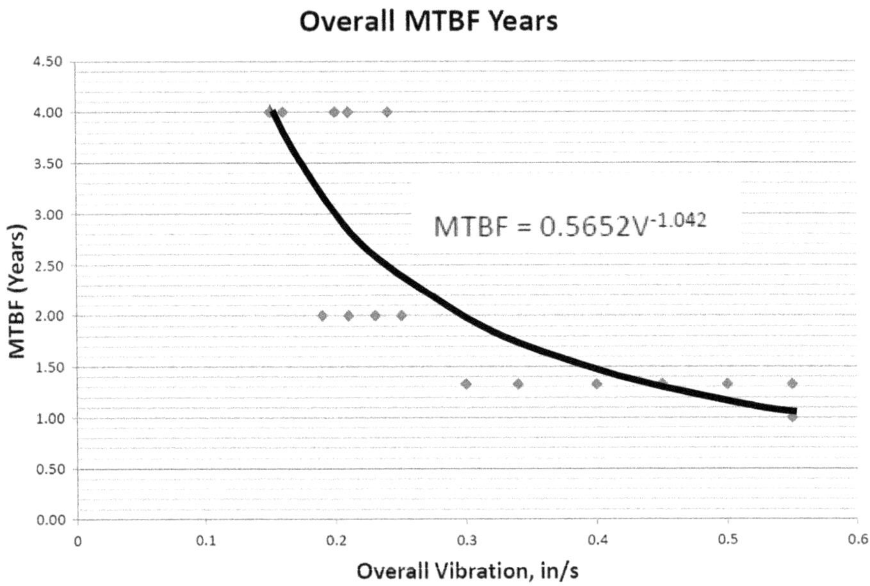

FIGURE 9.2 MTBF versus pump vibration.

force or forcing function generated by the machine itself or the process it is operating. Vibration analysis and other predictive tools may help diagnose active defects such as bearing defects (inner race, outer race, ball spin, cage) or pre-failure conditions such as misalignment, unbalance, looseness, vane pass, cavitation, and flow turbulence.

Since pumps will exhibit some vibration as normal operation, all pump vibration is not necessarily harmful. Some non-damaging vibration might be called benign vibration such as motor line frequency (60 Hz or 50 Hz), vane pass, gear mesh, and flow turbulence. This type of vibration is likely to be present anywhere the components are used. More serious vibration that is damaging might be bearing defects, misalignment, unbalance, significant cavitation, elevated vane pass or flow turbulence, resonance, and looseness.

There are several measurements of vibration such as displacement, velocity, and acceleration. Displacement is the distance traveled from a reference and usually expressed in mils (0.001″ = 1 mil). Velocity is the rate of change of displacement that the part is vibrating with units typically expressed in in/s or mm/s. Acceleration is the rate of change in velocity as it is moving from peak to peak and generally expressed in units of gravity such as gs. There are three terms to describe the vibration such as amplitude, frequency, and phase. The amplitude is simply the in/s or gs or mils of vibration measurements. The frequency is how often the event occurs such as cycles per second (Hz), rpms, and orders.

With respect to vibration frequencies, the vibration may be either synchronous, non-synchronous, or sub-synchronous. Sub-synchronous is less than 1 rpm frequency with defects such as belt defect, rubbing, cage, oil whip, and oil whirl. Non-synchronous is a frequency greater than run speed (1X) with defects such as electrical faults, another machine vibration, roller bearing defects, resonance, and flow turbulence. Synchronous vibration occurs at the running speed (1X) or at integer multiples of the running speed (2X, 3X, $X, etc.). Synchronous defects may be unbalance, misalignment, looseness, gear mesh, vane pass, bent shaft, etc. Peaks with the same harmonics are typically caused by the same defect.

VIBRATION SEVERITY

Even if a specific type of defect is unknown, the overall vibration trend can still be a useful tool. While a specific defect generates vibration at a particular frequency, the overall vibration of the pump is the sum of vibrations occurring at all frequencies. Frequency bands allow alarms to be set for different defect types instead of just an alarm for overall problem as shown in Figure 9.3.

Determining the severity of a pump vibration is not an exact science and there are many factors to consider. Factors such as the type of defect causing the vibration, velocity magnitudes, acceleration (impacting) magnitude, historical vibration trend/history, and many details of the pump design (bearings, sealing, shaft, impeller, etc.) affect the severity of the vibration being seen in any given pump design.

There are many different standards for equipment vibration. The hydraulic institute has a standard specifically for centrifugal pump vibration in ANSI/HI 9.6.1. Without getting into specifics of the standards, a general acceptable pump vibration for end suction foot-mounted pumps is typically in the 0.12–0.018 in/s for velocity. For end suction paper stock pumps, the acceptable vibration is a little higher between 0.14 and 0.22 in/s. The larger the pump (HP) the higher the acceptable pump

Percentage of Overall Value for each Narrow Band Alarm				
Sleeve Bearing – Simple				
Band Name	Band No.	Percent	Band Fault	
Sub. Synchronous	1	15 – 20%	Oil Whirl	
1 x rpm	2	80 – 90%	Imbalance/Misalignment	
2 – 4 x rpm	3	60 – 67%	Misalignment/Looseness	
5 – 20 x rpm	4	30 – 40%	Bearing Wear/Vane Passing	
Roller Bearing – Simple				
Band Name	Band No.	Percent	Band Fault	
Sub. Synchronous	1	15 – 20%	Cage Defect	
1 x rpm	2	80 – 90%	Imbalance/Misalignment	
2 – 4 x rpm	3	60 – 67%	Misalignment/Looseness	
5 – 20 x rpm	4	30 – 40%	1ST Bearing Band	Note:1
21 – 50 x rpm	5	20 – 30%	2nd Bearing Band	Note:2
Note: 1	Fundamental Bearing Defects Outer & Inner Race.			
Note:2	Harmonics for Bearing Defects Outer & Inner Race			

FIGURE 9.3 Vibration defects – parameter band ranges.

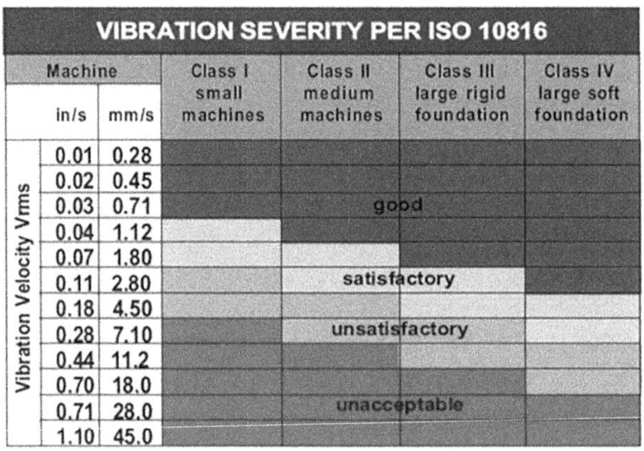

FIGURE 9.4 Vibration severity based on velocity and size machine.

vibration as the hydraulics of the pump will drive the overall vibration levels. I have seen some plants or industry picking a single amplitude of 0.25 in/s as a single limit for acceptable vibration across all equipment regardless of size, speed, or type. Alternatives for what is acceptable vibration may also match that shown on the vibration chart in Figure 9.4.

Another factor influencing vibration is how it changes over time as the pump is monitored. For pumps that normally operate at a known vibration level, a change of 25% or more is a concern that should be evaluated further.

The severity of vibration should also take into account not only the overall vibration but also the amplitude of the specific defect frequency. For some defects to increase in sideband harmonics can indicate increasing severity. For bearing defects, further understanding the stage of failure the bearing is in also factors into the severity of the vibration condition. The acceleration of vibration and vibration mass will also determine the impacting forces on the assembly. For example, a 20 lb rotor with 10 g acceleration has a much different impacting force than a 100 lb rotor with the same 10 g acceleration as Newton's law $F = ma$ is applied. At the end of the day, the impacting forces are what cause damage to the rotating assembly.

MECHANICAL-INDUCED VIBRATION

There are many sources or causes of pump vibration. Pump vibration may occur from mechanical or hydraulic reasons. Figure 9.5 shows some of the mechanical sources of pump vibration.

BEARING DEFECTS

Bearing defects is one of the common issues evaluated using vibration analysis. Which plane the vibration defect shows up best will depend on the type bearing or type load at a certain position. The axial reading on the drive end of an end suction centrifugal pump is likely to be more informative as the resultant shaft thrust is in the axial direction on the drive end angular contact bearings. The pump wet-end bearings typically show higher vibration in the radial plane, as they carry radial loads and are configured as radial bearings.

Bearing defects may be ball pass frequency inner race (BPFI), ball pass frequency outer race (BPFO), ball spin (BSF), or cage frequency or fundamental train frequency (FTF). Bearing vibration data is more accurate when taken as close to the bearing as possible and in the correct plane as the bearing load. For example, a cylindrical roller bearing only takes a radial load so taking vibration data in the radial plane is necessary and the same goes for a thrust bearing. There are detail bearing geometry equations that go into determining the bearing fault frequency which also vary if the bearing has inner race or outer race rotation. There are some general expressions for inner race rotation that are within 20% of frequency shown below. I generally am more concerned with inner race defects as it is already under shaft fit hoop stress so surface fatigue or other defects can accelerate much faster.

$$\mathrm{BPF}O = 0.4 \times N \times S$$

$$\mathrm{BPF}I = 0.6 \times N \times S$$

$$\mathrm{BSP} = 0.23 \times N \times S \ (N < 10)$$

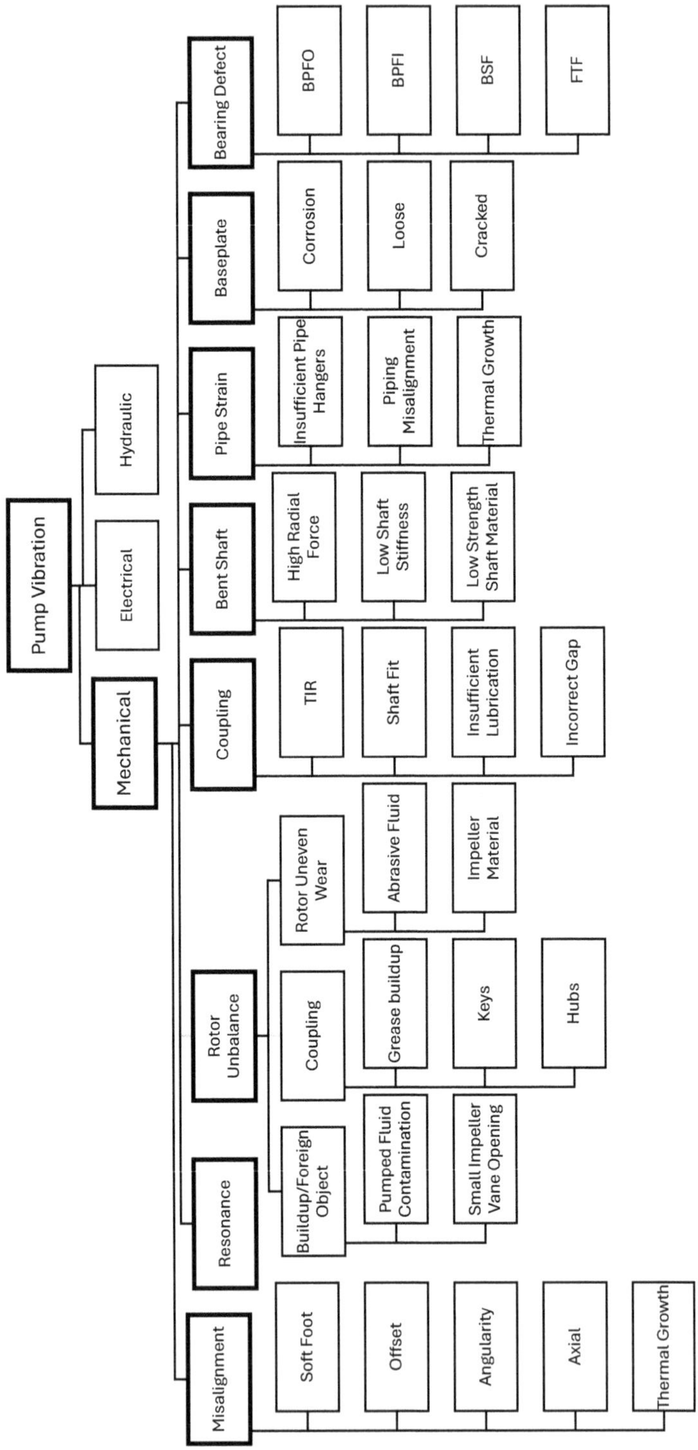

FIGURE 9.5 Basic mechanical sources of pump vibration.

$$BSP = 0.18 \times N \times S \ (N > 10)$$

$$FTF = 0.4 \times S$$

where:
N = # of rollers
S = shaft speed

There are typically four stages of bearing failure. Stage 1 is the earliest stage of failure and appears in ultrasonic frequency ranges (250,000–350,000 Hz). As condition worsens, the frequency drops to 20,000–60,000 Hz. Stage 2 is where the bearing defect begins to ring the bearing natural frequency which are in the 30,000–120,000 cpm range. Stage 3 bearing failure is when harmonics begin to show up and sidebands around these bearing natural frequencies. Stage 4 sees a progression of stage 3 amplitudes and even begins to affect the 1X run speed frequency and run speed harmonics. Stages 1 and 2 defects are normally not a cause for immediate response in a centrifugal pump but gives maintenance time to plan and get on a schedule. If stage 3 or 4 is the first stage identified, then a more urgent response might be necessary.

MISALIGNMENT

Misalignment can show up in any plane depending on if it is offset or angular in the horizontal or vertical plane. Angular misalignment is typically magnified in the axial plane. Offset misalignment is typically magnified in a radial plane (horizontal or vertical), depending on the type of offset misalignment. Misalignment will typically show up in some form of 1X, 2X, and 3X run speed frequencies as shown in Figure 9.6 from a misaligned pump with vertical offset misalignment. If phase analysis is used, a 180° phase shift will typically occur across the coupling with misalignment.

Pipe strain can also impact alignment and pump vibration. Figure 9.7 shows a before and after vibration on a high speed (3,600 rpm) horizontal multistage pump that had the pump suction and discharge nozzles supporting the pipe loads. In this case, the 2X vibration was reduced by 5X by adding pipe hangers on the suction and discharge piping.

Keep in mind that the velocity spectrum shows all the vibration frequencies, and it illustrates that multiple forcing function are causing vibration on the pump, with misalignment being certainly one of them. One of the first decisions when evaluating pump vibration is to determine which vibration to try and solve or improve, especially when there are many different vibrations and causes. Several of these will not be solved or eliminated so which ones are targeted for improvement or which ones are problematic to pump reliability?

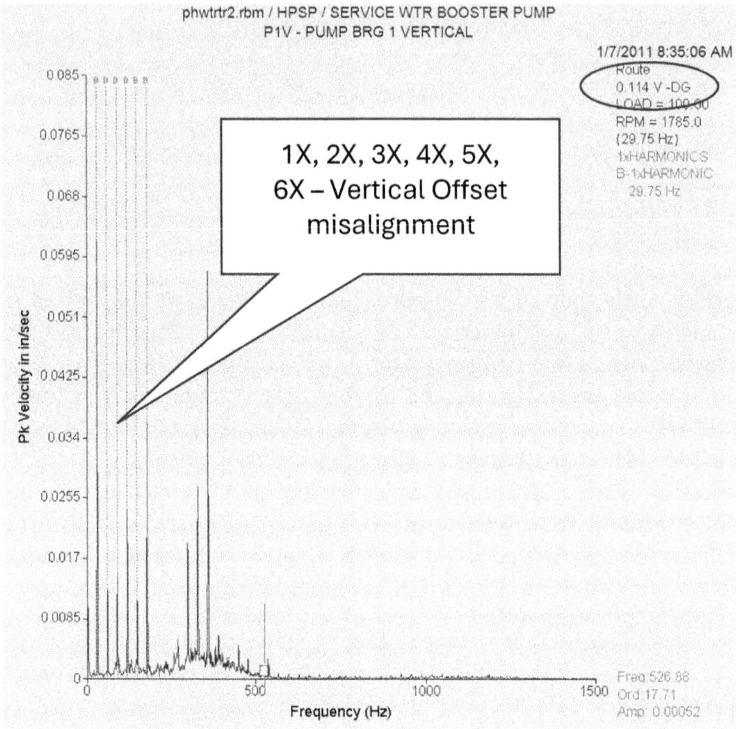

FIGURE 9.6 Typical pump vibration with vertical offset misalignment.

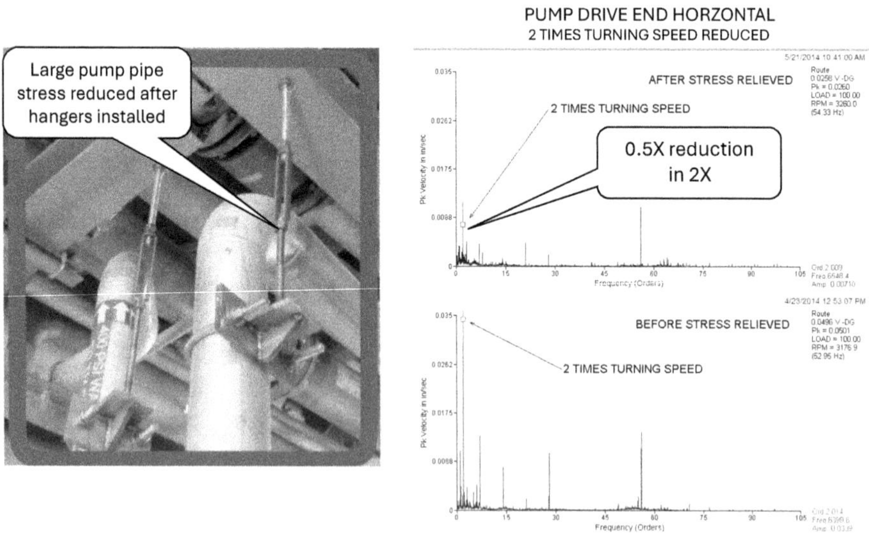

FIGURE 9.7 Pipe strain effect on pump vibration.

VANE PASS

Another question to decide is whether the vibration problem is mechanical or hydraulic in nature? Some issues are just mechanical, some just hydraulic, and some can be either. For example, a high 1X rpm frequency may be mechanical unbalance of the assembly, or it could be a hydraulic unbalance (one vane is plugged with a foreign object). Vane pass could be hydraulic or mechanical if something is rubbing the vane during each revolution. Usually this will require some trial and error, mechanical inspection, or additional methods to sort through the possible causes.

Vane pass vibration is caused by several conditions but at low amplitudes, vane pass is a normal vibration in centrifugal pumps. The impact of fluid forces on the impeller vane causes normal vane pass vibration and is a function of pump speed and number of impeller vanes as shown in the formula below. Figure 9.8 shows vane pass frequency in the velocity spectrum as part of normal operation on a self-priming sump pump. Note the two-vane impeller and the high amplitude at 4X due to the inlet vane hitting the discharge vane. Figure 9.9 shows two examples of a normal vane pass vibration and a high vane pass when the pump was operating in a dead head condition and the amplified vane pass frequency and waveform energy increase.

$$\text{VPF} = \text{Ps} \times V$$

where:

VPF = vane pass frequency
Ps = pump speed
V = # impeller vanes

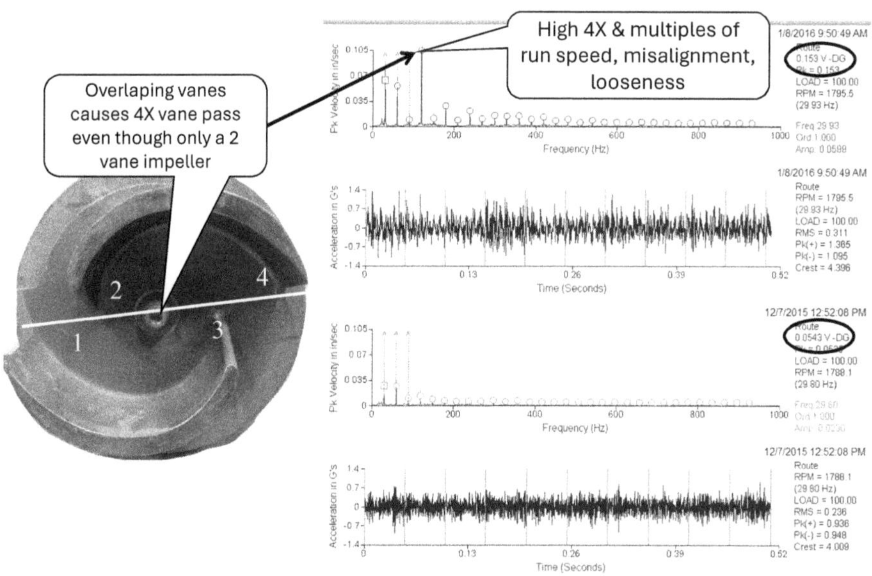

FIGURE 9.8 Typical vane pass frequency pump vibration.

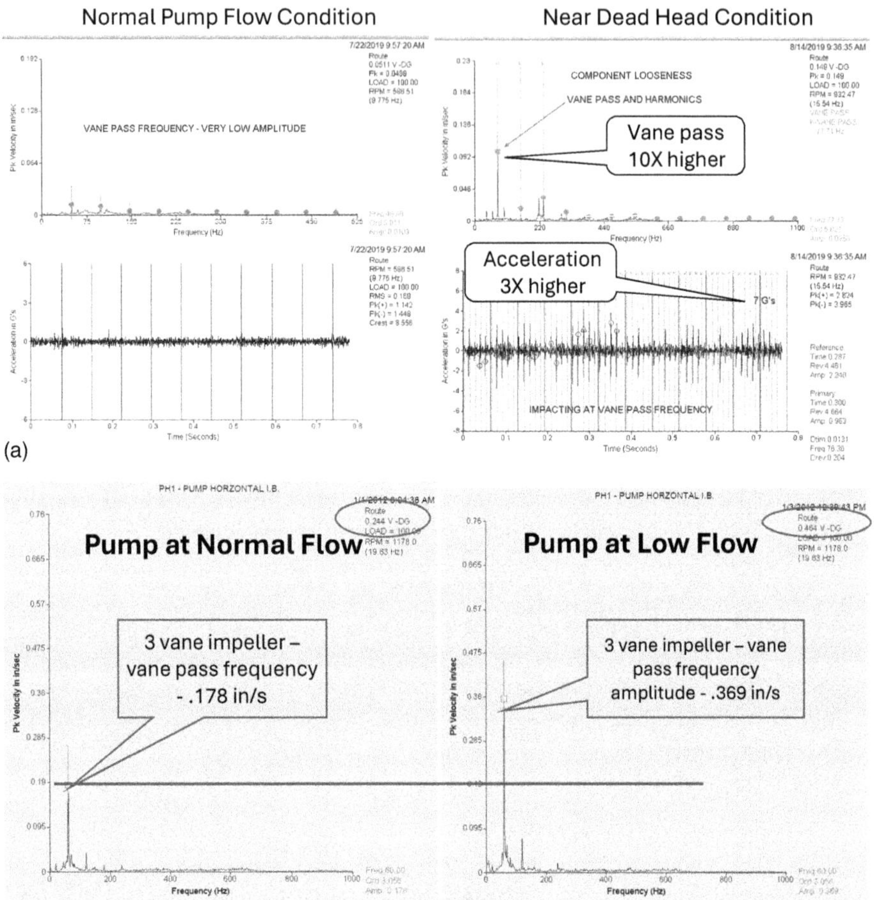

(b)

FIGURE 9.9 (a) Normal vane pass vibration and (b) near dead head condition vane pass vibration.

Vane pass may show up easily in the velocity spectrum, but some other views may also show a good picture of vane pass. A circular plot of the waveform with autocorrelation as shown in Figure 9.10 gives a very plain picture of the three-vane impeller. The autocorrelation shows the repeatable energy from the vibration. In this figure, the repeatable energy was 92%, which is why the picture was so pronounced at the three-vane. In this case, there is certainly a major issue at vane pass frequency for the repeatable energy to be this high.

A max size impeller will amplify vane pass vibration as the impeller vanes passes by the cutwater creating a pressure pulse. In the pump sizing process, if the impeller is designed with a slightly increased cutwater gap, it will minimize the natural vane pass vibration. Of course, the hydraulic efficiency is decreased, and recirculation

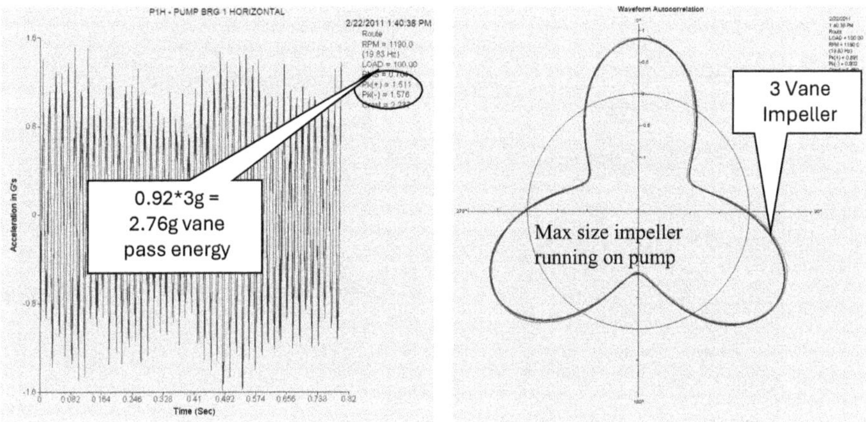

FIGURE 9.10 Vane pass vibration waveform and autocorrelation orbital plot.

may increase. A rule of thumb can be used to target an impeller diameter to reduce vane pass vibration from small cutwater gaps as shown below.

$$Dt = D\max - 0.25(D\max - D\min)$$

where:

Dt = target impeller diameter
$D\max$ = max impeller diameter for pump
$D\min$ = min impeller diameter for pump

For a double suction pump vane pass vibration causing flow pulsations can be reduced by using staggered vanes on each half of the impeller. A double suction pump with five vanes on one half and five on other will have a total of ten vanes showing up in vibration. The five-vane vibration also shows up, but the amplitude of the vibration is reduced compared to a full width ten vane alignment.

Vane pass frequency can have several causes such as a throttled pump where internal recirculation is high. A simple check when vane pass is to throttle or shut the discharge valve. If the vane pass vibration does not improve then the issue could be operating the pump outside the allowable operating region (AOR). If the vane pass improves, it could indicate a cavitation issue. NPSHR decreases as the flow decreases. Cavitation also can cause vane pass as the micro implosions impact each impeller vane.

One thing to watch for is when a bearing fault frequency is close to another fault frequency like vane pass as shown in Figure 9.11. This pump was initially called a bearing defect, which had a defect frequency at 4.887 orders. The peaks lined up at 4.954 orders, but after further investigation a five-vane impeller was in pump and lined up with basically the same frequency. The signature was nothing more than normal vane pass vibration during normal pump operation.

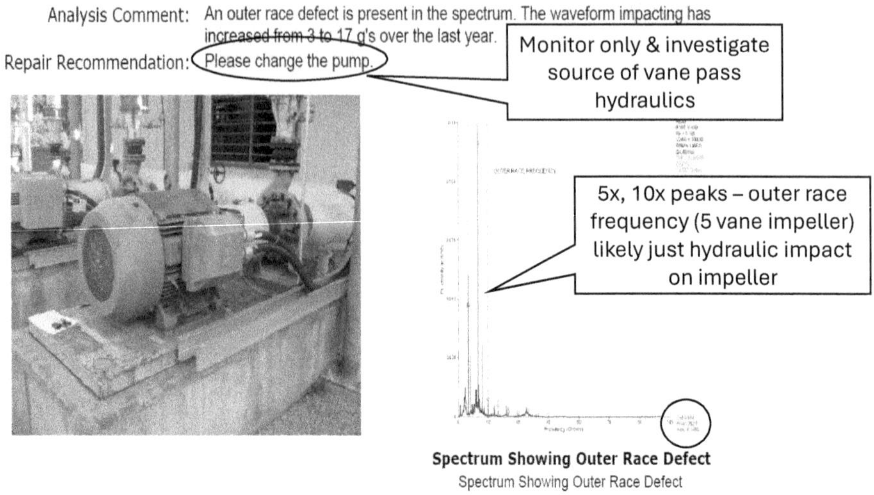

FIGURE 9.11 Similar defect frequency – vane pass and bearing frequency.

MECHANICAL LOOSENESS

Mechanical looseness can not only be a common defect but in some cases can be very serious and lead to catastrophic failure if not corrected. There are several types of looseness that vibration analysis tries to differentiate – types A, B, and C.

Type A looseness is typically a 1X vibration and an indication of structural looseness at machine feet, baseplate, or foundation. The normal response is vertical when measured at the affected area. A 90–180° phase shift on vertical between hold down bolts and baseplate can often be observed. The vibration signature and trend in Figure 9.46 shows the column bolts loose on a vertical turbine pump. The condition was increasing in severity as the failure progressed and the vibration amplitude increased. The vibration signature in Figure 9.12 shows a loose pump motor foot.

Type B looseness shows vibration at 0.5X, 1X, 2X, and 3X, which can indicate cracks in frame structure or bearing pedestal; this response can be very similar to misalignment and hard to distinguish between the two. This can be another case where misalignment is the symptom, but the root cause is a cracked frame. Type B looseness can have elevated 2X vibration.

Looseness, misalignment, and variations of unbalance can all be tricky at times. Figure 9.13 shows a pump vibration where misalignment was showing up. The pump and motor were re-aligned but didn't show any major improvement. Despite the issue persisting, phase analysis was used to help identify a loose baseplate. The baseplate had an additional anchor installed as shown in Figure 9.14 to ground the loose area and resolve the vibration. The tricky part here was that vibration signature showed misalignment, which was real. The misalignment was a symptom or result of the loose baseplate. As the baseplate flexed under load, it created a shaft misalignment condition, which was identified on vibration analysis. This pump and motor set operated at 3600 rpm as well.

Later the vibration changed with more unbalance with an overall vibration amplitude of 0.627 in/s. After the loose baseplate was corrected with another anchor

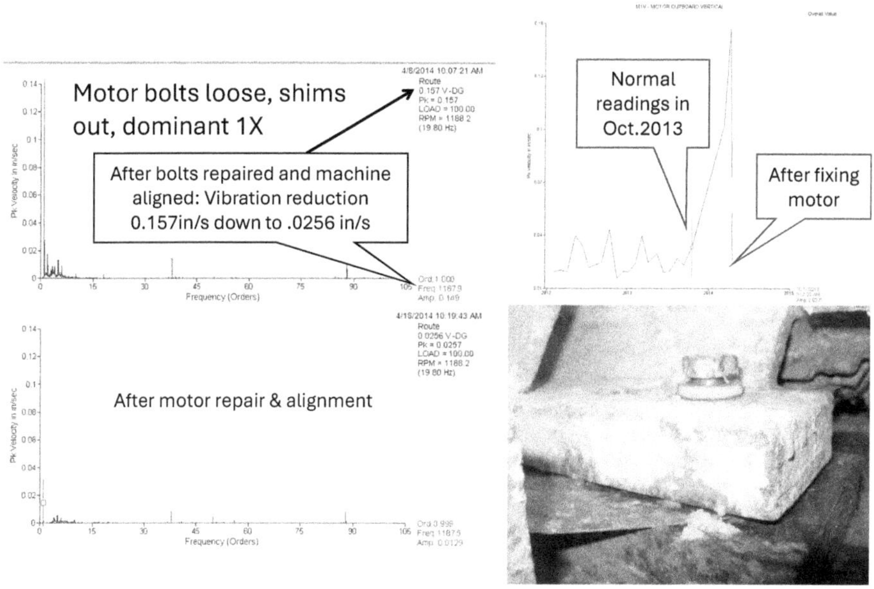

FIGURE 9.12 Loose bolts on pump motor.

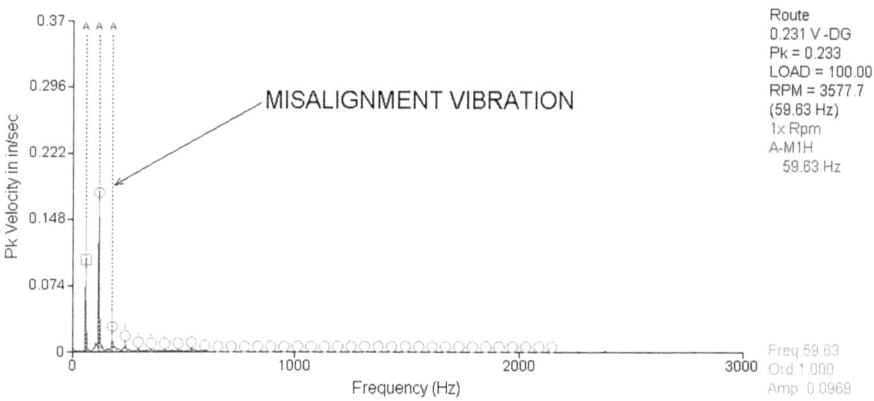

FIGURE 9.13 Vibration signature velocity spectrum with misalignment.

bolt in the loose area, the overall vibration dropped by 80% down to 0.127 in/s as shown in Figure 9.15.

Type C looseness shows vibration at 1X, 2X, 3X, 4X, etc is generally created by improper fit between machine parts, such as a bearing fit. This type of looseness is very concerning as failure can be imminent and very impactful. An inner race losing its fit on the shaft can deteriorate into a catastrophic failure very fast. The improper fit will cause many harmonics out in the spectrum as high as 10X or more run speed peaks. Figure 9.16 shows a pump where it had completely lost its bearing fits on the rotating assembly and type C looseness was dominant. Higher peaks at the three and six vane

FIGURE 9.14 Loose baseplate and anchor addition to solve looseness.

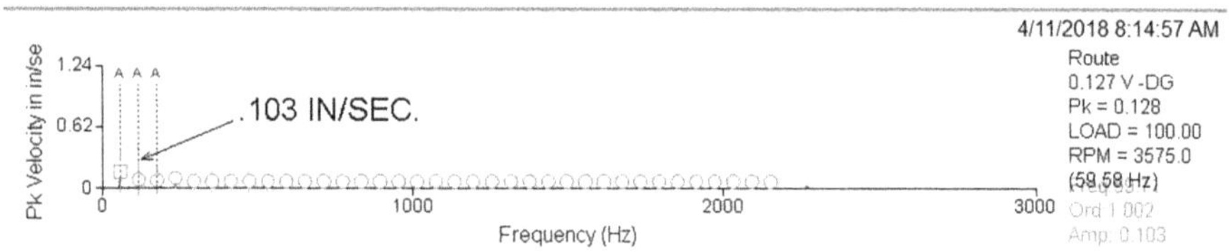

FIGURE 9.15 Vibration signature velocity after looseness correction.

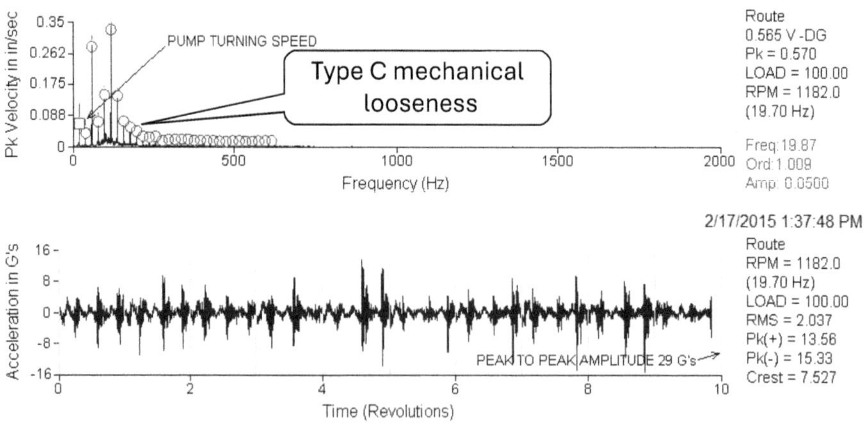

FIGURE 9.16 Type C mechanical looseness in pump vibration velocity spectrum.

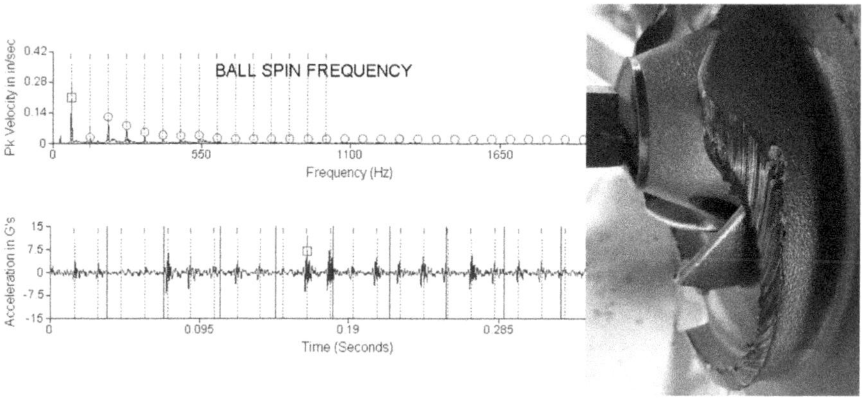

FIGURE 9.17 Impeller rub vibration signature.

frequencies were also present in the vane pass vibration. As with most machine vibration, there is typically more than one fault present, making it challenging to determine which one is dominant or more critical. Type C looseness will always become the primary concern when it shows up in the velocity spectrum.

Rotor rub is similar to mechanical looseness with run speed peaks every 0.5X rpm so 0.5X, 1X, 1.5X, 2X, 2.5X, etc. The pump vibration in Figure 9.17 shows a pump impeller that had an impeller rub. This is usually from a bearing failure allowing the shaft to drop into the wear plate or something has gone through the pump and caught between impeller and wear plate.

UNBALANCE

Pumps can have mechanical or hydraulic sources of unbalance. Most pump balancing is centered around the impeller as that is the large mass in rotation on the assembly. Impellers may be statically or dynamically balanced. Only small, slow turning impellers should be statically balanced. Most should be dynamically balanced to some level of balance quality. Mass may be removed or added to achieve impeller balance. With today's improved casting quality, balance correction should involve relatively small mass changes. I prefer small amounts of mass removed versus adding mass. Adding mass can create flow turbulence on the impeller, which is certainly not desired.

The question of how much unbalance or level of quality of balance is necessary is debated in industry all the time. There are several standards and opinions on the subject of pump impeller balance quality. The amount of rotor unbalance is defined by the ISO 1940 standard for balance quality grades from G0.4 to G4000. The lower the number the lower the unbalance or better the balance quality. Unbalance is expressed as mass multiplied by the distance from the center of rotation, typically in units such as g-mm or oz-inches. Figure 9.18 shows the ISO 1940 balance grade list.

Impeller or rotor unbalance is defined by the mass of unbalance at some radius from the center of rotation as shown in Figure 9.19. The same unbalance can result

Machinery types: General examples	Balance quality grade G	Magnitude $e_{per} \cdot \Omega$ mm/s
Crankshaft drives for large slow marine diesel engines (piston speed below 9 m/s), inherently unbalanced	G 4000	4 000
Crankshaft drives for large slow marine diesel engines (piston speed below 9 m/s), inherently balanced	G 1600	1 600
Crankshaft drives, inherently unbalanced, elastically mounted	G 630	630
Crankshaft drives, inherently unbalanced, rigidly mounted	G 250	250
Complete reciprocating engines for cars, trucks and locomotives	G 100	100
Cars: wheels, wheel rims, wheel sets, drive shafts Crankshaft drives, inherently balanced, elastically mounted	G 40	40
Agricultural machinery Crankshaft drives, inherently balanced, rigidly mounted Crushing machines Drive shafts (cardan shafts, propeller shafts)	G 16	16
Aircraft gas turbines Centrifuges (separators, decanters) Electric motors and generators (of at least 80 mm shaft height), of maximum rated speeds up to 950 r/min Electric motors of shaft heights smaller than 80 mm Fans Gears Machinery, general Machine-tools Paper machines Process plant machines Pumps Turbo-chargers Water turbines	G 6,3	6,3
Compressors Computer drives Electric motors and generator[...] speeds above [...] Gas turbines and steam t[...] Machine-tool drives Textile machines	G 2,5	2,5
Audio and video drives Grinding machine drives		1
Gyroscopes Spindles and drives of high-precision systems	G 0,4	0,4

General Pump balance target G6.3

Higher speed motors don't have hydraulic effects of pumps so tighter unbalance specs

FIGURE 9.18 ISO 1940 balance grade.

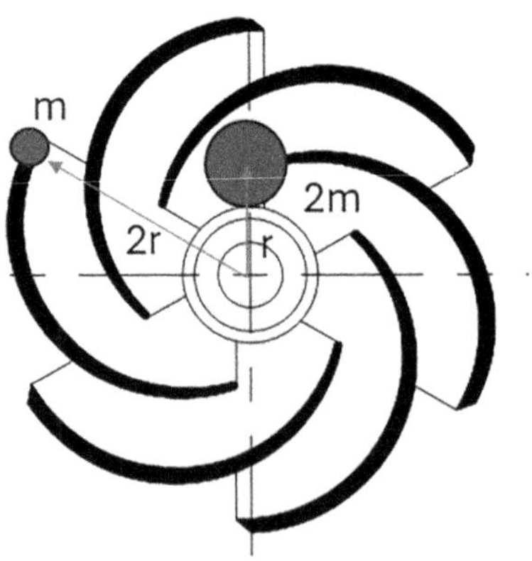

FIGURE 9.19 Impeller unbalance comparison.

from a mass at a distance of 2r from the center or a mass of 2 m mass at a distance. Unbalance is given in mass-distance units.

$$Uper = 1000\frac{[(eper\ \Omega) \times m]}{\Omega}$$

where:

$Uper$ = numerical value of the permissible unbalance in g-mm
$(eper\ \Omega)$ = numerical value of balance quality grade in mm/s, i.e. G1
Ω = numerical value of the angular velocity in rad/s, Ω~rpm/10

Note that for most pumps, the target balance grade is G6.3. For a pump rotor (shaft and impeller) that weighs 32.65 Kg (87 lbsm) operating at 1,800 rpm then the amount of unbalance would be 1,143 g-mm or 1.4465 oz-in. At 202 mm (8″) impeller diameter, which would be 5.79oz (0.4 lbm) of unbalance at that diameter. Look at the table in Figure 9.20 to compare these numbers with what a G1.0 balance grade would have.

With more focus on precision maintenance in the past 40+ years, many have focused on targeting tighter balance grades by balancing every pump rotor to G1.0. In a vacuum or with the rotor operating in air and no other variables, balancing every pump rotor to G1.0 might be good and beneficial; however, when the variables of hydraulic external forces are factored in, the case to balance every pump rotor to G1.0 might be an overkill.

The reason for potential overkill in mechanical balance is for many reasons. First, there are many potential mechanical reasons that might cancel out a G1.0 impeller balance. Other mechanical problems to be addressed in achieving a balanced assembly are listed below.

Condition 32.65Kg (87lbm) Rotor@1800rpm	G6.3 Balance Grade	G1.0 Balance Grade
Unbalance, g-mm (oz-in)	1,143 (1.4465)	181 (0.23)
Unbalance mass at 8" diameter of impeller oz (lbm)	5.79 oz (0.4 lbm)	0.92 oz (0.069 lbm)
Hydraulic water mass per vane per revolution for 5 vane impeller, 3000gpm, 1800rpm	2.78 lbm/vane If only 14% of water volume is affected in one vane = same as G6.3 balance grade	2.78 lbm/vane If only 2.5% of water volume is affected in one vane = same as G1.0 balance grade

FIGURE 9.20 Comparing G6.3 to G1.0 balance grade.

- Coupling balance: For most couplings < 3,600 rpm, an AGMA class 8 balance is standard balance grade which is a max 4,000 μ inches displacement. Beyond that an AGMA 9 (component balance) or AGMA 10 (assembly balance). This does not even account for a lubricated coupling and the potential for grease-induced unbalance within the coupling.
- Keylength balance: The keylength sticking through the coupling hub serves no purpose but to improve shaft balance (hub portion of key to assist in torque transmission is complete). A mill study on pump and motor keylength unbalance revealed that over 70% of keylength unbalance issues were on the motor shaft. This is due to motor shafts typically being larger, keys larger, and keyways being longer. To calculate the proper keylength, follow the graphic shown in Figure 9.21. Key phase at coupling assembly is also important to keep keyways 180° apart from each other to reduce couple unbalance.

Since the unbalance force is proportional to the square of the speed, a higher speed application is much more impactful than a low speed, as shown by the equation below. As an example, a keylength with A = 5.375″ and B = 3.875″ has a full-length key. The unbalance force at 3,600 rpm is ~94 lbf, while the same keylength situation operating at 1,800 rpm is ~24 lbf.

$$F = ma = \frac{m \times r\left(\left(\frac{N}{60}\right) \times 2\pi\right)^2}{gc}$$

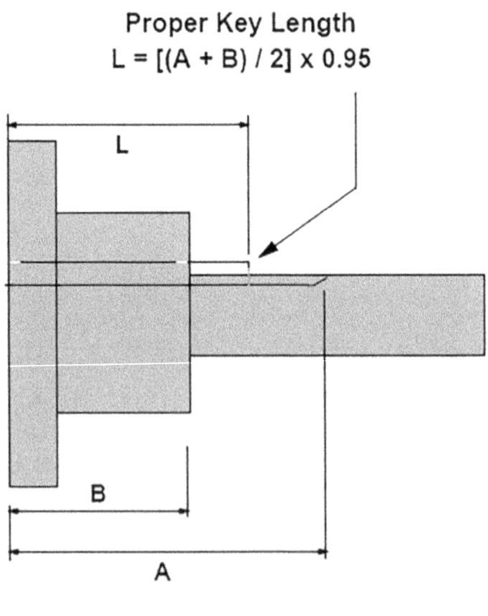

FIGURE 9.21 Keylength unbalance.

where:

m = unbalanced mass of key (too long or too short)
r = radius of keyway
N = operating speed, rpm
gc = gravitational constant

The biggest reason for pump unbalance vibration might be hydraulic unbalance issues. Hydraulic imbalance in the pump and impeller can occur for several reasons listed below.

- Non-uniform pressure or fluid around impeller.
- Uneven vane spacing or wear.
- Bad castings.
- Air entrainment or cavitation causing random masses passing through impeller vanes.
- Pump handling solids that are not uniform in each vane.
- Plugged impeller vane as shown in Figure 9.22. A foreign object on one vane not only chokes pump flow output (hydraulic unbalance) but also contributes to mechanical mass unbalance.

For example, if the same pump in Figure 9.20 is pumping 3,000 gpm with a five-vane impeller, the mass of water per vane per revolution is 2.78 lbm. For reasons listed above for hydraulic unbalance, the potential can be significant. A G6.3 unbalance is only 14% of the mass of water in one vane area. A G1.0 unbalance is only 2.5% of the mass of water in one vane area so any non-concentric mass of fluid in the impeller vane passage area can wipe out a G1.0 balance of the rotor.

This is just one unbalance defect and does not include other pump vibration elements such as flow turbulence, cavitation, and all the other mechanical causes. This is likely one reason that centrifugal pumps generally have a higher target

(a) (b) (c)

FIGURE 9.22 Plugged impeller vanes by various foreign objects. (a) Cigarette lighter stuck in enclosed impeller, (b) recycle paper stock wire and plastic, and (c) wood plugged in open impeller.

unbalance grade of G6.3. The other pump vibration factors (mechanical and hydraulic) just drown out a tight balance grade. There are applications where many pump manufacturers will recommend a G1.0 impeller/rotor balance. High speed applications are one. A pump with speed > 3000 rpm should be balanced to G1.0 balance quality. In a clean fluid application, good pump hydraulic operating point, and an enclosed impeller, I would consider a G1.0 at a lower speed even down to 1,800 rpm.

In another example, a large plastic pipe clamp had gotten inside a large 8×10×19 end suction pump. The foreign object was approximately 3.5″×8″×3.5″ in size and it created a significant noise and vibration in the pump. The vibration exhibited a high 1X and 2X vibration and flow turbulence as shown in Figure 9.23. Due to the size of the object, it was believed that the vibration was mostly hydraulic and impacting the impeller and not the impeller/rotor imbalance. Once the foreign object was removed, the vibration returned to normal-looking signature. The overall velocity vibration dropped from 0.346 in/s to 0.100 and the acceleration dropped from 10 gs to 3 gs.

As shown in other case studies, a number of conditions may show symptoms of unbalance. Another example might be a bent shaft. In a bent shaft, there is actually an unbalanced mass outside the center axis of rotation. A G1.0 balanced impeller will not do well if the shaft is bent. Another example is shown in Figure 9.24, where the pump coupling had a loose fit on the shaft, which created a high 1X vibration on this 3,600 rpm application.

It is always better to balance the entire rotor (impeller and shaft) as an assembly if possible. I always preferred removing some metal to achieve balance instead of adding balance weight. Adding a balance weight can create flow turbulence and can come off. Use specific locations when removing material for balance and always ensure sufficient structural integrity of the impeller. The next case study is about comparing various balance grades of pumps and the resultant vibration.

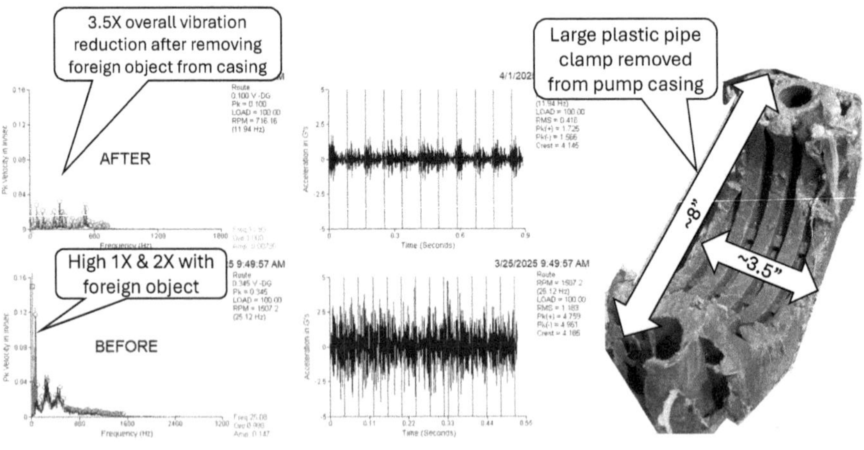

FIGURE 9.23 Foreign object effect on pump vibration.

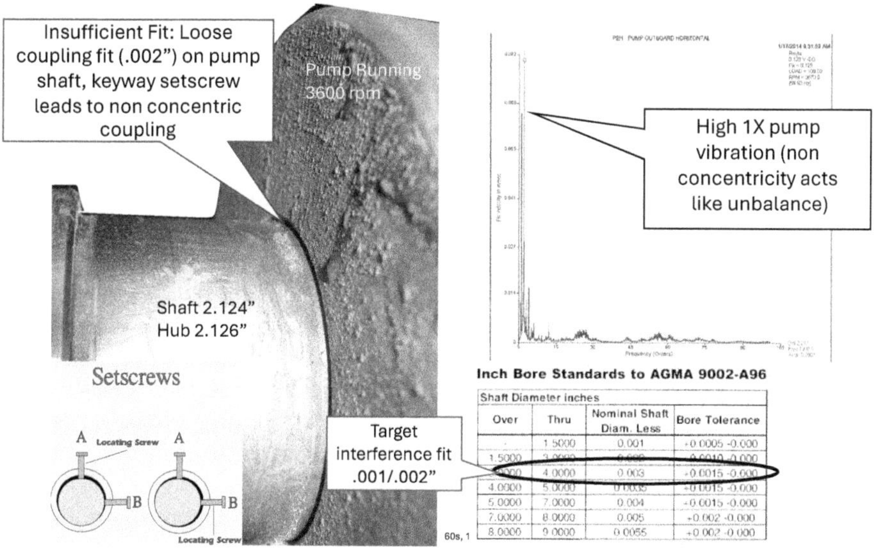

Shaft Diameter inches			
Over	Thru	Nominal Shaft Diam. Less	Bore Tolerance
-	1.5000	0.001	+0.0005 -0.000
1.5000	[illegible]	[illegible]	[illegible]
[illegible]	4.0000	0.003	+0.0015 -0.000
4.0000	5.0000	[illegible]	+0.0015 -0.000
5.0000	7.0000	0.004	+0.0015 -0.000
7.0000	8.0000	0.005	+0.002 -0.000
8.0000	9.0000	0.0055	+0.002 -0.000

FIGURE 9.24 High vibration from coupling fit issue.

CASE STUDY 49 IMPELLER BALANCE QUALITY IMPACT ON PUMP VIBRATION

The discussion around what impeller balance grade to specify when purchasing a pump has been a topic of much conversation. I had never really seen any hard data comparing balance grades on an actual pump in operation. The only comparisons that had been made were looking at unbalance and vibration with the rotor in a balance stand. Mechanically, there is no question G1.0 is better than G6.3 and G6.3 is better than G16 and so on; however, pumps do not operate in air on a shop test stand.

There are so many more variables to consider when evaluating the total pump unbalance during actual process operation besides just the mechanical unbalance. It becomes one of feasibility and ROI for the extra investment for a reliability improvement. For it to improve pump reliability, it must reduce the pump vibration, as supported by many sources and previously discussed data.

Field trials were conducted on several pump systems regarding impeller unbalance. A couple of different pumps were selected and comparisons were made between them, allowing some general trends to be identified. One pump was a standard end suction stock/process pump. The other pump was a high solids handling (vortex, recessed impeller) pump. There are many other types of pumps that could also be studied but these two provided a great opportunity to compare in real-life application. The end suction pump was the same model, operated with two different impeller balance quality grades. The recessed impeller pump consisted of two different pumps with similar hydraulic characteristics, with only a few minor differences. While the trails where not lab conditions with precise measurements, the field examples were very telling in their results.

The recessed impeller pump used in the vibration study was a 4×4×12 end suction pump. Each pump was brand new and operated in the same type of process and stock cleaning system. The differences in the two pumps were that one pump had a 0.25″ smaller diameter impeller and one pump ran on a VFD, while the other pump operated with a fixed speed pump. The pumps were very similar in hydraulic design, so the hydraulic effect are minimized in the comparisons. Both pumps were new when this data was taken so no general wear and tear on other mechanical components should have been a factor in the vibration data. The 1X run speed was a primary focus of the vibration data since that is the unbalance frequency of focus around balance grade. That would remove many other variables.

The process streams were not scientifically controlled as this was run on actual process equipment and conditions. There were no major variables observed between the two systems which may have presented flawed results when the vibration operating data was taken. That is not to say that one of them may not have had a unique condition such as higher trash load to contribute to the results. If it did, that is essentially the point of taking actual operating data on process pumps: process variables drive pump vibration, especially when evaluating very tight levels of unbalance compared to G1.0 balance quality.

In Figure 9.25 is the G1.0 and G16 impeller pump curves. The pump had a 12″ impeller and a design point of 300 gpm at 122 ft and 1,600 rpm. The pump operated on a VFD, so the speed was not at the exact design speed when the vibration data was taken. Looking at the two curves, it can be observed even though there are a few differences, the two pumps are very close in design with the impeller balance being one of the main differences.

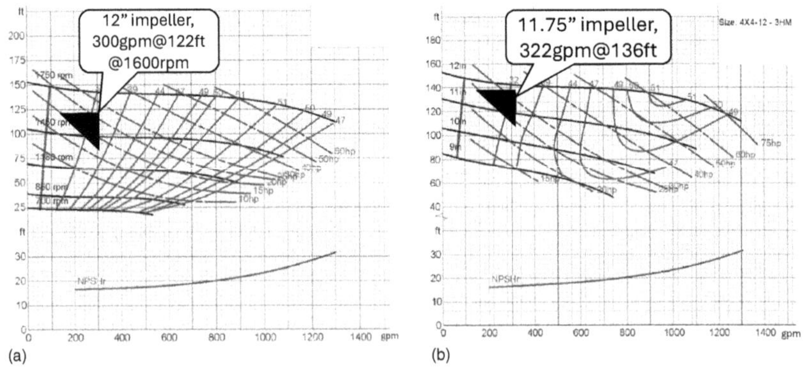

FIGURE 9.25 Pump curve recessed impeller pumps: (a) G1.0 balanced impeller and (b) G16 balanced impeller.

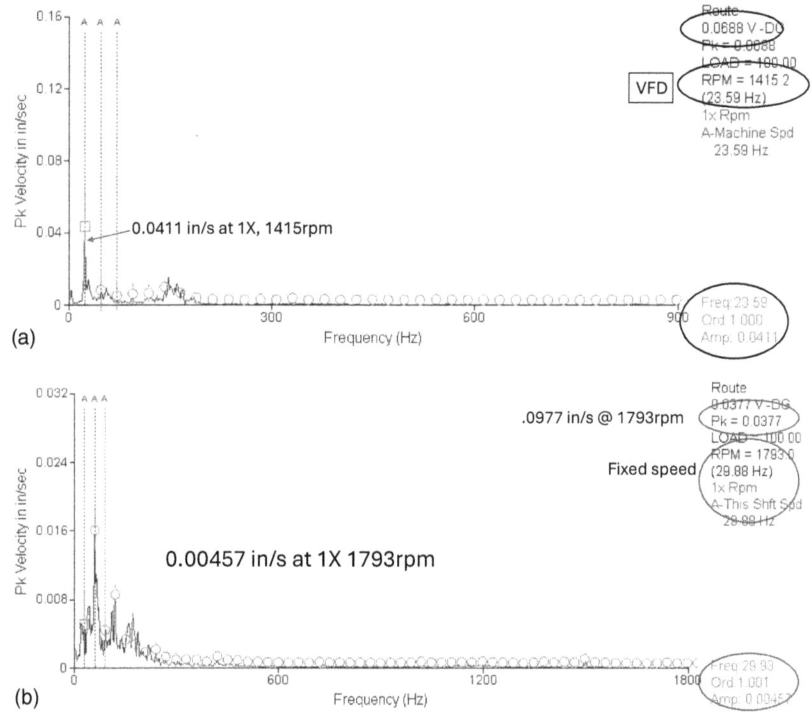

FIGURE 9.26 Velocity spectrum vibration signature for (a) G1.0 pump and (b) G16 pump.

The results of the vibration signature weren't what was expected but interesting as seen in Figure 9.26. The G1.0 balanced impeller had a vibration amplitude (0.04 in/s) at 1X frequency that was 10X higher than the G16 balance grade impeller vibration (0.004 in/s). Not only that but the G1.0 impeller was also operating at a lower speed so it should have been lower. The overall amplitude (all frequencies) was 0.07 in/s for the G1.0 impeller, while the G16 overall vibration was 0.10 in/s. This was understandable since it was operating at a higher speed but still wasn't grossly higher.

With the recessed impeller pump, the operating data and vibration did not support that the G1.0 impeller delivered lower vibration than the target impeller balance grade of G16. The reasons could be many, as the high solids handling pump yields large process variables that can impact pump vibration and negate any precision balance quality. The G1.0 pump typically had a higher solids/trash load than the G16 pump, so the results were even more surprising.

The other pump evaluated was a typical end suction stock pump used in paper process. The pump was an $8 \times 10 \times 19$ pump that operated off a VFD. This pump study involved the same pump and the same system with different

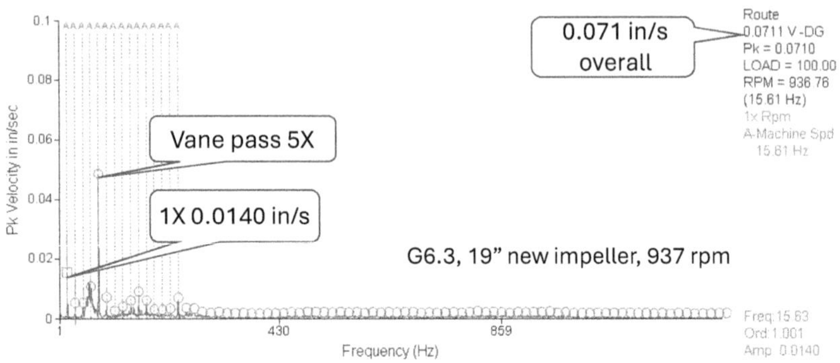

FIGURE 9.27 G6.3 balanced impeller velocity spectrum vibration signature.

impellers and some varying pump speed and hydraulics for each but very close to the same speeds were also obtained so a better comparison of vibration with only one variable (impeller balance quality) could be easily seen. This was also on a production pump and system. The signature in Figure 9.27 is the G6.3 impeller with a new full size 19″ impeller. The G6.3 impeller is the target balance grade for this type of pump from the pump manufacturer.

The next impeller installed was a 17.5″ G1.0 balance grade impeller. The impeller size change was part of a pump system improvement project and also provided an opportunity to compare the impeller balance grade against the previous G6.3 impeller. These grades are two levels apart, representing a significant difference in balance quality. Figure 9.28 shows the G1.0 impeller vibration signature.

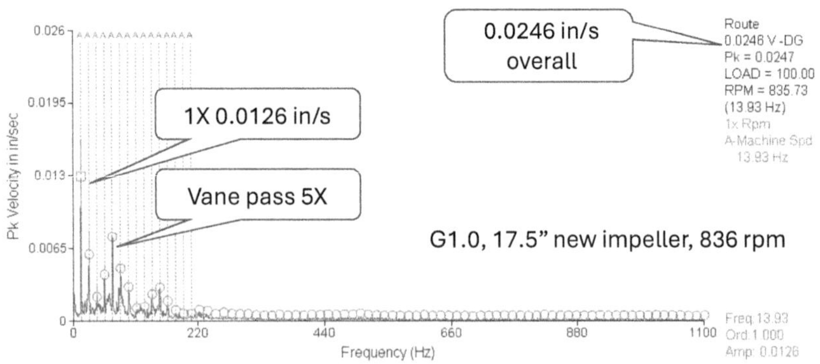

FIGURE 9.28 G1.0 balanced impeller velocity spectrum vibration signature.

Impeller Diameter	Balance Quality	Speed, rpm	Unbalance 1X Vibration, in/s	Overall Vibration, in/s (all frequencies)	Vane Pass, 5X Vibration in/s
16" Old Impeller	**G6.3**	831	**.0029**	**.029**	.016
16" Old Impeller	G6.3	1200	.0187	.048	.008
16" Old Impeller	G6.3	1800	.0443	.127	.022
19" New Impeller	G6.3	937	.0140	.071	.048
17.5" New Impeller	**G1.0**	836	**.0126**	**.025**	.008

Speed drives vibration

FIGURE 9.29 Summary of different impeller balance grades on end suction pump.

There were several other conditions monitored on pump vibration with an old impeller which would likely not have been at G6.3 balance quality since there was uneven wear on it but the data is there for further comparison. That older impeller was also a smaller diameter. The summary of impeller vibration is shown in Figure 9.29.

Several conclusions can be drawn from this data. These conclusions may include the following.

- Increase speed – increases 1X and overall vibration. This is not surprising as it is typical for most machine applications regardless of balance grade.
- Maximum 19″ impeller exhibited higher vibration levels. Especially at vane pass, which is not surprising as the impeller cutwater gap was the smallest and excited that vibration frequency. The smaller 16″ impeller with a much larger impeller cutwater gap running at twice the speed as the 19″ impeller had a +50% lower vane pass vibration but higher overall and 1X vibration maybe recirculation part of the reason.
- Impeller balance quality did not appear to drive pump vibration at 1X frequency. The G6.3 impeller was only ~10% higher 1X, while operating at ~15% higher speed. The overall vibration was higher but mostly driven by the vane pass vibration which again on that datapoint was driven by the hydraulics of the max size impeller.

At the end of the day, the trial data for targeting G1.0 impeller or rotor balance grade is not one that should be applied across the board that provides a return on investment every time. A top balance grade will usually come at an add-on price as much as $1,000–1,200 per impeller which as shown in the end not really produce a lower unbalance vibration due to the pump application and hydraulics of the system or process fluids being pumped. With no improvement in vibration, there will not be any substantial gain in pump reliability.

Case Study 49 Learning – Understand the application before paying a premium for a higher balance quality grade, there may be no return for the balance grade investment.

RESONANCE

Pump resonance is not typically an issue on most pumps with good installation practices; however, when field conditions change, resonance can become a factor in pump vibration and ultimately pump reliability. As with any machine, resonance is maybe one of the most damaging vibration conditions that exists. Resonance is not the source of vibration but is a magnifier of vibration. It is many times the source of extreme vibration amplitudes.

Resonance can occur on any part of equipment (stationary or rotating). Resonance is the natural vibration frequency of the machine or component. The critical speed is the speed at which the natural frequency is excited on the rotating element. Critical speed is a single speed, and most rotating equipment is designed so that the operating speed is nowhere near a critical speed or resonance frequency. Resonance is a function of mass and stiffness as shown below.

$$Fn = \frac{1}{2\pi}\left(\frac{k}{m}\right)^{0.5}$$

where:

Fn = natural frequency (resonance)
k = stiffness
m = mass

There are a number of indicators for suspecting resonance such as the following:

- High amplitude run speed (1X) or multiples of run speed vibration peaks.
- If vibration amplitude changes drastically with speed change especially around 1X.
- 180° phase shift as speed goes through resonance.
- Ratio of 3:1 or more of 1X from horizontal to vertical.
- Perform a static bump test on the pump or machine to identify resonance frequencies.

For a pump and other process equipment, it is good to remember that resonance is the rotor in air. If a pump is full of fluid, its resonance frequency will be affected, as the fluid will dampen the rotor's natural frequency. There are a number of potential fixes for a pump or machine operating in resonance as listed below.

- Change machine speed, operating at least 15–20% away from the resonance frequency. A bode plot may be used on some machines to plot forced vibration and natural frequencies verses operating speed. Intersecting lines indicate possible resonance conditions. This is only an option if operating on a VFD. Changing impeller size to operate a different speed would allow similar hydraulic performance but operating at a different speed away from resonance.

- Change natural frequency of the machine by adding mass or stiffness. This is not easily done.
- Add dampening to the system to reduce resonance effect.
- Reduce root vibration source – vane pass, or adjacent vibration exciting pump resonance frequency.

CASE STUDY 50 STOCK PUMP ELEVATED VIBRATION

A stock pump was experiencing very high vibration much higher than normal. The pump was a 10×8×21 end suction pump which was designed with a 21″ diameter open impeller. The original design point was 3,500 gpm at 172 ft with an operating point just to the left of BEP. The pump was driven by a 250 HP, 1200 rpm fixed speed motor. The pump system had a pressure control recirculation loop and a level control valve to the stuff box on the machine as shown in process diagram in Figure 9.30.

The trend had been coming up for almost a year. The normal vibration was even higher than desired much of the time but the increase in vibration went well beyond the upper alarm limit as shown in Figure 9.31. While trending doesn't solve the issue or get specific about what the problem is, overall trending can still be a powerful tool on the front end of an issue as shown. The basic vibration amplitude trend had been steady for about 12 years and then shot up. Something significant had changed, as indicated by the vibration trend. The vibration amplitude had been somewhat high most of the time

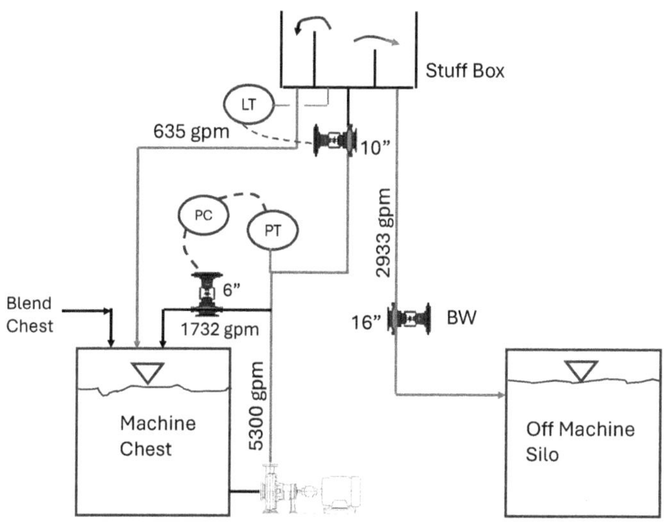

FIGURE 9.30 Process diagram machine chest pump system.

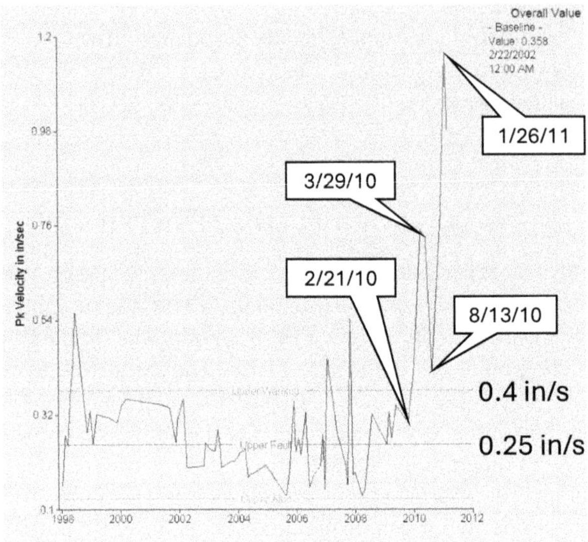

FIGURE 9.31 Overall vibration trend on machine chest pump.

above 0.25 in/s and many times going over 0.4 in/s. The pump did operate in a variable process flow condition so that was obviously a factor as the control system affected pump condition.

Looking further at a typical vibration signature, the pump has always had a little higher vane pass vibration than desired as shown in Figure 9.32. Part of this is the fact that the pump had the largest size impeller (21″ diameter) installed in the pump. The pump had a discharge level control valve which operates 35% or lower in operating position. The vane pass vibration was 3X run speed with a three-vane impeller which is 59.5 Hz and was very high in the horizontal plane at 0.86 in/s. The overall vibration was over 1 in/s which was very excessive for a 1,200 rpm machine. Autocorrelation of the waveform in a circular plot clearly showed that 92% of the vibration energy was repeatable at the vane pass frequency so very little of the energy resulted from other vibration sources (unbalance, flow turbulence, etc.).

With this information, one of the first actions was to install a trimmed impeller (1″ smaller with a 20″ diameter) to reduce this vane pass energy and evaluate the results. The control valve had plenty of range to reduce the system pressure with a smaller impeller. Inspection of the piping system revealed some pipe hanger supports missing from the discharge piping, so this was also corrected. After this was corrected it did ground some of the vibration energy and reduced the overall vibration by 25%, but the root cause remained unresolved.

With the amplitude of the vibration, one of the suspected causes was system resonance. Resonance isn't a cause of vibration but amplifies an existing

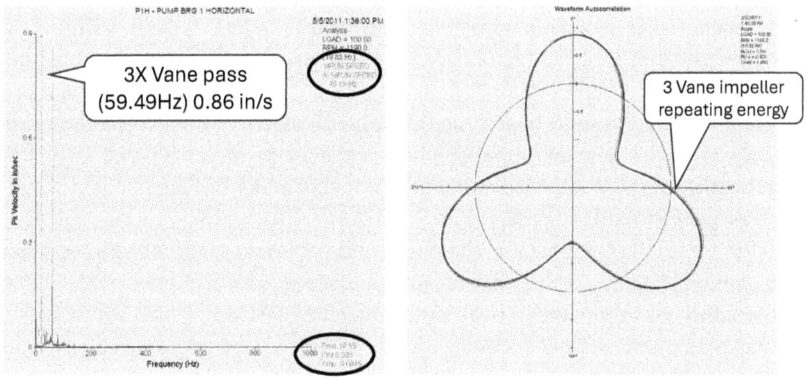

FIGURE 9.32 Typical pump velocity spectrum and waveform autocorrelation.

vibration. In this case, the existing vibration is the normal vane pass frequency, but it is amplified by a change in the system stiffness. We did some resonance testing by conducting a bump test on the pump and system. What we learned was a few potential resonance frequencies as shown in Figure 9.33. There were two large frequency spikes – one at 62.5 Hz and another at 89 Hz. The run speed or vane pass frequency should be at least 20% away from any forcing function frequency. The vane pass frequency was 59.5 Hz, while one of the resonance frequencies was 62.5 Hz. The safe frequency away from resonance would be ~50 Hz–74.5 Hz which the vane pass frequency fell almost directly on one of these resonance frequencies.

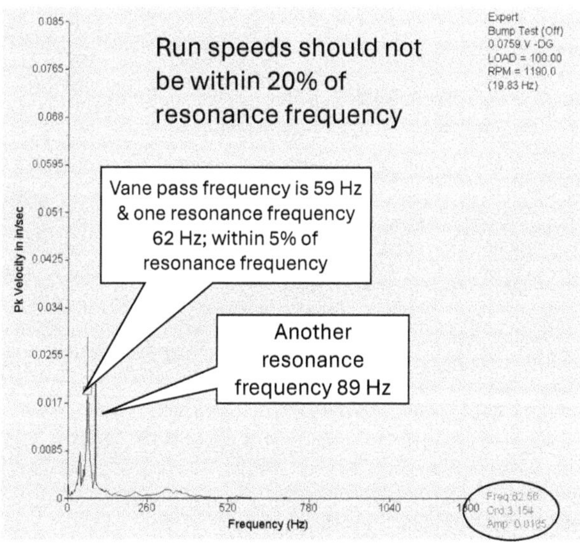

FIGURE 9.33 Bump test resonance testing.

Knowing that resonance involves speed, mass, and system stiffness, a closer evaluation of these factors are required. Speed had not changed as this is a fixed-speed pump system so that can't be a contributing factor. The mass of the pump and system hasn't significantly changed. We changed the mass slightly with the smaller impeller but not the main lever. So that only left system stiffness. We adjusted some pipe hangers and while that helped some, the root problem still existed.

Now we are looking back at the pump and system to see if we can see what else changed the system stiffness to make this resonance frequency the same as the vane pass frequency. The pump baseplate was very corroded in spots, but no major issues were recognized on the first investigation. The rust and process stock had covered up a major compromise on the pump support as shown in Figure 9.34. Once the pump and baseplate were cleaned up knocking all the residual rust out of the way, the issue was obvious. The suction-side pump foot baseplate support was completely gone. A temporary repair was put in place on the pump to support the pump suction end to the baseplate as shown in Figure 9.35. The overall vibration dropped from over 1 in/s to 0.39 in/s and the vane pass dropped to 0.343 in/s after making the temporary pump support correction. This put the pump vibration back down to an upper limit where the pump had operated previously for many years. The change in pump stiffness on the foundation was the root cause of high vibration and increased failures.

FIGURE 9.34 Pump baseplate corrosion damage.

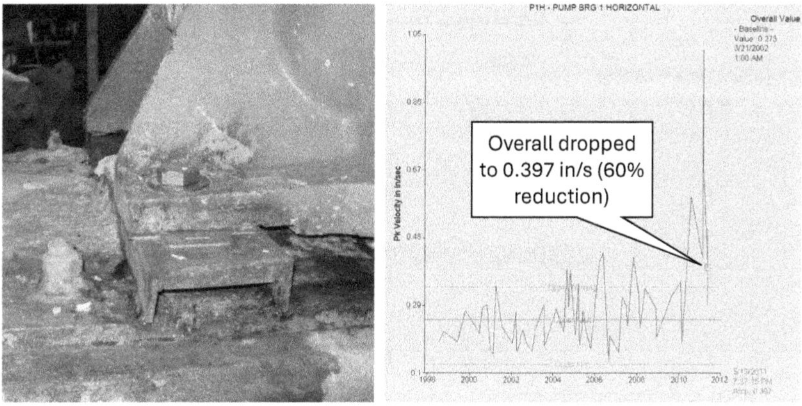

FIGURE 9.35 Temporary repair and resulting vibration reduction.

The last part of this pump improvement involved a pump system redesign and upgrade to improved pump system reliability and reduce energy consumption. Since the pump needed the baseplate replaced, a considerable amount of investment was already needed; it created an opportunity for more improvement to the overall system.

From a reliability end, this pump system had a spare installed pump. This arrangement from a system standpoint increased the overall system reliability. The actual reliability showed two replacements on one unit in 2.5 years and three changes on the other. This resulted in a pump reliability of 0.45 and 0.3. With the pumps operating in parallel, the equivalent system reliability was 0.61.

In a proposed system, if the installed spare was removed and the pump reliability remained at 0.45, then the equivalent system reliability would drop to 0.41. However, it has been seen with a pump upgrade and VFD, with some confidence the new single pump can achieve a 0.9 reliability at one changeout every ten years. The equivalent pump system reliability then would be 0.81, which is 20% improvement from the original system, with only half the equipment footprint and investment. This is summarized and shown in Figure 9.36.

The proposed pump process would involve removing the header pressure control valve and loop as well as the stuff box level control valve loop as shown in Figure 9.37. A pump VFD would be installed on a new pump which would control from stuff box level control. A tremendous reduction in pump size, motor size, and energy usage would be realized. Also, a reduction in two control valve loops and associated instrumentation would be removed.

Due to the process efficiency improvement, 80% of the life cycle cost savings on this project proposal was energy. The existing pump operation was around 5,300 gpm at 140 ft, while the new pump was 3,500 gpm at 78 ft. There was also a small pump efficiency improvement. The motor went from a

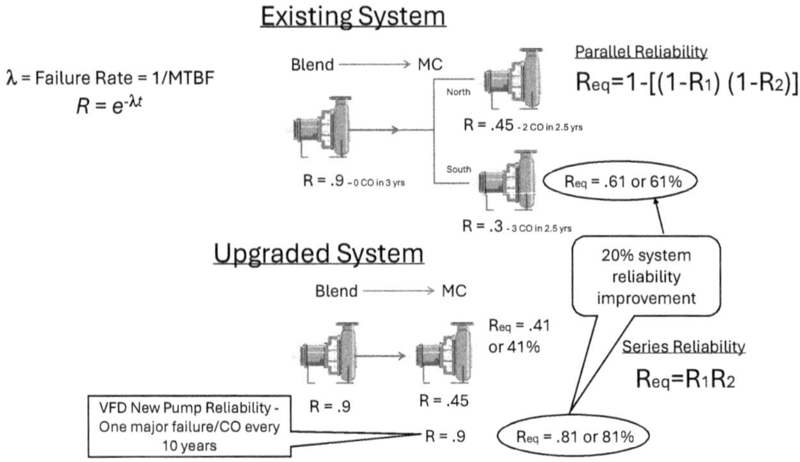

FIGURE 9.36 Existing and proposed pump system reliability.

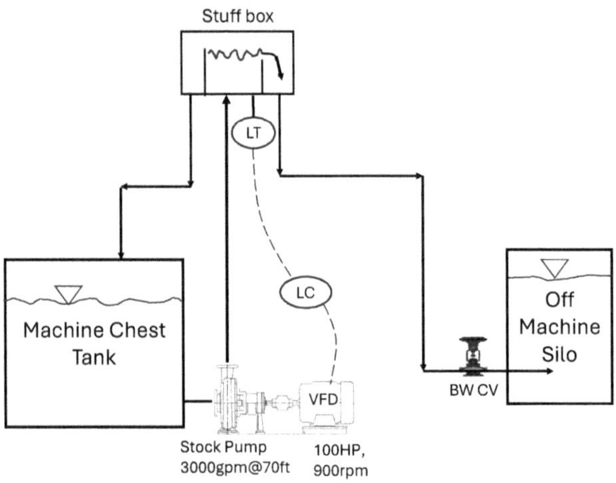

FIGURE 9.37 Proposed process diagram for machine chest pump system.

250 HP to a 100 HP motor. The total life cycle cost comparison is shown in Figure 9.38. In ten years, with the existing design, total life cycle cost would be ~\$1.16 million while the proposed new pump with VFD system would only be ~\$426,000. Energy was the largest life cycle cost benefit, but maintenance cost was also a significant contributor.

The project went in as designed and it provided the benefits as expected. The pump vibration reduction was significant even from the corrected

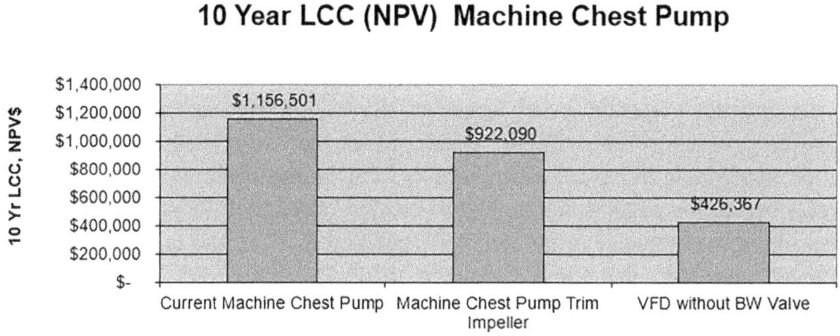

FIGURE 9.38 Ten-year life cycle cost of energy savings on pump options.

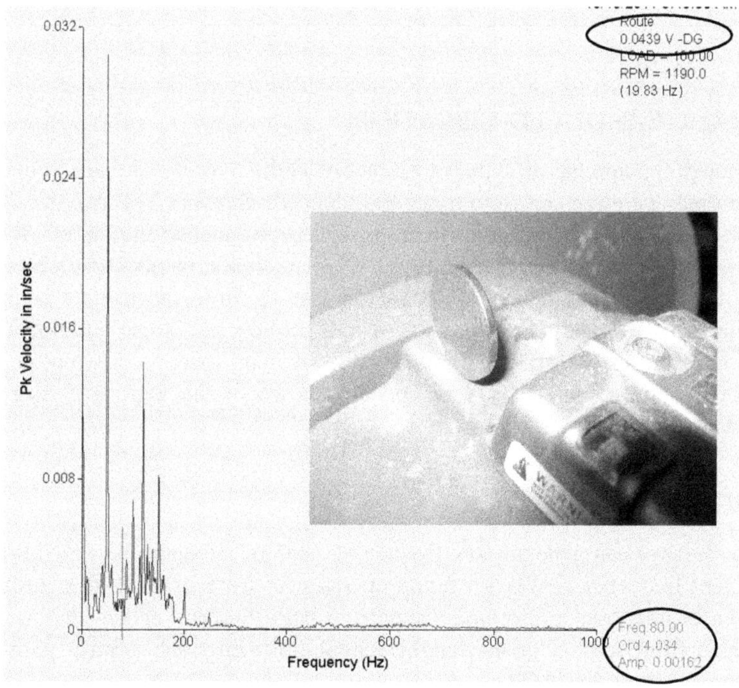

FIGURE 9.39 New pump velocity spectrum vibration signature.

resonance issue with the baseplate issue repaired. The new VFD pump system had a vibration reduction of 90% from an overall vibration of 0.34 in/s down to 0.043 in/s as shown in Figure 9.39. It was so smooth; I had to try the old trick of standing a nickel on its end on the running pump and it never moved (see Figure 9.39).

The change in the pump system stiffness was due to improper maintenance of the pump baseplate. Pump baseplate deterioration led to the resonance condition which caused elevated pump vibration. This was clearly a pump issue and not a system issue; however, a very good pump system upgrade project was executed with an improved design and made even greater improvements in pump system reliability. This pump operated for almost five years before the paper machine had to be shut down, but the pump never had any major maintenance or failures performed on it. The extra cost of having to replace and redo the entire baseplate made the possibility of the project upgrade possible as the capital cost could be offset with a smaller pump and motor.

Case Study 50 Learning – Any changes in a pump system structure (stiffness or mass) can have big impact to pump system vibration and reliability.

HYDRAULIC-INDUCED VIBRATION

Even though a pump has all types of vibrations in the overall spectrum, some pump vibration may be dominant from hydraulic-induced vibration. This may be due to a number of reasons, such as various causes of flow turbulence and various forms of cavitation. Flow turbulence can be forms of recirculation or poor suction entrance flow path. Some vibration symptoms even can be from mechanical sources or hydraulic such as vane pass vibration. It will usually take some additional information to determine which type of vibration is the main source of a symptom. Some of these hydraulic sources of vibration are shown in the logic tree diagram in Figure 9.40.

FLOW TURBULENCE

Flow turbulence can generate low frequency, random vibration but may also show up at other frequencies. When a pump operates far from its BEP, flow turbulence increases as flow vectors become irregular in both direction and magnitude. Flow turbulence from suction or discharge recirculation can result from a poor pump operating point outside AOR.

One example of this is an overall vibration trend on a pump which had a suction vortex breaker. Figure 9.41 shows this vibration trend with the vortex breaker, after vortex breaker was removed and then the same pump running at a lower speed. The pump with the vortex breaker had an overall vibration around 0.3 in/s over many years. Removing the vortex breaker reduced the suction flow turbulence and lowered the vibration by almost 40%. Slowing the pump down also lowered the pump vibration by another 50%.

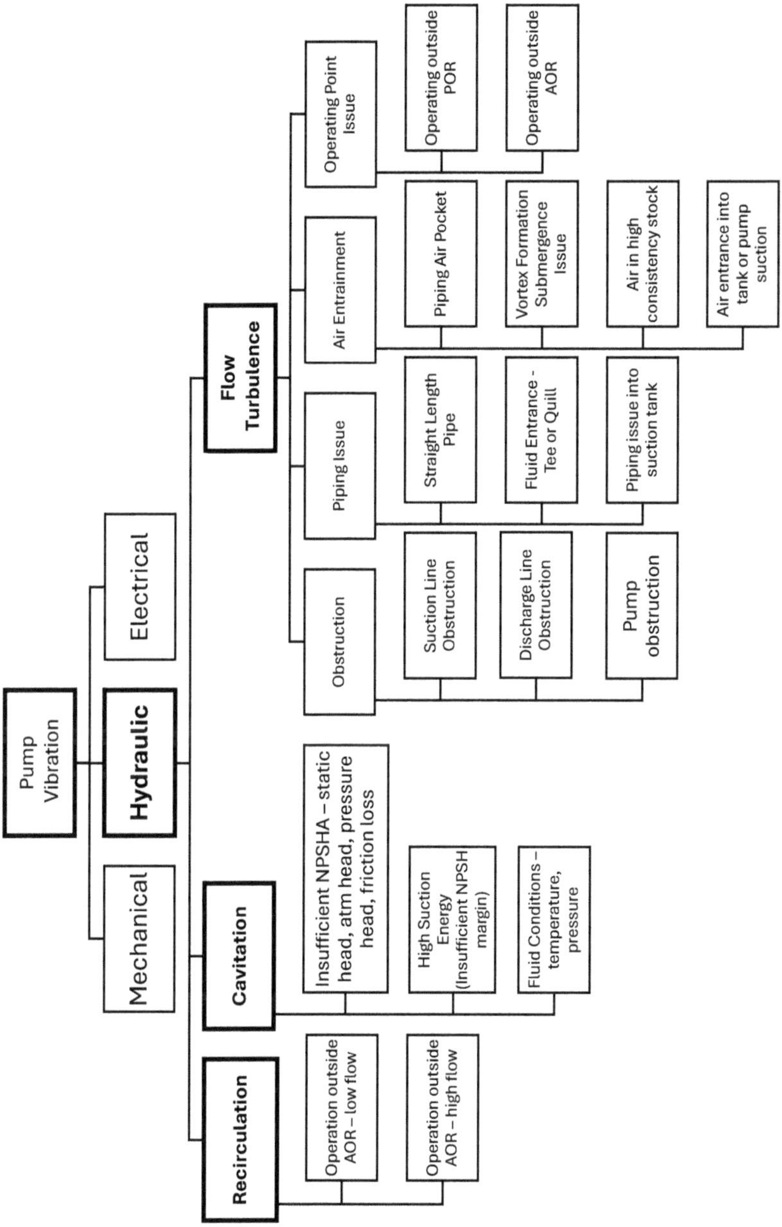

FIGURE 9.40 Basic hydraulic sources of pump vibration.

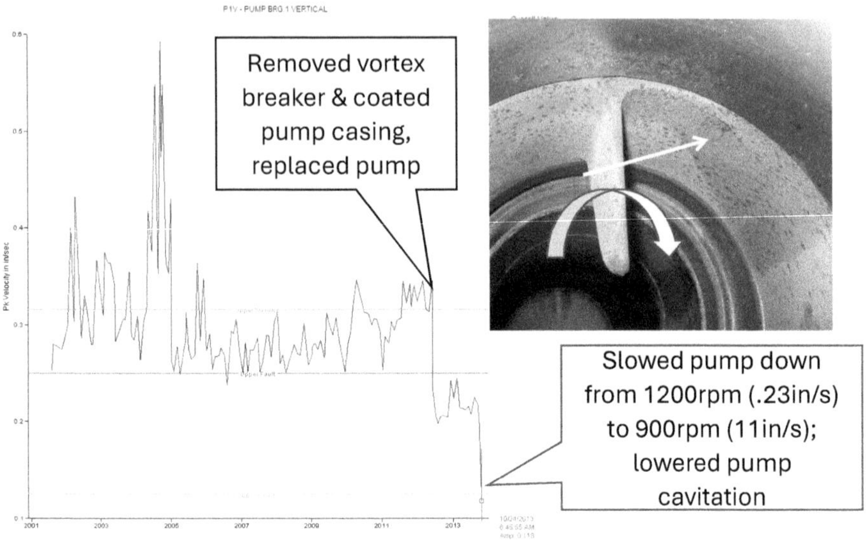

FIGURE 9.41 Vortex breaker suction nozzle vibration impact.

CAVITATION

Cavitation energy is usually random, high frequency broadband energy, and many times incorporates harmonics of blade pass frequency as the implosions occur in the impeller vanes. Cavitation energy is typically in the 10–120 kHz range as shown in Figure 9.42. Severe cavitation may also be audibly heard with a sound

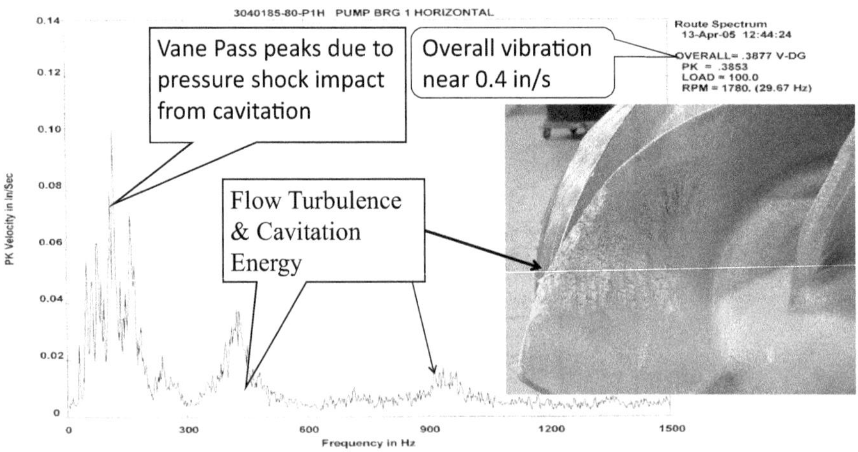

FIGURE 9.42 White water pump cavitation – vane pass.

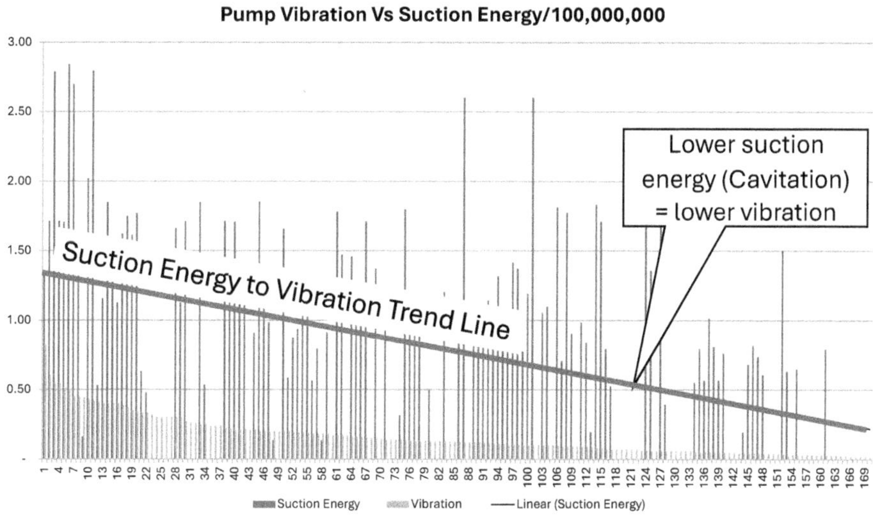

FIGURE 9.43 Suction energy trend versus pump overall vibration.

like gravel traveling through the pump and piping. Cavitation can also be observed using ultrasonic measurements with the contact probe on the pump suction and subsequent piping. Using ultrasound is especially helpful to identify early low-level cavitation but can also be clearly heard at high levels. If the ultrasonic reading gets above 50 db then damaging cavitation is possible. I've seen some very damaging cavitation on high suction energy pumps with cavitation noise at 67 db on the suction nozzle. The pump cavitation in Figure 9.42 also shows the elevated amplitude at the vane pass frequency as implosion on impeller vanes shows impacting.

Suction energy is not only a good indicator of cavitation, but also cavitation-related vibration. One study of nearly 140 pumps revealed the relationship between suction energy and pump vibration as shown in Figure 9.43. Of course, there are other causes of vibration but just taking these two parameters shows the clear relationship.

Pump vibration will also look very different at various operating points on the curve with reliability being closely connected. Figure 9.44 shows the vibration of the same pump and three distinct operating points – (a) off end of curve operation, (b) near BEP, and (c) dead head condition. It is clear that a dead head condition is the most violent vibration condition and one that a pump will not tolerate to run very long in that condition. A pump dead head exhibited almost three times the velocity spectrum vibration and over five times the acceleration/impacting energy.

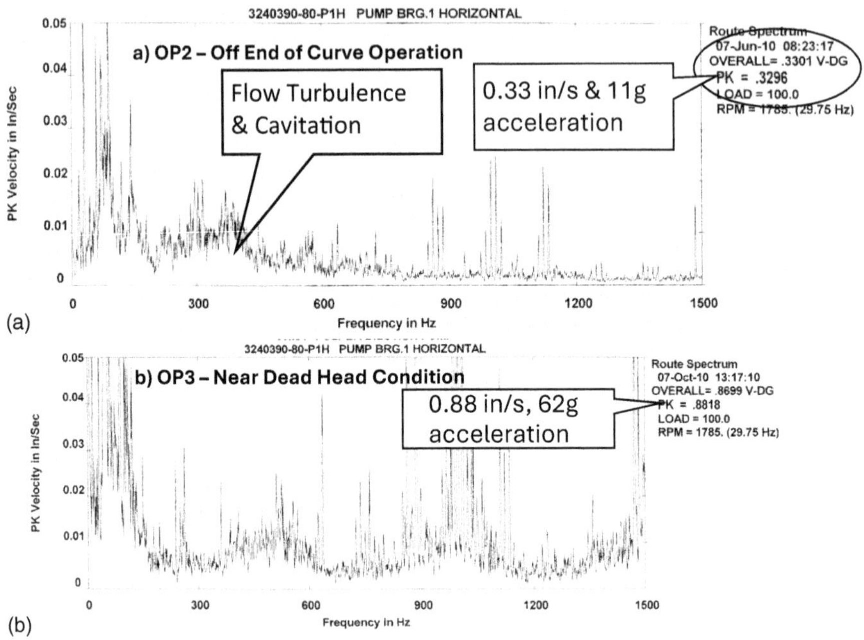

FIGURE 9.44 Pump vibration at various operating points: (a) high flow outside AOR operation and (b) low flow outside AOR operation.

OTHER PREDICTIVE TOPICS

Vibration Trending

A rising trend at any frequency may be telling. The rising trend in Figure 9.45 shows a 300% increase over 1.5 years. The 1X rpm was the dominant frequency as it was over 80% of the energy in this vibration spectrum. This was on a vertical turbine

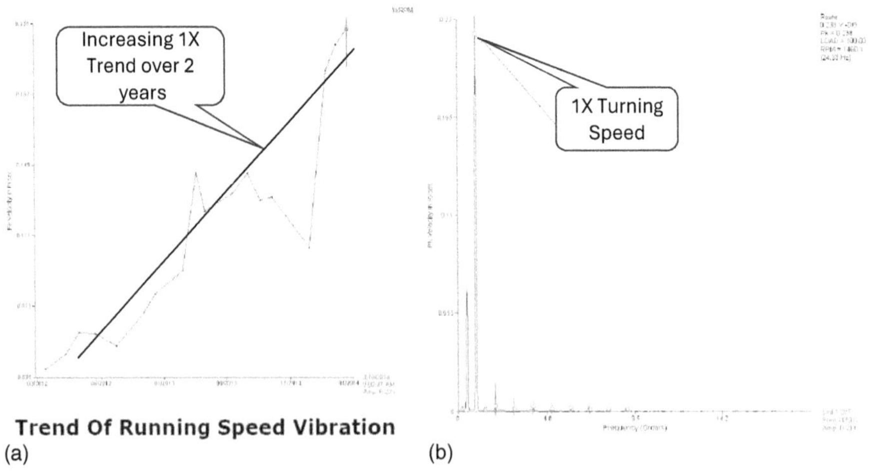

FIGURE 9.45 (a) 1X overall vibration trend and (b) 1X dominant peak in velocity spectrum.

FIGURE 9.46 Vertical turbine pump column failure.

three stage pump about 40 ft long with four column sections. The 1X is typical indicating unbalance and the assembly was unbalanced once the failure was discovered. The column bolts had failed which was whipping the assembly flopping around the shaft creating a large unbalance force showing up on the vibration signature. Figure 9.46 shows the pump column assembly as it was being removed from the sump. One other symptom that happened at the end was the pump lost flow. When the column became completely loose, the flow short circuited out of the pipe so no flow exited the pump.

Another good learning here is to differentiate between the vibration diagnosis as a symptom or the root cause. Unbalance might make one think that there was something wrong with balance in the pump impeller or shaft assembly, so the action would be to remove and balance assembly. In this case, unbalance was a symptom of another root cause – broken column assembly bolts. However, the physics of the vibration analysis told the truth in that there was unbalance in the rotating assembly as operating. High 1X is also a sign of Type A structural looseness which is what we had here. The vibration tools can't really determine the difference in symptoms or root causes but only give indication of how the pump is vibrating. Once the column bolts broke, the large pump column mass was not rotating around the shaft center, but outside which created the large 1X, unbalance-type vibration symptom.

Improvements to the vertical pump column design included upgrading the column bolts from grade 5 to grade 8 (improved bolt preload and joint clamp load) and installing double nuts to reduce the likelihood of loosening under vibration. It is believed that the pump vibration that loosened the column bolts was from the increase in overall pump vibration, as the pump bearings began to wear and shaft whip was occurring. The pump also had a shaft size upgrade and improved column bearings to help reduce this wear impact and increase shaft stiffness. The wear on this pump was mostly constant, with an established service life of five years for the pump and bearing assembly. Implementing a five-year preventive maintenance replacement interval, along with the condition vibration monitoring, was a good reliability measure.

Static Vibration

Some other pump vibration involves static failures. That is pump failure modes induced by pump vibration while the pump is not even in operation. For systems that have a spare pump installed (to improve system reliability), the spare pump may be vulnerable to vibration while it is idle. The foundations and piping all connect the idle pump to the operating pump. There will always be some residual vibration on the spare pump. Most of the time this vibration is not damaging, but sometimes it may be severe enough to create false brinelling damage on the pump bearings. Figure 9.47 shows a pump vibration on a spare pump which exhibited high vane pass even while not running.

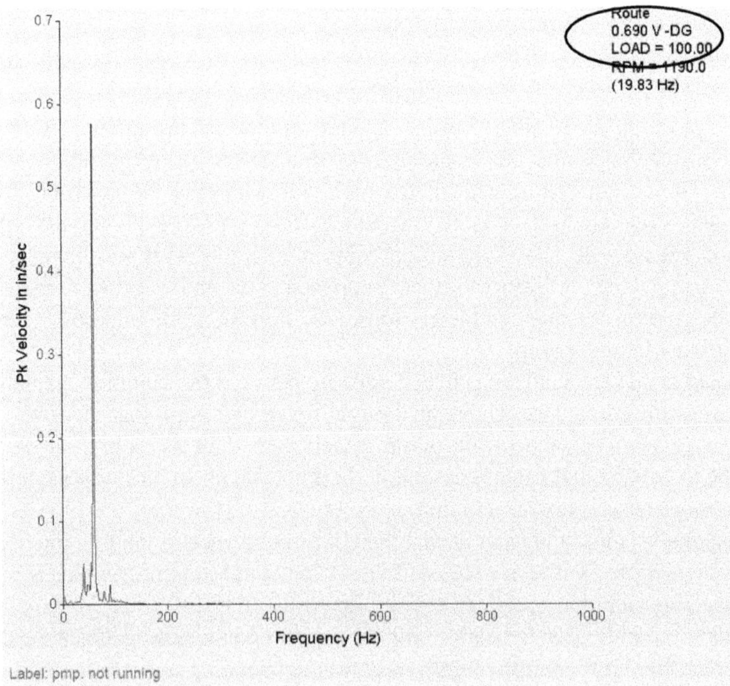

FIGURE 9.47 Vibration of idle spare pump.

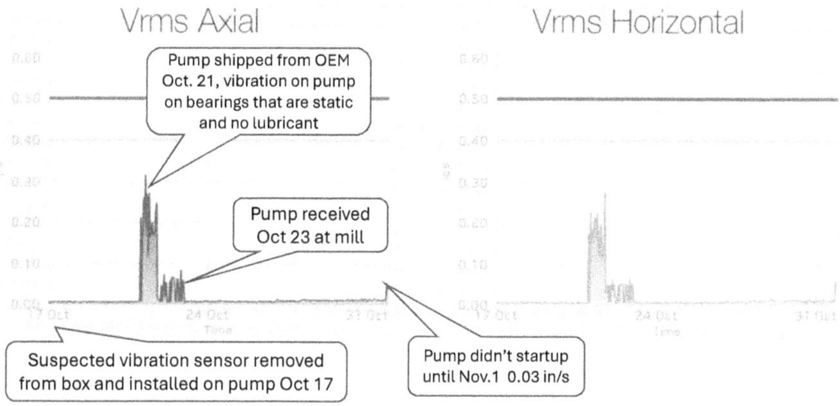

FIGURE 9.48 Pump vibration during shipping recorded by wireless continuous monitoring sensor.

Another potential static vibration issue that can also lead to bearing damage is vibration during shipping. During one new pump order, a wireless vibration sensor was activated by mistake as the pump was finishing assembly at the manufacturer. When the pump was received at the plant, it was observed that the vibration sensor was already on and probably got turned on during assembly process. Figure 9.48 shows the vibration trend from the time it was turned on in assembly with the high vibration peak during shipping.

Another challenge in pump system vibration can be troubleshooting the motor. An electric motor is an electro-mechanical device that can have electrical and mechanical defects. It can be hard at times to discern if the problem is an electrical issue or a mechanical issue. One action is to compare vibration levels during operation with those immediately after power shutdown.

One quick test that can be done is to take vibration data while the system is operating (motor on) and continue to take data as soon as the motor power is shut off. If the vibration is basically unchanged then the issue is purely a mechanical one. If the vibration changes or improves significantly then there is an electrical component contributing or at least participating in the vibration forcing function.

One pump motor had multiple issues showing up, such as 2X line frequency on electrical reading and elevated 1X as well as looseness on normal vibration readings. The pump motor was operating at 3,600 rpm on a fixed speed drive. The motor was also a 600 HP open frame motor. The normal vibration data was taken while the power was on the motor and with power shut off. This comparison is shown in Figure 9.49. The vibration reduction was significant (over 100X reduction) on the 1X vibration from 0.31 in/s down to 0.0035 in/s. This obviously tells us that the electrical element of the motor is a major factor in the vibration.

This led to thinking about the motor stator, rotor, and motor air gap effect on the motor operation as it affected vibration. As the motor current is shut off, the effect on the motor rotor is removed. Could there be an issue with the motor causing the issue? A more precision approach to pump and motor installation was undertaken after seeing this vibration change. While there were numerous small things found, the biggest thing

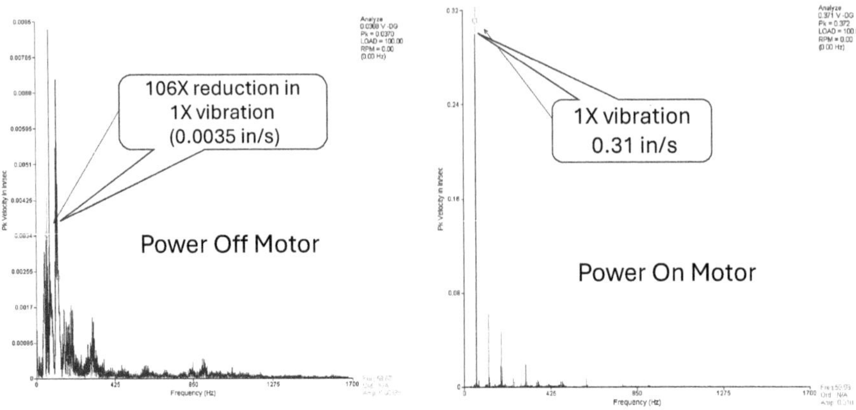

FIGURE 9.49 Pump motor vibration with power on and off.

was a 0.007″ soft foot on the motor to baseplate on opposite corners. The theory at this point was that the soft foot and resultant motor housing stress twisting the motor frame distorted the air gap, resulting in an electrical-mechanical effect.

To further test the theory of the soft foot causing this elevated motor vibration, one of the motor feet with the soft foot was loosened to remove the stress from the motor frame. The result reduced the vibration by ~65% from 0.31 in/s to 0.125 in/s as shown in Figure 9.50. Later the motor base was machined to improve the flatness and remove the soft foot. When this happened, the motor vibration stayed at the reduced vibration level, resulting in good motor reliability.

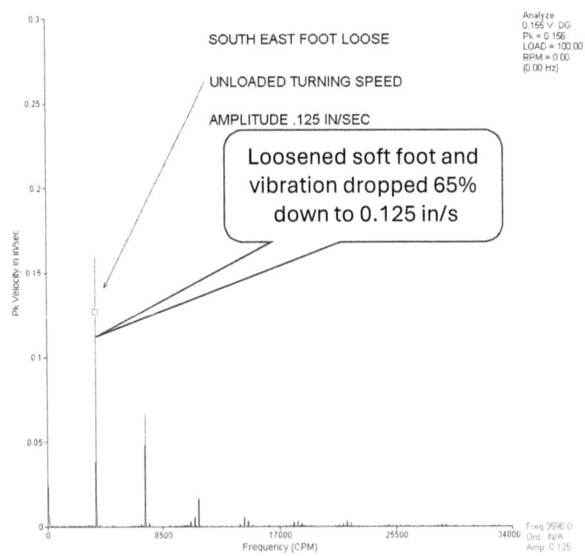

FIGURE 9.50 Vibration response with stress removed from soft foot.

INFRARED (IR)

Infrared or IR can show a host of issues in pump system applications from plugged lines, hot bearings to coupling issues. Figure 9.51a shows a pump system on water which had built up a temperature of nearly 200°F due to a dead head condition. Note how hot the pump is relative to the discharge line just past the pump.

A spacer gear coupling has two flex half gear couplings and will float/oscillate axially depending on how much misalignment is present. This axial movement is what creates wear when gear coupling lubrication is not adequate. Figure 9.51b shows this pump where the coupling was found dry and locked up creating a high axial load on the pump bearings. Figure 9.51b also shows the same pump with a good coupling properly lubricated and the normal pump thrust bearing temperature.

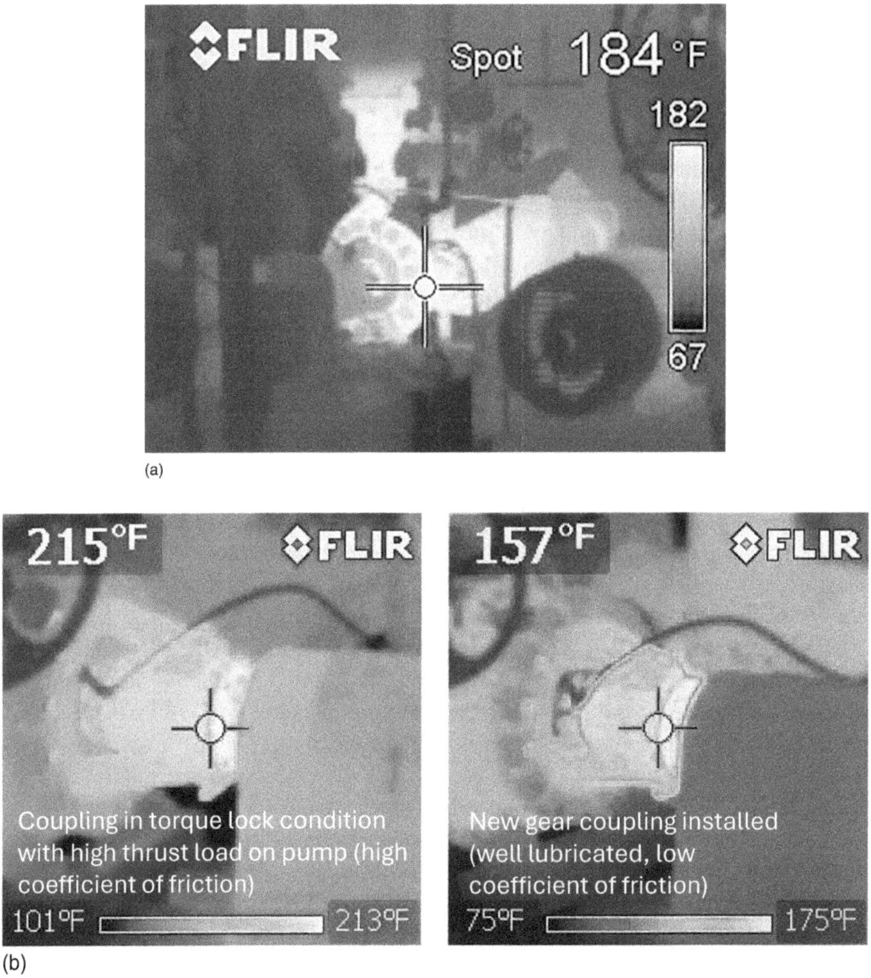

FIGURE 9.51 IR picture: (a) pump dead head condition and (b) failing coupling – high bearing temperatures.

10 Pump Reliability Summary

Key Elements

Reliability in general can be traced back to three main areas – design, operation, and maintenance of an asset. The pump and system that has a reliable design, operated and maintained correctly will result in good reliability. Every case study of unreliability in this book has a root cause that falls into one of these three areas. While just knowing the category the failure root cause is in doesn't solve anything, it can help you focus your energy on a certain area to bring about resolution and improve reliability results.

The results of the pump and pump system failures of case studies from this book are shown in Figure 10.1. Poor design was the biggest issue with over 60% of pump system problems. Maintenance was second, causing 25% of pump system issues. This is usually shocking to many in management who will believe that maintenance is responsible for most pump reliability issues. Every equipment has some level of reliability in its DNA from the original design. Design issues cannot be corrected through maintenance or operation. Only redesign can resolve the problem. This will require much know-how and understanding of pump systems and their failure. It will also take some patience, investment, discipline, and perseverance from company management.

Pump system issues arise from both the pump and the system, with many failures resulting from their interaction. In the case studies discussed in this book, the breakdown of issues were ~44% pump issues and ~56% pump system issues. Pump reliability can many times be traced back to its operating point on the pump and system curves. This can be shown on the pump curve and identify the different regions of operation and the impact on pump reliability.

Looking at Figure 10.2, the allowable operating region, AOR, (1) is set by the pump manufacturer for that specific pump. The pump in general should not be operated outside of this region. Keeping the pump operation within the AOR is not considered to compromise the service life of the pump. Smaller pumps typically have a larger range on AOR. While bearing life may vary over the range of AOR, within the AOR it is generally expected that the bearing service life will at least be two years. There are many factors which pump manufacturers may consider when determining the AOR for a pump, such as vibration, noise, suction recirculation, temperature rise, slope of head/flow pump curve, NPSHmargin, shaft fatigue, bearing life impact, and seal life impact. Some special considerations may also be considered for slurry- or solids-handling pumps.

DOI: 10.1201/9781003729525-10

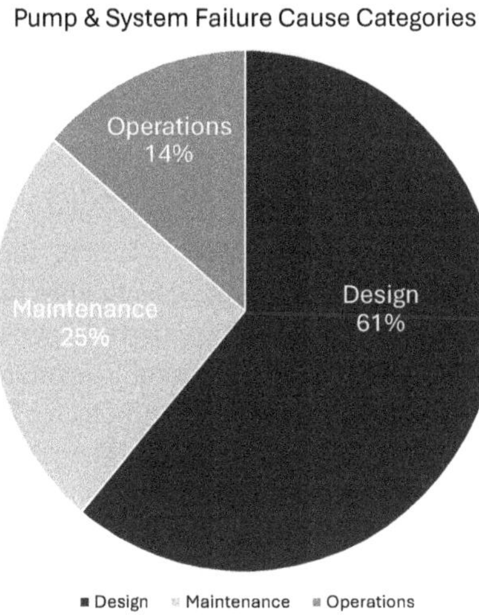

FIGURE 10.1 Pump and system failure cause categories from case studies.

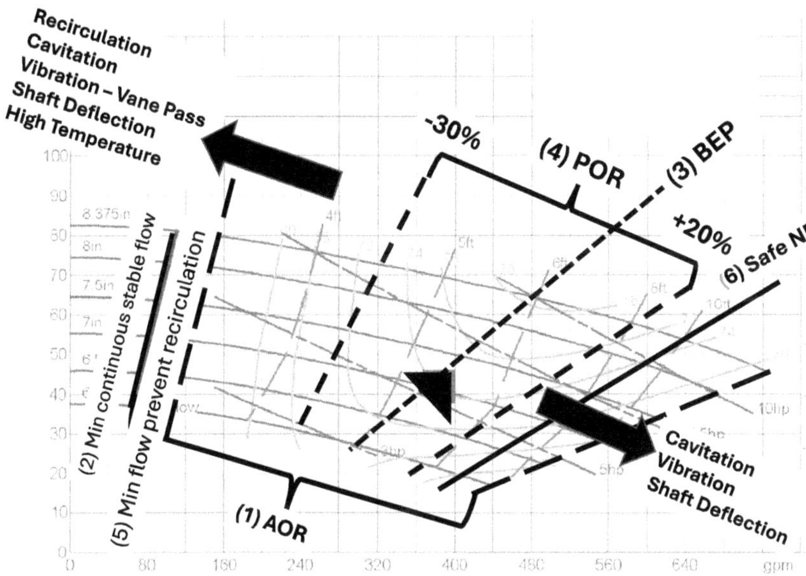

FIGURE 10.2 Pump reliability conditions on pump curve.

Pump system low flow operation (off AOR operation) has many negative consequences such as the following:

1. Increased discharge pressure
2. Increased axial and radial loads on pump shaft and bearings
3. Increased vibration

4. Increased operating temperature of bearings and process fluid
5. Increased recirculation inside the pump
6. Decreased pump efficiency
7. Low flow output

Most of the other operating conditions that may exist on the pump curve are centered around the pump best efficiency point (BEP) (3). For the fixed speed pump in Figure 10.2, the BEP runs through the range of impeller diameters. BEP in many pumps is about 80–85% of shutoff head. The preferred operating region (POR) (4) is an operating range recommended by the Hydraulic Institute (HI) centered around hydraulic efficiency. While reliability was not the focus for POR, pump reliability in many ways tracks along with hydraulic efficiency. The farther away from the BEP, the less reliable the pump is likely to operate.

The POR for pumps with a specific speed of $<$ 4,500 is -30% to +20% of BEP flow or 70–120% of flow at BEP, as shown in Figure 10.2 POR (4). For pumps with specific speed of $>$ 4,500 the POR is -20% to +20% of BEP flow or 80–120% of BEP flow. If possible, operating the pump in the POR region is a preferred operating condition for optimum efficiency and reliability.

The two regions of pump operation outside the POR and even more extreme AOR to the right or left of BEP will result in significant reduction in not only pump performance but also pump reliability. Going the left of BEP produces several consequences such as suction recirculation and related cavitation, increased vibration, increased bearing loads and resultant shaft deflection, and lower hydraulic efficiency. As a result, fluid temperatures can increase and ultimately poor pump reliability. Going to the right of BEP produces similar negative results with increased bearing loads, shaft deflection, vibration, and suction side cavitation. Right of BEP and NPSHR increases significantly.

There are many other areas where pump reliability is determined but evaluating the pump curve and the actual operating point is a good way to estimate how reliable your pump system may operate. The farther away from the BEP, the worse the reactions will be on pump system reliability.

The minimum flow to prevent recirculation marked by line (5), in Figure 10.2, is typically just inside the low flow AOR. A discharge restriction in the pump system causes recirculation in the impeller eye. This minimum flow can be estimated by the relationship between the suction-specific speed and the operating point as percentage of BEP flow, shown in Figure 10.3. Note that the higher the suction-specific speed, the higher the flowrate needs to be to prevent suction recirculation. BEP pump efficiency is ideal flow vector angles in and out that match the impeller to minimize the hydraulic losses. Altering pump flows can cause inlet and outlet angles to misalign with the impeller vanes, resulting in lower pump efficiency and other issues. When the pump is throttled severely, the flow vectors can reverse out of the pump (suction recirculation), which can create cavitation and/or choked flow conditions.

In Figure 10.3, a pump with a BEP flow of 8,000 gpm and suction-specific speed of 11,000, the lowest flow to prevent recirculation would be 47% of BEP flow or 3,760 gpm. Another factor which can be problematic for a double suction pump

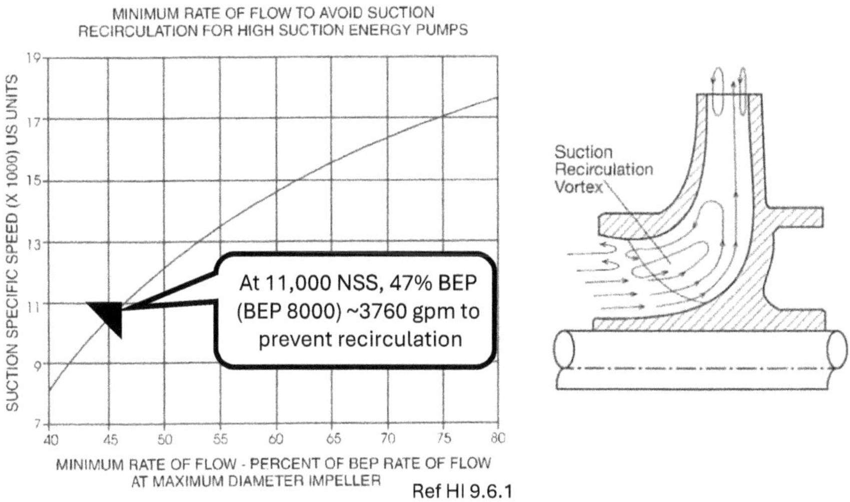

FIGURE 10.3 Minimum flow rate to avoid excessive suction recirculation (Ref HI 9.6.1).

reliability is suction recirculation caused by throttled flow or other disturbances. When this happens, the double-suction pump rotor can become unbalanced, and the rotor can shuttle back and forth within bearing clearances. This small shuttle movement can cause havoc on a mechanical seal on double-suction pumps, especially split seals, which are often used on these pumps due to easier maintenance (can install seal without removing bearings). If the flow issue can't be resolved, many times the pump sealing may have to be converted back to packing.

A choke ring on the pump suction can be used to redirect some of this recirculation in cases where the pump is running too far from BEP. The choke ring will increase the impeller eye velocity and reduce recirculation. A choke ring is not necessarily a permanent solution to a pump that is operating far from BEP where recirculation is an issue. The choke ring can also increase NPSH3, so this must also be considered. A smaller pump to fit the system conditions or recirculation line to increase pump flow is a better choice to optimize pump system reliability.

The safe NPSH line (6) is a point where safe operation of the pump limits significant cavitation damage on the pump. This can be set by the system designer based on system design conditions. The minimum continuous stable flow, MCSF, (2) or just called minimum flow, represented on the curve, defines the lowest flow at which the pump should operate, which is the lower APR limit. Outside that line, the flow recirculation becomes high and the pump flow may become unstable.

The minimum continuous thermal flow (MCTF) is the flow rate at which the pump should not operate below to avoid maximum fluid ΔT. The allowable maximum ΔT is the difference between the saturation temperature of fluid at suction pressure and suction temperature. Some suggest to take an additional 15°F margin on top of that number. The basic equation for minimum continuous thermal flow is:

$$T = \frac{P \times 5.1}{Q \times Cp \times SG}$$

$$\Delta T = \frac{Q \times T}{Qx}$$

$$\text{Allowable } \Delta T = T\text{sat} - T1 - T\text{margin}$$

where:

T = °F/min
Q = volume of pump casing, gallons
Cp = specific heat
SG = specific gravity
P = horsepower at actual flow condition
Qx = flow in gpm through the pump
ΔT = temperature rise across pump at some flow, Qx
Tsat = saturation temperature of fluid
$T1$ = starting temperature of fluid
Tmargin = 15°F design margin

To satisfy a minimum continuous thermal flow for a pump system, there are several methods to address. The most basic one involves a continuous bypass from the pump discharge. Typically, this will involve a fixed orifice plate or valve, which allows flow to pass continuously while the pump is operating, regardless of whether it is needed. This is easy to implement, low initial cost, almost no maintenance cost if designed correctly, and will provide pump thermal protection. This constant flow option does waste energy (but typically small) and must include enough pump and driver to handle the additional flow system demand.

Another option is an automated flow control line on the pump discharge. This could be a flow control valve or an automated relief-type valve. This is energy-efficient as it only operates and relieves when the pump demand flow is very low, and recirculation flow is needed to control the thermal condition of the pump. The initial cost is higher and has some maintenance with long-term care.

CASE STUDY 51 MINIMUM CONTINUOUS STABLE THERMAL FLOW

For example, a fire water pump system runs monthly tests at various flow conditions. The pump even runs a period of time in a dead head condition checking system pressure. The pump must have a minimum flow line to prevent damage during the no-flow condition, so a minimum flow line must be used for monthly flow tests. What is the minimum continuous thermal stable flow for the fire pump?

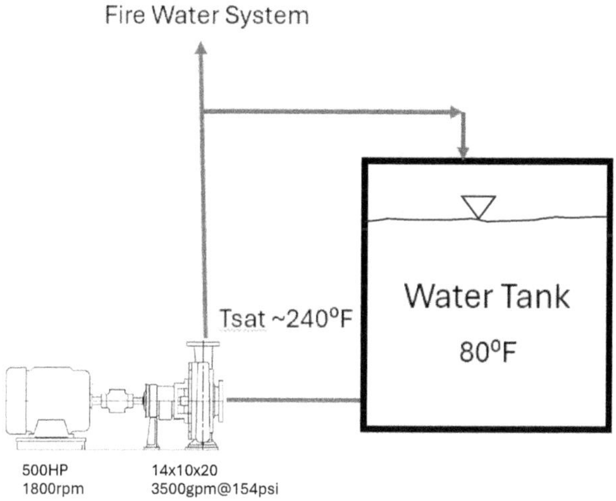

FIGURE 10.4 Process diagram for fire water pump system.

The fire water system pulls water from a water tank and has a recirculation line back to the tank with the main flow feeding the fire water header, as shown in Figure 10.4. The pump is a $14 \times 10 \times 20$ double suction pump with an 18.9″ diameter impeller. The pump has a design point of 3,500 gpm at 154 psi. The pump is driven by a 500 HP, 1,800 rpm fixed speed motor.

The fire water pump had many minimum flow points noted, as shown in Figure 10.5. The minimum flow to prevent recirculation is ~2,500 gpm. The AOR minimum flow is ~1,500 gpm. The minimum flow recirculation line flow is ~60 gpm. Is the minimum flow line enough flow to satisfy the minimum continuous thermal stable flow (MCSF)?

Using the equations for fluid temperature changes in the pump, the temperature rise for the pump at dead head is 20.4°F/min. The dead head power is 220 HP and pump casing volume is ~55 gallons. The fluid properties are $SG = 1$ and $Cp = 1$ for water. The allowable ΔT for the pump fluid is ~145°F. A ΔT of 145°F/20.4°F/min = 7.1 minutes of safe operating time as the maximum time the pump could be run without a minimum flow line. The pump did have a minimum continuous flow line, which was sized and provided ~60 gpm of flow continuously when the pump was operating. Rearranging the equation and solving for flow through the pump, Qx, with a $T =$ 20.4°F/min, allowable $\Delta T = 145$°F and $Q = 55$ gallons, the minimum continuous stable thermal flow through the pump is ~8 gpm. Therefore, with a 60 gpm continuous minimum line flow, the MCSF is satisfied, and safe operation is achieved.

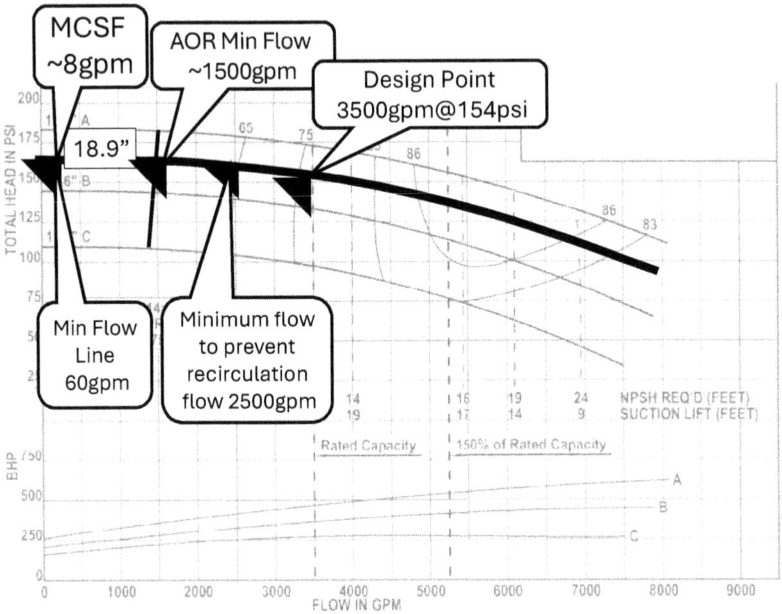

FIGURE 10.5 Fire water pump performance curve with operating points.

Case Study 51 Learning – Design minimum continuous stable thermal flow for pump systems which will or easily may run near dead head for any period of time.

Many of the next several case studies will look at various pump systems and the operating points that generated certain reliability results. These illustrate how the actual operating point, relative to the pump's AOR and POR, impacts pump and pump system reliability. These case studies will show various reliability results based on pump operating condition, as shown in Figure 10.2. The pump curve and system curves will determine what kind of overall pump reliability can be expected. The pump operating point is one of the major determining factors in predicting and solving pump reliability issues. A summary of these case studies is listed below.

Case Study 52 – Pump Operating Outside AOR (Low Flow) – Cavitation Damage
Case Study 53 – Pump Operating Outside AOR (High Flow) – Cavitation Damage
Case Study 54 – Pump Operating Outside AOR (High Flow) – System Change
Case Study 55 – Pump Operating Outside AOR (High Flow) – Speed Change
Case Study 56 – Pump Operating Outside AOR (Low Flow) – Shaft Fracture

Case Study 57 – Pump Operating Outside AOR (Low Flow) – High Vibration
Case Study 58 – Pump Operating Outside POR (Low Flow) – Low Seal Life
Case Study 59 – Double Suction Pump – Mechanical Seal Issues
Case Study 60 – Pump Operating Inside POR (At BEP) – Best Reliability

CASE STUDY 52 PUMP OPERATING OUTSIDE AOR (LOW FLOW) – CAVITATION DAMAGE

A pump dilution system consisted of a pump providing dilution water to over 18 destinations in a pulp operation, as shown in Figure 10.6. The pump is an $8 \times 10 \times 16$ end suction pump with a 13.6″ diameter impeller and a design point of 3,000 gpm at 127 ft. The pump was driven by a 200 HP, 1,800 rpm fixed speed motor. The pump had an NSS = 9,420. The system had over 15 control valves serving different destinations, most of which operated at less than 30% open. There were a few control valves that operated at over 75% open. The pump system also had a header pressure control recirculation valve (3″). Due to a few control valve positions, the pump system operating pressure had a minimum setpoint of around 60 psi to keep enough dilution to all destinations.

The pump and pressure recirculation valve operated with poor reliability. The 3″ header pressure recirculation control valve had a very high pressure drop and very high fluid velocities, which created significant vibration. The vibration

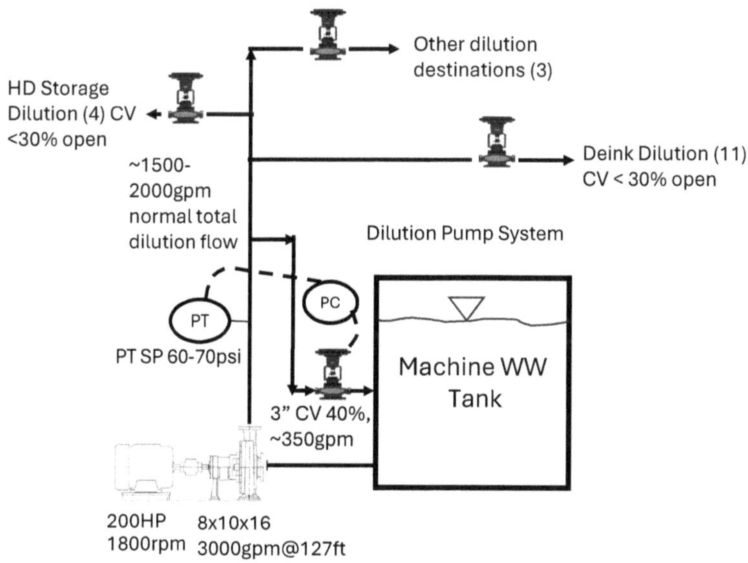

FIGURE 10.6 Process diagram for dilution pump system.

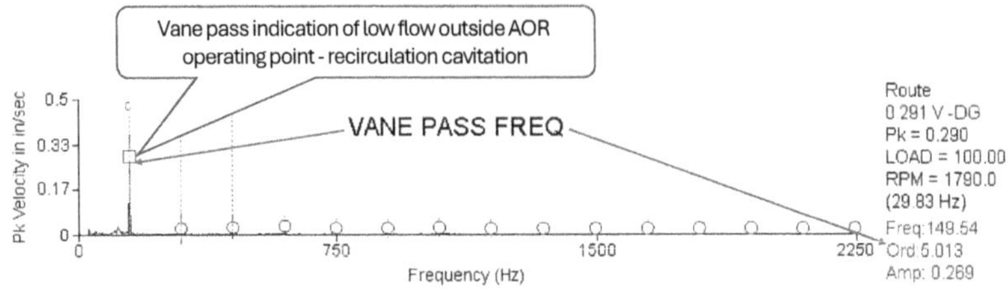

FIGURE 10.7 Typical vibration signature velocity spectrum.

usually caused cracking in the piping and tank connection. The pressure control valve typically operated 90–100% open to try and get the header pressure down to setpoint but many times it was unable to achieve. The control valve was essentially undersized to control the target system pressure. The pump also exhibited significant vibration, especially at vane pass frequency at 0.269 in/s, as shown in Figure 10.7.

The physics of failure for the pump and components was a focus to determine the root cause of failure and poor reliability. The pump experienced several different failure modes such as short life on the rotating assembly (bearings), seal life, and casing life. The cast iron pump casing typically would become pitted in less than 18 months, as shown in Figure 10.8. The suction wear plate area was the focal point of the pitting damage. Since this wear plate

FIGURE 10.8 Pump casing pitting damage on suction wear plate.

was integral to the pump casing, the entire pump casing had to be replaced every two years or so.

Also, the pump hydraulic efficiency was drastically reduced when the pitting became severe to the point that with the pressure control valve shut, the system pressure would drop 60 psi below, affecting process dilution controls on several end points. As the header pressure dropped, control valves would have to open and cause several to operate near 100% open.

So, what was the issue? The pump operating point was verified with pressure readings and motor load. The TDH of the pump measured around ~69 psi (159 ft). The pump motor load measured and calculated to be ~141 HP. These two different pump operating characteristics were very close on the pump curve, as shown in Figure 10.9, giving high confidence that they represent the actual pump operating point. As shown in the pump curve in Figure 10.9, the pump was operating outside the AOR at a low flow condition at ~1,700–1,800 gpm. In this region, the pump will have recirculation cavitation, high vane pass vibration, high radial loads, and shaft deflection. All this matched the pump symptoms and failure modes that were seen with the pump.

There were two related issues with the pump and system. First, the pump was too large for the demand in the system. The pump needed an additional ~1,500 gpm to be operating in or near the POR for good reliability. Second, there was at least one destination where the control valve ran out of control range and the system pressure could not be lowered past 60 psi. To fix the issue, flow demand had to be increased on the pump system and the control valve station had to have a larger control valve installed.

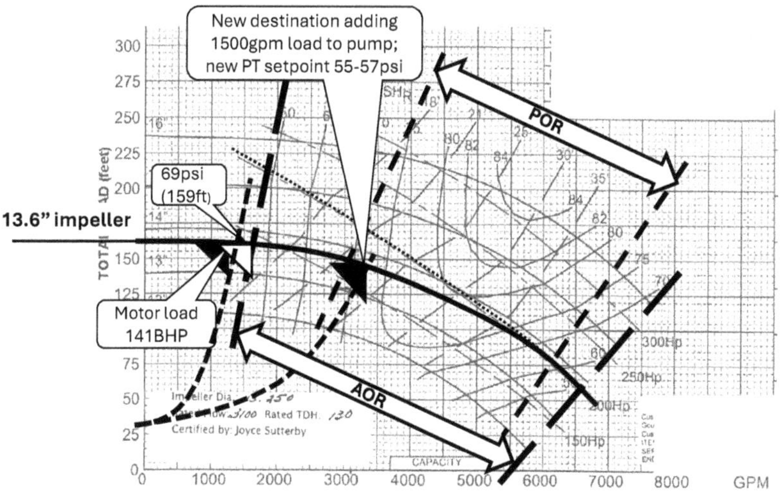

FIGURE 10.9 Pump and system curves with operating points.

In the process of developing corrective actions for the system, a new project was being installed, and an additional 1,500 gpm was designed to be added onto this system which would satisfy the pump operating point. During the same project, the problem dilution application had its demand significantly reduced to where the existing control valve could operate at the lower header pressure ~55 psi and stay in control range.

This pump system reliability issue had a root cause category of design in that the design did not match the process demand and operating point. There was something missed from the original design and where the current process was operating. A process change or pump change would be required to get the pump system operating within the AOR for the pump.

Case Study 52 Learning – Never operate pump system outside AOR (low flow) as poor reliability will result.

The next case study is one where the pump was also operating outside the AOR but in a high flow condition.

CASE STUDY 53 PUMP OPERATING OUTSIDE AOR (HIGH FLOW) – CAVITATION DAMAGE

A dilution pump system was experiencing chronic failures and pump casing leaks in a very short run time. The dilution pump system was a system pulling out of an atmospheric tank and feeding dilution water to multiple sources, as shown in Figure 10.10. The pump was controlled by a 150 HP, 1,800 rpm VFD motor on header pressure control. The pump was an $8 \times 10 \times 15$ end suction pump with a 13.5″ diameter impeller and a design point of 2,600 gpm at 160 ft. The pump had a NSS = 12,050. Typically, the header pressure was set high enough for the pump to operate at 100% speed continuously.

The pump failure mode was typically severe, pitting on the casing integral suction wear plate area and on the stuffing box, as shown in Figure 10.11. The MTBF for the pump casing was about 18 months. The casing or stuffing box will then begin to leak dilution water. The list of possible causes for the severe pitting damage is shown below.

Possible Causes

- Impeller rubbing wear plate and stuffing box.
- Abrasive material in dilution water
- Corrosion-related – material selection, pH, material defect, etc.
- Air entrainment/vortex – submergence
- Recirculation cavitation operating low flow outside AOR
- Suction cavitation – insufficient NPSHA

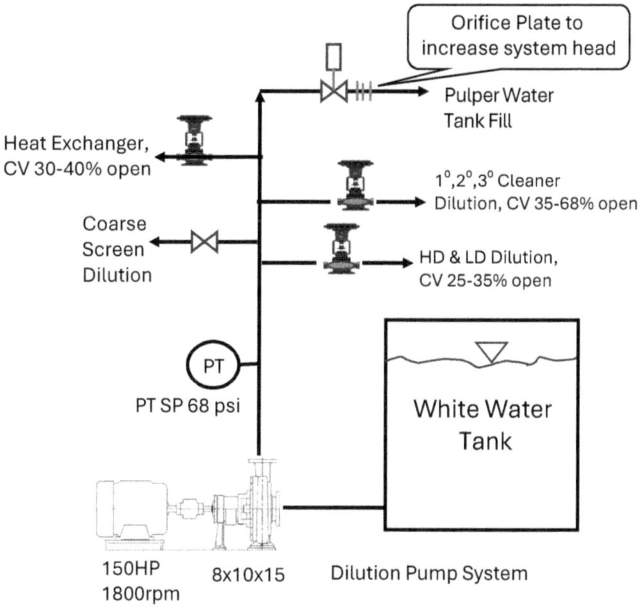

FIGURE 10.10 Process diagram for dilution pump system.

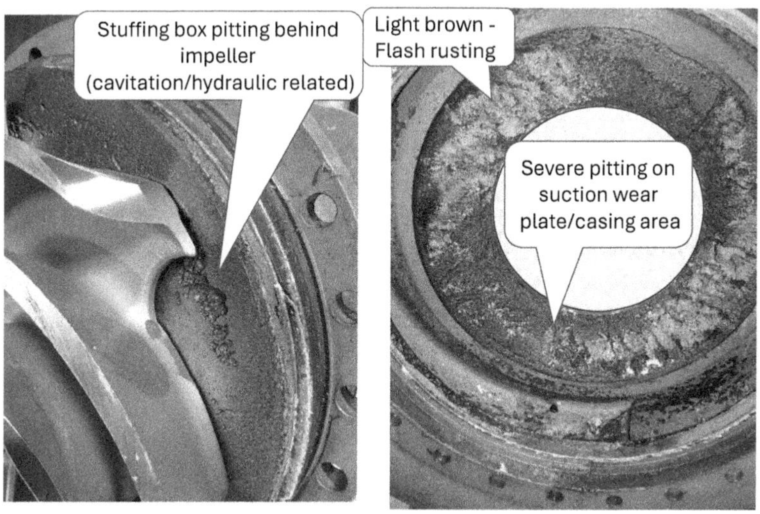

FIGURE 10.11 Pump casing and stuffing box pitting damage.

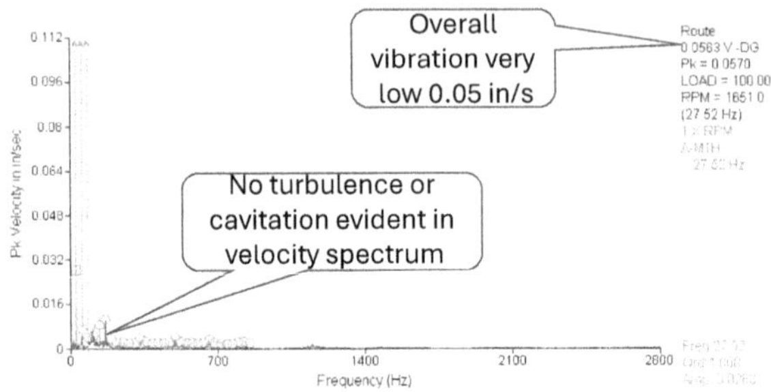

FIGURE 10.12 Vibration signature velocity spectrum.

The pump was observed to be operating very smoothly during normal operation. Vibration data was collected monthly and other on the run inspections never saw the pump operating with any unusual noise. The overall vibration was less than 0.06 in/s, as shown in Figure 10.12. The 1X was the highest peak but still low at 0.026 in/s and the vane pass was very low at 0.012 in/s. There was no significant flow turbulence or cavitation energy showing up in the vibration either.

The pump impeller clearance was never found significantly out of specification and the impeller was never found rubbing the pump wear plate or casing. The dilution water quality was verified and did not have any heavy abrasive solids that might impact pump wear.

The pump material was a standard material for dilution water with ductile iron for the casing and 316SS for the impeller. The cast iron is a suitable material for a fluid with a pH in the 6–9 range. The process fluid was verified and tested at a pH of 7.3, therefore, the material choice is sufficient for the application.

The pump suction was evaluated for submergence and potential for vortex air entrainment. The minimum submergence level was found to be between 41 and 51″ for the range of operating points observed. The low tank level for the pump suction was never below 60″ (5 ft), with most of the time the tank level being near 8 ft.

The pump pressure was verified at 68 psi, with a pump TDH of ~153 ft. The motor load was also checked at 132 HP. The normal operating point (OP1) for the pump was found to be in the POR for the pump, as shown in Figure 10.13. The normal operating point, OP1, was even right at the BEP. For all observed pump operating points, a suction recirculation and cavitation-type issue do not appear to be a problem.

Further investigation into the pump system operation revealed that there was at least one other operating point that occurred every so often. Looking back at process diagram in Figure 10.10, there was an operating condition where the process control valves were shut and the fill valve to the pulper water tank operated 100% open. The system curve on this condition created a very low system head, which led to a pump operating point (OP2) that was well outside the AOR in a high-flow situation as shown in the pump and system curves in Figure 10.13. Since the pump is controlled with a VFD on header pressure control, the low system head will force the VFD to speed up the pump to 100% speed to try and achieve the target header pressure.

Under this high flow condition, a cavitation condition now exists. The NPSHA for this system was calculated at 31 ft. With the suction energy of this pump, an NPSHmargin of 1.5 is recommended. With the NPSHmargin, the new NPSHR is ~39 ft. The normal operating point (OP1) has an NPSHR < NPSHA so cavitation is not an issue; however, at the alternate operating point (OP2) the NPSHR > NPSHA. Therefore, the alternate operating point (OP2) creates a cavitation condition which is damaging the pump casing.

What is the solution to correct the alternate operating point where significant cavitation is damaging the pump? One option is to adjust the VFD to run at a lower speed during the pulper fill operation. This could be done manually from the pump controls or lower the header pressure setpoint. With the system curve, the pump would need to be slowed down to around 900 rpm which would only yield ~1,500–1,700 gpm. This will work from a pump reliability standpoint but now the tank fill process will take twice as long and affect process production rates.

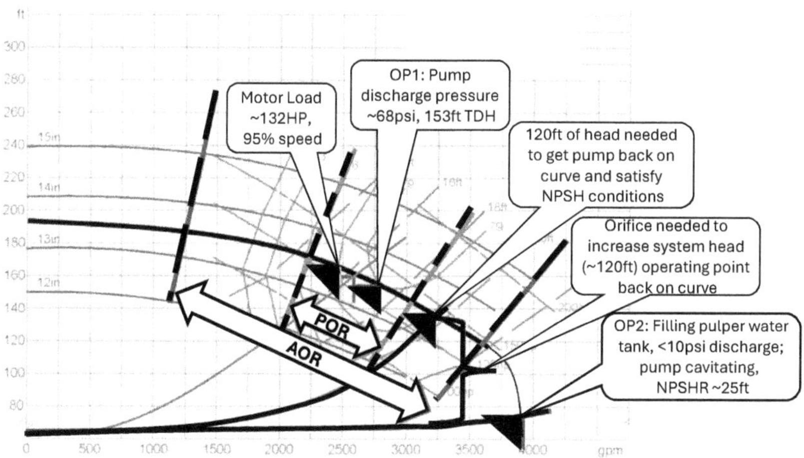

FIGURE 10.13 Pump and system curves with various operating points.

Parameter	OP1	OP2	OP2Rev
Submergence, Inches (Min tank level 5', 60")	40	51	44
Operating Point - Flow	2750 gpm	3900 gpm	3200 gpm
Operating Point - Head	153ft	25ft	140ft
NPSHA	31ft	31ft	31ft
NPSH3	17ft	26ft	20ft
NPSHR (NPSHmargin=1.5)	**25.5ft NPSHA, 31'>NPSHR, 25.5' No Cavitation**	**39ft NPSHA, 31'<NPSHR, 39' Cavitation**	**30ft NPSHA, 31'>NPSHR, 30' No Cavitation**
MTBF (annual cost)	10 Years ($12,000)	1.5-2 years ($12,000)	10 Years ($12,000)

FIGURE 10.14 Summary of hydraulic pump conditions for three operating points.

The best option is to increase the system head so that the pump operates on the curve within the AOR but also produces good flow to support the process. This can be accomplished by adding an orifice plate in the piping system to the pulper water tank. This is shown in Figure 10.13 of the pump and system curve. A summary of the three different operating conditions evaluated for the pump system are shown in Figure 10.14. The revised operating point (OPRev) with the orifice now has improved cavitation conditions and NPSHR < NPSHA. This modification was completed and to date the pump has run ~8 years with no failures or reliability issues. The pump and system are now operating in a reliable condition for all operating points. The pump is now happily married to the pump system.

Operating the pump way outside the AOR for even a short amount of time can have catastrophic effects. This pump probably only ran in the other operating condition less than 1% of the time. This pump system had a root cause of failure due to system design for one of the operating conditions. The system was not designed for the pump to operate within the AOR for all operating conditions. A system design modification would be required to correct this issue.

Case Study 53 Learning – Never operate pump outside AOR (high flow) even in intermittent conditions as poor reliability will result.

Operating a pump outside the AOR has significant impact on pump reliability by destruction of the pump components (from cavitation-type pitting failure modes) as seen from the previous case studies. However, other pump failures can occur by operating outside the AOR. The next case study will investigate one of the other potential outcomes of pump operation outside AOR.

CASE STUDY 54 PUMP OPERATING OUTSIDE AOR (HIGH FLOW) – SYSTEM CHANGE

A sump pump system was experiencing very low reliability. The pump was an 8×8×14 self-priming lift (trash handling) pump operating in a batch-type sump operation. The pump would cut on and off as the level reached a high and low. The original design point from the pump manufacturer was 1,400 gpm at 50 ft. The pump had a 13.75″ diameter, three-vane semi-open impeller that operated with a direct drive of 40 HP, 1,200 rpm fixed speed motor. The pump had a NSS = 6,320. The process diagram is shown in Figure 10.15.

The pump experienced all types of failures leading to very short run times with a mean time between failure (MTBF) of ~12 months. Some of the failure modes of various pump failures include coupling failures, seal failures, and bearing failures. The discharge piping also experienced some fatigue cracking due to pump and system vibration. Bearing failures often include much more secondary damage to the impeller and wear plate as clearance is not held due to loss of geometry from bearings failing. The internal check valve inside the self-priming lift pump also randomly fails. The pump always exhibits high vibration which is a big factor in poor pump reliability.

The pump always exhibited high vibration with elevated vane pass (3X run speed) vibration. Cavitation energy also is typically present, which also manifests itself in the vane pass. Figure 10.16 shows typical vibration signatures with vane pass and cavitation-type indications.

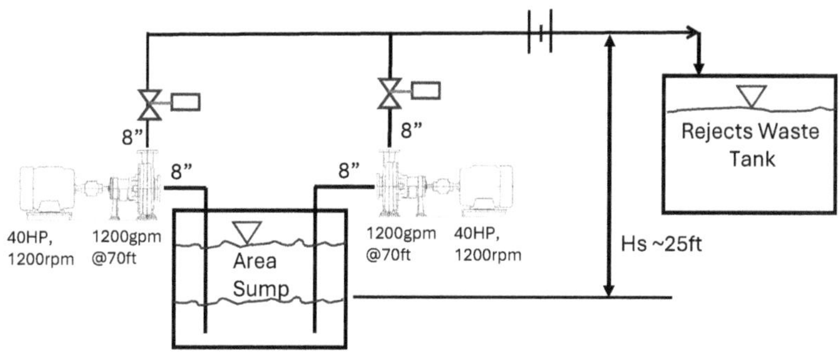

FIGURE 10.15 Process diagram for sump pump system.

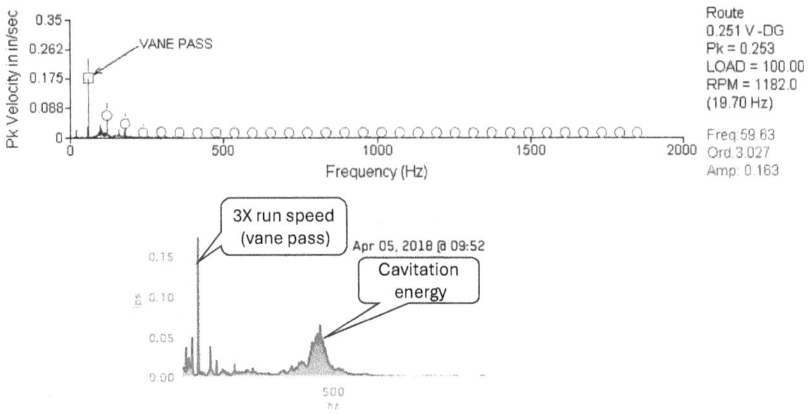

FIGURE 10.16 Typical vibration signature velocity spectrum.

An evaluation of the hydraulic condition of the pump and system was the next thing executed. The sump levels for operation had the pump operating between 30 and 45%, which was low and created several issues. First, the pump's minimum submergence was around 42″, while the low level provided about 24″ of submergence, leaving it about 18″ short. The 15% sump level indicated the pumping duration, resulting in about eight starts per hour, which is the maximum allowed for a 40 HP, 1,200 rpm motor per NEMA.

Pressure gauges were installed on the pump to measure the operating head. The pump TDH ended up at ~25 ft compared to an original design of 70 ft resulting in a flow of ~2,200 gpm. With a static head of 15 ft, this left about 10 ft of friction head in the system. The pump system was a very short run of piping and a very small static head to a nearby wastewater tank, so the low head made sense. There must have been some errors in the original engineering for this pump system.

The NPSHA was calculated at ~24 ft. The NPSH3 for the current operating point was ~23 ft. When the suction energy and NPSHmargin of 1.2 was added, the NPSHR is ~28 ft so cavitation is a potential issue as the NPSHR > NPSHA which matches up with vibration indications.

With the pump head, the system curve, and pump curve were evaluated as shown in Figure 10.17. The pump had an operating point well outside the AOR, as it operated in a high flow condition off the end of the curve. This also has a steep increase in NPSH3 at the high flow. One of the first things to correct on the pump system is get the pump operating back within the AOR. Since the system has low head, the TDH (friction head) must be increased to operate in the AOR. One option was to install a system orifice in the discharge line (don't ever install an orifice or pressure drop element in a pump suction line). A 5.25″ orifice would increase the head to ~50 ft and get the pump back within the curve AOR. This also lowers the NPSH3 down to about 15 ft, so now the NPSHR is at 18 ft. The new pump condition would have NPSHA > NPSHR, so the cavitation potential is low.

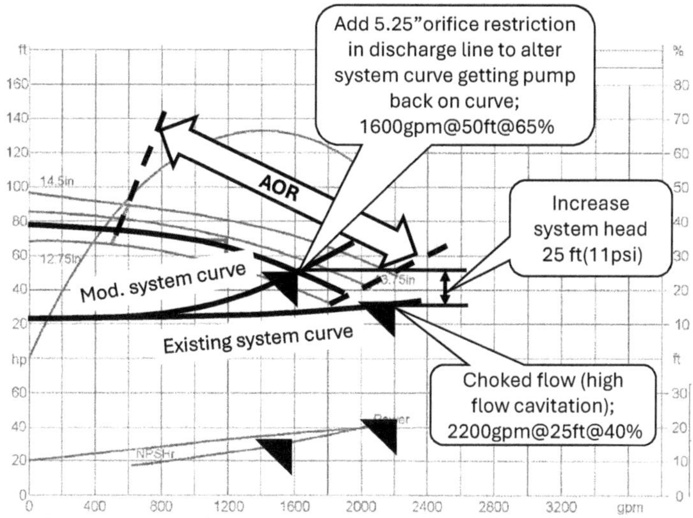

FIGURE 10.17 Pump and system curves for sump pump system.

Since the system head is low, a look at the impeller size was also evaluated. The 13.75″ impeller was not required, so a 13″ impeller was selected. This lowers the pump head on the curve by another 10 ft, better matching the system and moving the operating point further into the AOR and within the POR. With this move, the pump flow decreases which isn't a big issue on the batch process level control. It just means the pump will run a little longer. Combine this with adjusting the sump level controls so the pump starts at 55% and stops at 35% results in around four starts per hour, which is well below the NEMA maximum of eight starts per hour. The submergence also lowers to 34″ while increasing the minimum sump level, ensuring that the required submergence is also maintained.

A summary of some of the pump improvements are shown in Figure 10.18. Although the TDH increased with the smaller pump impeller, the improved operating point in the POR and higher pump efficiency led to a slight reduction in BHP of about 4 BHP. As a result of these improvements, the MTBF increased to over three years. Figures 1.1–1.4 from Chapter 1 shows the remarkable improvement in repair work orders from this pump and system improvements in MTBF after pump and system modifications were completed. The pump also saw a big improvement in pump vibration. The vane pass velocity amplitude dropped from 0.24 in/s to 0.02 in/s (92% reduction) and the overall pump vibration dropped from 0.276 in/s to 0.161 in/s (40% reduction).

There were a few other system improvements that were also done. The sump pumps had air operated on/off gate valves on the pump discharge. These were removed and replaced with swing check valves. The on/off valves had

Parameter	Existing	Proposed	Differential
Impeller Diameter	13.75"	13"	-.75"
Operating Point - Flow	2200 gpm	1600 gpm	-600 gpm
Operating Point - Head	25ft	50ft	+25ft
NPSHA	24ft	24ft	-
NPSH3	23ft	15ft	**-8ft**
NPSHR (NPSHmargin=1.2)	**28ft** **NPSHA,** **24'<NPSHR, 28'** **Cavitation**	**18ft** **NPSHA,** **24'>NPSHR, 18'** **Low cavitation**	**-10ft**
BHP (Energy Efficiency) Only 16 minutes operation/hr	35 (40%)	31 (65%)	$400 annual savings
MTBF (annual cost)	12 months	3-5 years	$8,000

FIGURE 10.18 Summary of pump and system hydraulic conditions.

maintenance issues with proximity switches, controls, and air leaks. This eliminated another maintenance issue from the system. The swing check valves lasted almost eight years before they were replaced, indicating another good reliability result.

Switching to a 13″ impeller standardized this pump with another spare from a different sump system, allowing one spare to be eliminated from the storeroom. Only a single 13″ impeller pump is now available for both locations.

The self-priming lift pumps did not have an air purge line on the discharge to help purge air out of the system. This can be left on all the time as it is a small flow or opened as needed on pump startups if priming is taking a long time. The sump had some buildup, so it was scheduled to be cleaned out and placed on an annual sump cleaning schedule. The rotating assembly vent filter was upgraded from a 40 micron to a 10 micron to reduce contamination.

The main issue causing pump failures was pump system design as the system did not match the pump to allow the pump to operate within the AOR.

Case Study 54 Learning – Match pump with system curve so that pump operates within the AOR for good pump reliability. Modifying the system is one way to get the pump operating in the pump AOR.

The next case study had similar issues to the last case study, but the solution ended up being much different. A good reminder that problems and solutions are specific to each pump system and depends on how it operates.

CASE STUDY 55 PUMP OPERATION OUTSIDE AOR (HIGH FLOW) – SPEED CHANGE

Another sump pump system was experiencing very low reliability, high vibration, and various system failures. The pump was a 6×6×11.5 self-priming lift pump operating in a batch-type sump operation. The pump would cut on and off as the level reached a high and low. The pumps discharged into the same piping system. If the sump level reached a high level, the second pump would start, allowing parallel operation to lower the sump level.

The pump had a 11.5″ diameter, two-vane impeller that operated with a direct drive 40 HP, 1,800 rpm fixed speed motor. Neither the original pump performance curve, the process flow sheet nor the process and instrument diagram (P&ID) gave a design operating point for this pump system. An ideal design point would be 1,100 gpm at 78 ft. The pump had a NSS = 7,115. The process diagram is shown in Figure 10.19.

The pump system typically had elevated vibration and poor reliability. The failure mode would develop in this order. High vibration and pump conditions would cause the mechanical seal to fail and start leaking fluid into the pump power end. The self-priming pump leaked internally and was not obvious that the seal had failed by the external observer. After operating for a while with water in the rotating assembly and bearings, the bearings would begin to fail. At that point the assembly would begin to experience catastrophic failure. At the later stages of pump failure, the vibration was more severe. There had been times where pipe failures also occurred likely due to operating in excessive vibration. Base bolts have also gotten loose at times.

The vibration signature during the onset of this failure is shown in Figure 10.20. The vibration signature at this point had begun to fail the bearings and rotating assembly starts to show mechanical looseness.

The piping system only had ~25 ft of total static head. Discharge pressure readings revealed only ~15 ft of friction head on the system so the pump

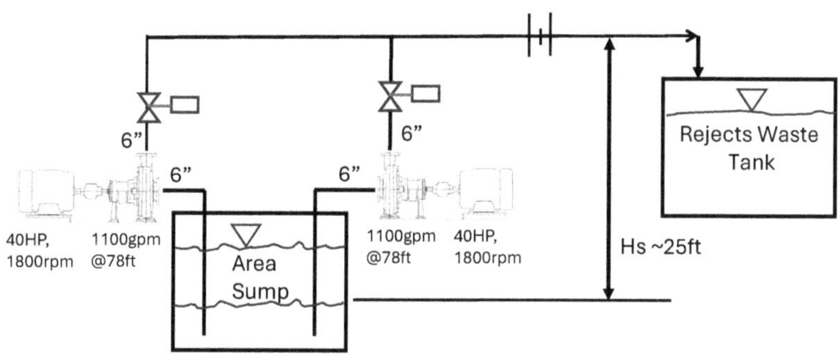

FIGURE 10.19 Process diagram for sump pump system.

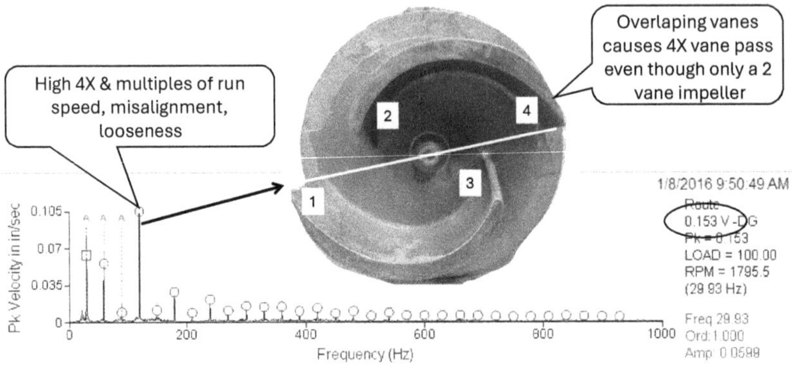

FIGURE 10.20 Vibration signature velocity spectrum for typical failing pump.

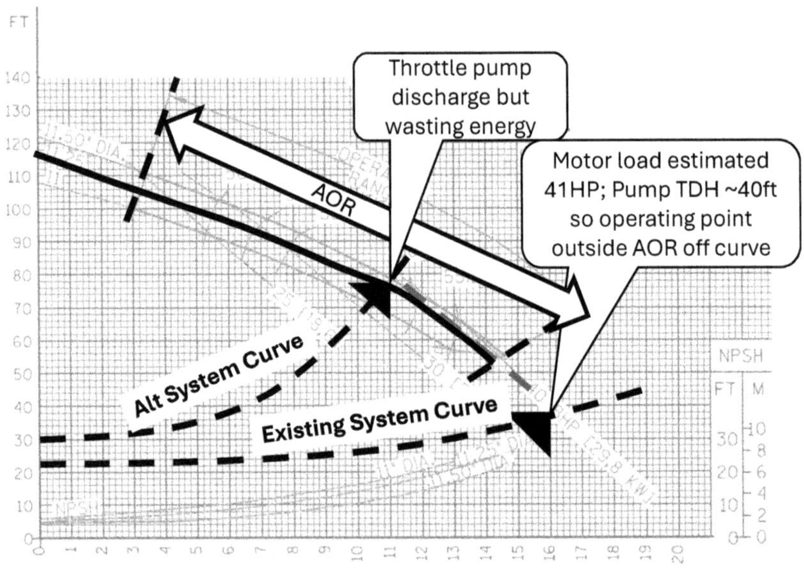

FIGURE 10.21 Existing pump and system curves at 1,800 rpm.

system TDH was at ~40 ft. Evaluating the motor load and utilizing the system pressure and head conditions, an accurate estimate of the pump operating point was determined. Looking at the existing pump and system curve in Figure 10.21, the pump clearly is operating outside the AOR. With this condition, there is high pump vibration, some likely cavitation, and shaft deflection that is causing the poor pump reliability. One quick solution to the operating condition would be to increase the system head similar to the last case study so the pump is operating within the AOR.

To get the pump and system in a good fit either the pump can be modified or the system. For the system modification, this could be done by installing an orifice or throttling a manual valve on the pump discharge. However, doing this would drop the process flow from ~1,600 gpm to ~1,100 gpm.

To consider any alternate pump or system modification to correct the pump operating outside the AOR, the question becomes, will the modification support the process to maintain level in the sump? To determine this, a mass balance must be completed and understood. In evaluating the sump system in a batch on/off operation, the pump must discharge more than the inlet flow into the tank in order to pump the level down. A new system operating point at a lower flowrate (to get pump operating back within the AOR) must be carefully evaluated. How much the pump pumps will determine how long the pump runs and how many times it comes on/off in every hour.

The sump fill rate was determined by measuring the trench flow into the sump. Using the weir flow formula or table shown in Figure 10.22, the typical weir flow into the sump was ~450 gpm. While there are times when higher flow conditions may arise, the typical or average flow was used based on random sampling. The sump pump system is configured so that the second pump comes on in parallel if a high, high level reaches and the higher friction head on the system improves the pump operating point closer to AOR.

$$Q = 3.33 \times (L - 0.2 \times H) \times H^{1.5}$$

where:

Q = ft^3/s water
L = length of weir, ft (typical L = 4–8XH)
H = head on weir, ft (measured 6 ft back from opening)

The fill time is simply the operating volume in the sump divided by the fill rate. With the current operating levels and sump operating volume of 1,050 gallons and 450 gpm fill rate, the fill up time is 2.3 minutes. For the proposed condition with a larger sump operating volume of 1,950 gallons, the fill time is 4.3 minutes.

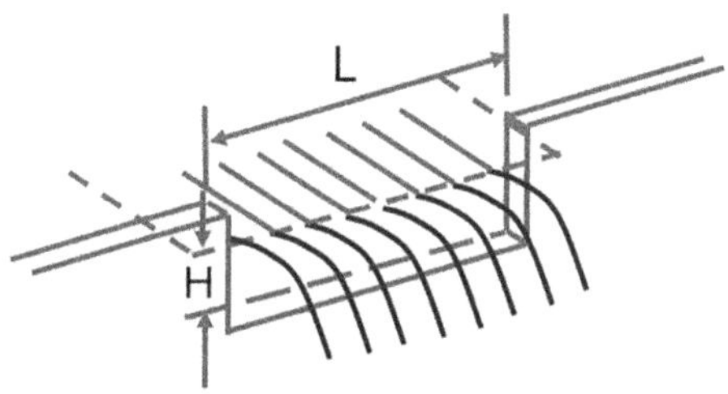

FIGURE 10.22 Weir flow rate calculation.

The pump out time depends on the pump flow, sump inlet flow, and how much volume is removed from the sump. The pump out time can be calculated by the formula below. For the current sump condition which has a very high pump flow and low sump volume reduction, the pump out time is very low at 0.9 minutes. With these variables, adjustments can be made to try and attain the pump starts per hour below the NEMA allowable for this pump system.

$$PT = \frac{SV}{Q - W}$$

where:

PT = pump out time, min.
SV = sump volume being pumped out, gallons.
Q = pump flow, gpm
W = Sump inlet (weir) flow, gpm

The pump motor starts per hour is not so much a factor with the pump reliability but is a factor with the motor reliability as inrush current can build up excessive heat in the motor. Figure 10.23 shows the NEMA allowable motor starts per hour for NEMA A & B motors. For the current sump operating conditions, the pump motor starts 19 times per hour, which is well beyond the allowable of 7.4.

	2 POLE			4 POLE			6 POLE		
	A	B	C	A	B	C	A	B	C
1	15	1.2	75	30	5.8	38	34	15	33
1.5	12.9	1.8	76	25.7	8.6	38	29.1	23	34
2	11.5	2.4	77	23	11	39	26.1	30	35
3	9.9	3.5	80	19.8	17	40	22.4	44	36
5	8.1	5.7	83	16.3	27	42	18.4	71	37
7.5	7.0	8.3	88	13.9	39	44	15.8	104	39
10	6.2	11	92	12.5	51	46	14.2	137	41
15	5.4	16	100	10.7	75	50	12.1	200	44
20	4.8	21	110	9.6	99	55	10.9	262	48
25	4.4	26	115	8.8	122	58	10.0	324	51
30	4.1	31	120	8.2	144	60	[illegible]	384	53
40	3.7	40	130	7.4	189	65	[illegible]	503	57
50	3.4	49	145	6.8	232	72	[illegible]	620	64
60	3.2	58	[illegible]	6.3	275	85	[illegible]	735	75
75	[illegible]	[illegible]	[illegible]	[illegible]	338	[illegible]	[illegible]	[illegible]	[illegible]
100	[illegible]	[illegible]	[illegible]	[illegible]	441	[illegible]	[illegible]	[illegible]	[illegible]
125	[illegible]	[illegible]	[illegible]	[illegible]	542	[illegible]	[illegible]	[illegible]	[illegible]
150	[illegible]	[illegible]	[illegible]	[illegible]	640	[illegible]	[illegible]	[illegible]	[illegible]
200	2.0	172	600	4.0	831	[illegible]	[illegible]	[illegible]	[illegible]
250	1.8	210	1000	3.7	1017	500	4.2	2744	440

Original Design 19 starts per hour > 7.4 NEMA limit

Lower speed pump, higher starts per hour

A = Maximum number of starts per hour.
B = Maximum product of starts per hour times load Wk^2.
C = Minimum rest or off time in seconds between starts.

FIGURE 10.23 NEMA allowable motor starts per hour.

Another good solution is to modify the pump by slowing the pump down. A 1,200 rpm motor is the next standard motor speed which could improve all the design issues with this pump and motor system. We could also adjust the sump pump operating window and increase the operating volume of the sump from the 40–60% operating range to 25–80% range. This will allow the pump to operate longer.

A lower speed pump will mean lower flow and lower BHP required. Using affinity laws, we can estimate the pump BHP and flow which was estimated to be less than 15 HP and 1,000 gpm. Keep in mind this is just an estimate and not as accurate as the actual pump and system curves. Figure 10.24 shows the actual pump and system curves for both the 1,800 rpm and 1,200 rpm pumps. The maximum flow is around 850 gpm, and the BHP is around 12.5 HP.

A 15 HP motor would give full curve runout at 1,200 rpm. Looking at the existing 40 HP, 1,800 rpm motor shows that it is a 324T NEMA motor frame size. Since a 25 HP, 1,200 rpm motor is also a 324T frame motor, this would be a good choice for this pump system as no modifications would be required for the motor/speed change and it also satisfies full curve runout BHP. The 25 HP motor is also a more standard and available motor size from the plant storeroom.

A summary of the existing pump system operating conditions and proposed conditions regarding the sump pump flow, pump time, sump fill time, and motor starts/hr are shown in Figure 10.25. Sump conditions on the left are current conditions, while the conditions on the right are the proposed conditions. Modifications to the pump alone will not get the motor starts per hour within NEMA allowable. This is not achieved until the operating volume of

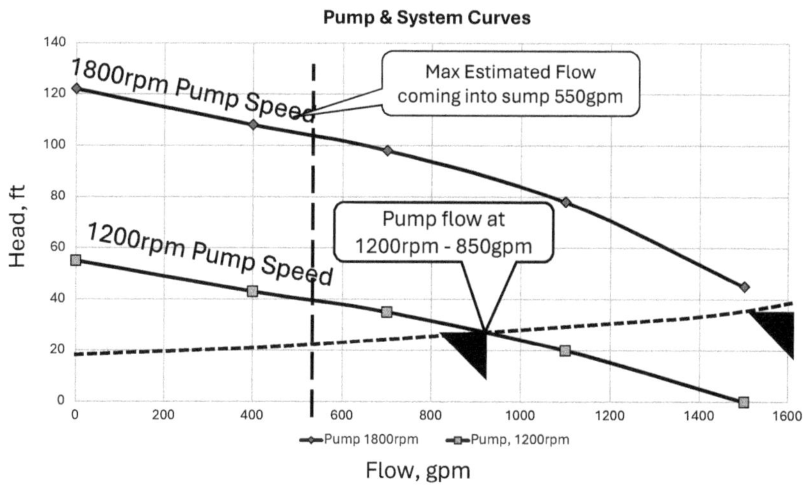

FIGURE 10.24 Proposed pump and system curve with new operating point.

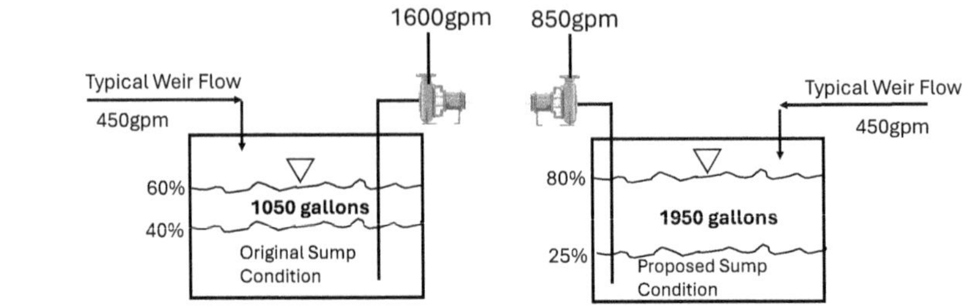

Original Sump Condition:

	Pump Flow, gpm	Pump Time, minutes	Fill time, minutes	Motor starts/hr
Current Flow	1600	0.9	2.3	**18.8**
Throttled Flow	1100	1.6	2.3	**15.3**
Lower Speed Flow	850	2.6	2.3	**12.2**

Proposed Sump Condition:

	Pump Flow, gpm	Pump Time, minutes	Fill time, minutes	Motor starts/hr
Current Flow	1600	1.7	4.3	**10**
Throttled Flow	1100	3	4.3	**8.2**
Lower Speed Flow	850	4.9	4.3	**6.5**

FIGURE 10.25 Summary of sump pump process conditions for current design and proposed design.

the sump is increased by changing on/off sump levels. The throttled flow condition is also summarized.

Other impact of executing the lower speed pump is better NPSH conditions, lower friction head (~10 ft), and improved pump efficiency (~58% versus ~40%). While the same volume of fluid must be pumped from the sump, the efficiency in which it does so has been improved which leads to some energy savings (~6 BHP, ~$2,200/yr) as a bonus.

Another example is a pump system designed to operate outside the AOR, which led to poor reliability. When a pump system, or any system, is designed with inherent unreliability, no amount of maintenance or operational effort can improve its performance. This problem was difficult to determine if it was a pump problem or system problem. Changes in either can solve the problem. The pump ended up being the better choice for modification.

Case Study 55 Learning – Match pump with system curve so that pump operates within the AOR for good pump reliability. Modifying the pump is another way to get the pump operating in the pump AOR.

The next case study will also show another outcome of operating the pump outside of AOR and some other factors in pump design which can create pump system unreliability.

CASE STUDY 56 PUMP OPERATING OUTSIDE AOR (LOW FLOW) – SHAFT FRACTURE

A paper machine shower pump system had a catastrophic failure where the pump lost flow and pressure. Further investigation revealed that the pump shaft had broken between the opposite drive end (ODE) bearing and the impeller (near the impeller fit on the shaft). The location of the shaft fracture is shown in Figure 10.26. The shaft fracture was at the shaft step where the impeller is installed on the shaft and starts to enter the stuffing box.

The shaft fracture failure mode was defined as rotating bending fatigue with severe stress concentration, as shown in Figure 10.27. Bending fatigue started at the shaft step, stress concentration as noted by the ratchet marks around the

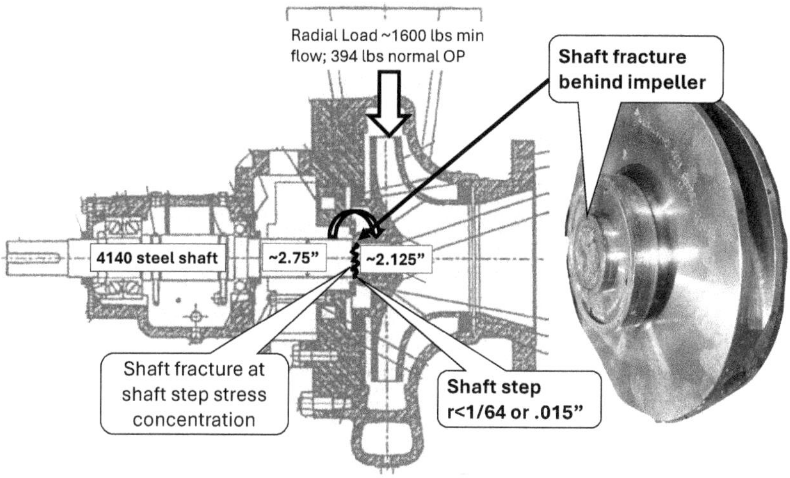

FIGURE 10.26 Pump shaft fracture location.

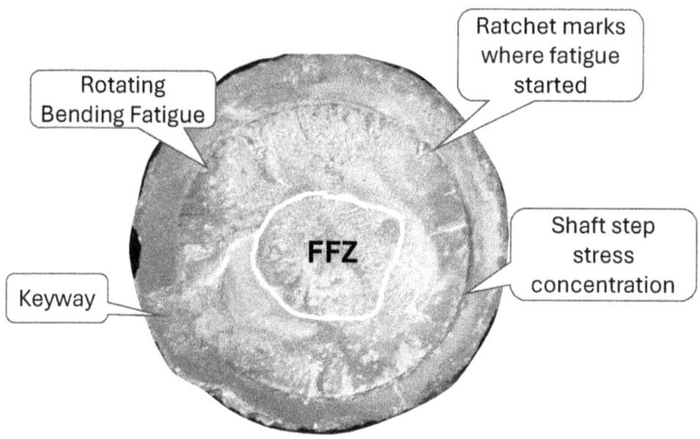

FIGURE 10.27 Shaft fracture surface fractography.

fatigue surface. The fast fracture zone (FFZ) was only ~30% of original diameter, so the fracture stress was not excessively high at the time of fracture. The shaft step was verified with a radius gauge and appeared to be 1/64″ (~0.015″). The original pump shaft design shaft step was a 0.031″ radius (1/32″) instead of the 1/64″.

The other interesting observation was the fact that another area where shaft cracking was initiated was at the corner of the keyway radius, as shown in Figure 10.28. Also, the keyway corner stress concentration intersected with the shaft step stress concentration which really made the shaft vulnerable to a fatigue crack initiating. One easy design fix would be to shorten the shaft keyway by 0.200″ so it is no longer in the shaft step area. Also making the shaft step one smooth undercut radius instead of the compound radius creating a higher stress concentration.

Another discovery was the surface appeared to have some machine marks at the shaft step radius. This will decrease the surface fatigue factor and thus reduce the endurance limit of the material. It was estimated a potential of 10–15% reduction based on the roughness of the shaft that fractured. The shaft machine surface should be ground and polished with no more than 32Ra surface roughness, preferably less.

While there were issues with the shaft design and even manufacture of it, this had been the case for many years and no shaft fractures, so these findings alone did not cause the failure but were certainly a contributing root cause. For fatigue failures there are two areas which usually lead to failure – part design and cyclic force. The other question is, what was the event(s) that increased the cyclic forces on the pump shaft that when combined with the shaft defects caused the rotating bending fatigue failure?

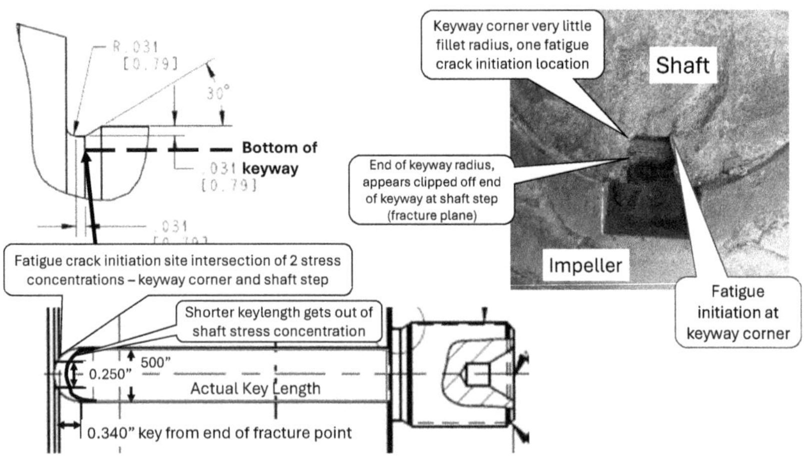

FIGURE 10.28 Shaft fracture location detail.

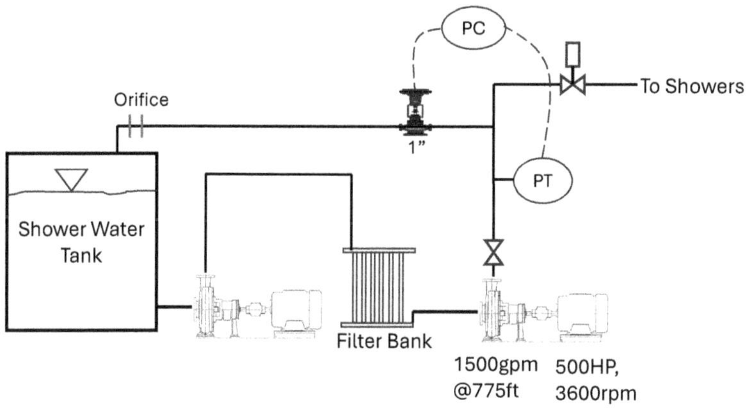

FIGURE 10.29 Process diagram for shower pump system.

This pump system consists of an end suction, overhung pump with enclosed impeller. The pump was a 6×8×16 size API-style designed pump driven by a 500 HP, 3,600 rpm direct drive fixed speed motor. The pump had an original design point of 1,550 gpm at 850 ft with a 14.5″ diameter impeller. The pump has a NSS = 11,225. The pump had some previous hydraulic re-rate on the impeller and was operating a 14″ diameter impeller. The pump was operating a shower system on the paper machine and was fed by another feed pump in series. The system had a recirculation pressure control valve with a flow orifice for minor control of header pressure as shown in Figure 10.29.

The system had been operating reliably for many years with no major failures of any kind. The API design had the pump casing with center mount so thermal growth was negligible as well as other improvements from the API standard. With the 14″ diameter impeller, the pump's typical operating point was in the ~1,800–2,050 gpm flow range, keeping the pump operated within the AOR. The lower flows were even in the POR range for the pump.

However, there was another operating condition that was more recent and intermittent that was learned of during the failure investigation. The shower water on/off valve in the system had experienced a failure of the machine controls which inadvertently was shutting this valve with the pump still running. The pressure recirculation loop was not sized to handle large flows for this condition and the result was the pump operated well outside the AOR on the low flow end at ~250 gpm, as shown in the pump curve in Figure 10.30. This resulted in very high shaft radial loads. For relative comparison, the estimated radial shaft load at POR operating point was ~400 lbs versus ~1,800 lbs at the low flow condition.

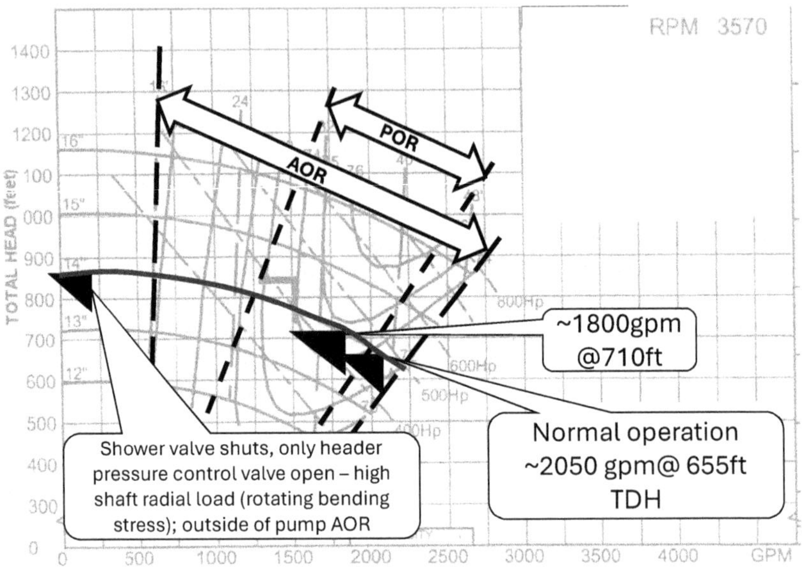

FIGURE 10.30 Pump and system curve with operating points.

This high cyclic load combined with the shaft design defects raised the actual shaft cyclic stress above the adjusted endurance limit for the shaft and rotating bending shaft fatigue was the result.

The physical root cause of the shaft fracture was due to two issues. First there was the shaft design condition. A stress concentration designed at the intersection of the shaft step (also shaft radius was smaller than design target) and keyway radius. The surface finish also contributed to surface fatigue factor reduction. This led to a lower fatigue endurance limit for the shaft. Another contributor was corrosion on the 4140-carbon steel shaft material. The second condition that was also part of the root cause was the pump operating outside AOR. This resulted in a high shaft radial bending stress which when combined with the shaft stress concentration produced a fatigue stress greater than the endurance limit for the shaft.

There were numerous potential action items to correct the failure. Some of those are summarized below.

- Increase shaft radius to ~1/16″ (0.062″) undercut from the original design of 1/32″ (0.031″) while removing the compound radius. A potential of ~25% fatigue strength improvement.
- Shorten the shaft keyway length ~0.200″ so keyway corners do not intersect with shaft step stress concentration.

- This is probably not necessary but could be done for an insurance design feature. Surface harden shaft (compressive surface stress) carburize or shot peen. 25% fatigue strength improvement.
- Modify shower controls so that if the shower valve shuts, the pump shuts down and does not run outside of AOR causing high radial shaft loads.

This pump failure had several layers as to the root cause category. It took all of them to be true for the failure to occur. First, there was the shaft design weakness or flaw which made the shaft vulnerable to fatigue. Then there was the secondary failure of a valve in which the pump operated outside the AOR at certain points which added the higher cyclic stress to the flawed shaft design. Then there was the risk or lack of attention to correct the valve issue by operations and continue to operate this way for an extended time. There were pump-specific issues and pump system issues which combined to cause the component failure.

Case Study 56 Learning – Failure may not occur on some design flaws until an additional pump operating flaw occurs. For some failures there are multiple root causes which must coexist to cause pump failure.

Sometimes there are clear warning signs of a serious pump system issue before functional failure occurs. These may be missed many times but if condition monitoring is functioning properly and maintenance and reliability professionals are paying attention, this can be corrected before failure. The next case study shows one example of this.

CASE STUDY 57 PUMP OPERATING OUTSIDE AOR (LOW FLOW) – HIGH VIBRATION

A pulper pump system was experiencing system issues such as pump and piping vibration, along with pipe and valve failures. The pump was an $8 \times 10 \times 19$ end suction pump with a 17.8″ diameter impeller driven by a 150 HP, 1,200 rpm fixed speed motor. The pump had a NSS = 9,379. The pump vibration velocity spectrum had an overall vibration of 0.309 in/s with vane pass being a dominant frequency (5X run speed with five-vane impeller) with a 0.17 in/s amplitude as shown in Figure 10.31. Component looseness was also high in the spectrum as the discharge valve handle nut had come loose and the valve handle was vibrating severely with pump in operation.

The machine pulper pump system operated all the time the pulper was in operation. The pump system had two discharge-level control valves operating in a split control arrangement, as shown in Figure 10.32. The pump system had

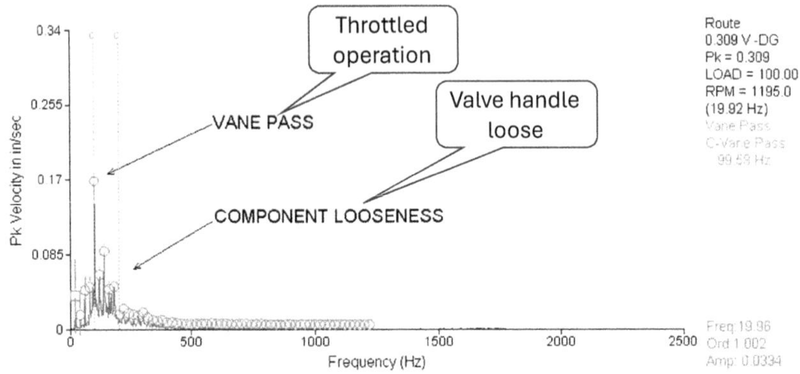

FIGURE 10.31 Pump vibration velocity spectrum.

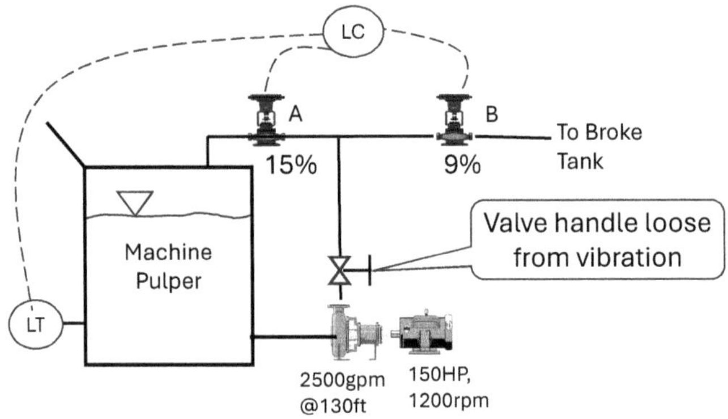

FIGURE 10.32 Process diagram for pulper pump system.

two main conditions for which to operate. One was a machine break where maximum pump out capacity was required. In this case, valve B would open more, and valve A would close more to pump the most volume from the pulper tank to maintain level. Normal running condition only had a little trim going into the pulper so recirculation mode with valve A opening more (typical was 50%) and valve B would be slightly open to maintain level and pump the trim out.

An understanding of the pump operating point was investigated at this point. The original design point was 2,500 gpm at 130 ft. There wasn't a pressure gauge on the pump so motor load was checked and the motor was pulling 86 amps at 475 volts. Motor nameplate full load amps, FLA, was 171 for the 150 HP motor. Nameplate motor efficiency was 95.8 and power factor was 85.5; however, the motor was only ~50% loaded so a reduced power factor was used of 0.68 to calculate motor HP. The existing motor BHP calculated to be ~62 HP. At the original design point, the BHP was ~105 HP. Looking at the pump curve in Figure 10.33, the current operating point is

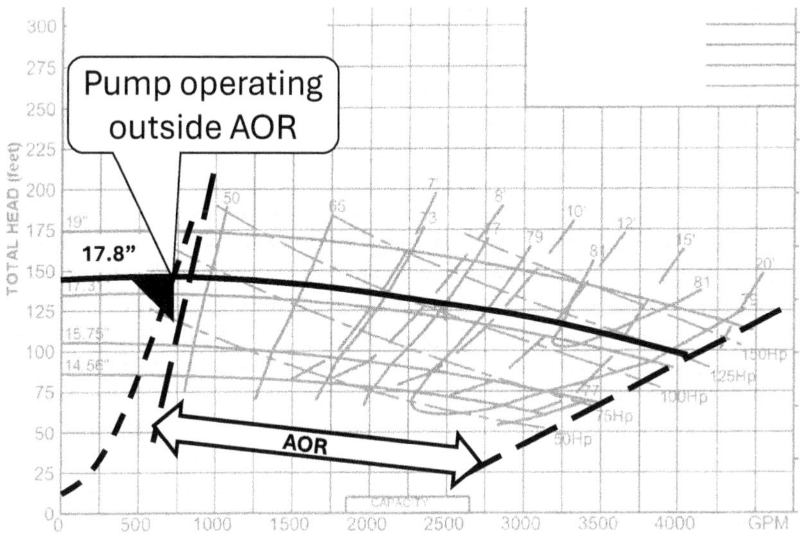

FIGURE 10.33 Pump and system curve with operating point.

only operating at ~750 gpm at 148 ft, which is outside the AOR for this pump. This operating point will cause the pump to have suction recirculation and high shaft loads causing shaft deflection which is likely the source of the high pump system vibration. What is causing this operating point?

Upon inspection of the pump system, both level control valves (A&B) were operating in a very closed position (A was 15% open, B was 9% open) in normal operating mode which was about 95% of the time with the machine only running trim to the pulper. Further investigation found the A valve was being operated in manual at 15% with the B valve in automatic.

Further questioning on the system operation revealed that there was a previous failure on the recirculation line back into the pulper which had not been repaired. The pump system was unable to operate in automatic due to the recirculation line pipe failure. Operating personnel were not aware that the new operating condition was creating reliability issues on the entire pump system. Once the recirculation line was repaired, normal pulper pump system operation resumed.

The main root cause of the pump issues was operation of the pulper pump system outside of AOR. Like some other case studies, people making the decisions on these did not understand the impact of operating in the alternate operating point. This is where it is good to have a management of change process so that knowledgeable resources can vet the process change before executing to avoid negative consequences. Deferred maintenance on the piping resulted in more needed maintenance on the rest of the pump system.

Case Study 57 Learning – Understand pump system and don't operate pump outside AOR for permanent system operation.

The next case study will look at some other reliability issues related to operating the pump outside the AOR. There were also many other elements of the pump mechanical condition which also added to the reliability challenges.

CASE STUDY 58 PUMP OPERATING OUTSIDE AOR (LOW FLOW) – LOW SEAL LIFE

A shower pump system was having several reliability issues that were leading to short pump life. Some of the issues consisted of mechanical seal failures, misalignment, high vibration, coupling failures, looseness, and unbalance. A typical vibration signature is shown in Figure 10.34, indicating some misalignment and high 1X vibration. From the vibration signature shown, a weld was found broken on the baseplate. The baseplate was a fabricated skid with only a piece of channel iron for the motor base. The channel was welded to a prefabricated skid and the weld had broken on one end. The pump system was a small unit but still lacked enough structural stiffness for the application.

The pump was a $1 \times 2 \times 10$ end suction low flow pump with a 7.75″ diameter radial vane impeller. The design point was 30 gpm at 300 ft. The pump was driven by a 15 HP, 3,600 rpm direct drive fixed speed motor. The pump had a NSS = 553. The pump system consisted of a pump and an installed spare being fed from a small tank where water and some chemicals were mixed. The pump had a couple of discharge filters and fed a chemical shower. A recirculation control valve was used to maintain constant pressure to the shower as shown in Figure 10.35.

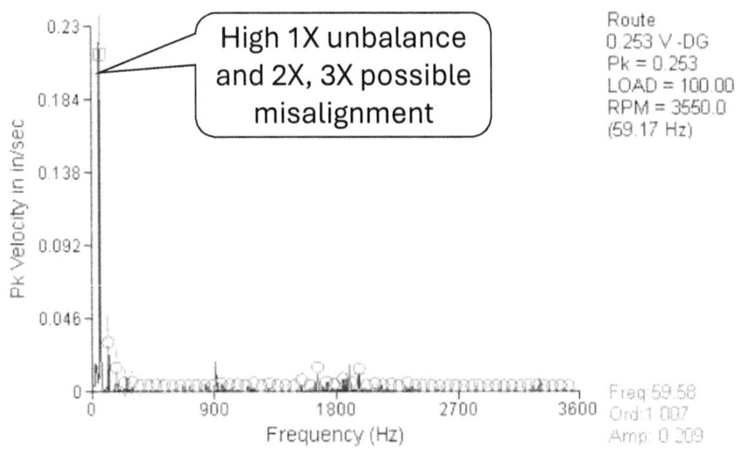

FIGURE 10.34 Pump vibration signature velocity spectrum.

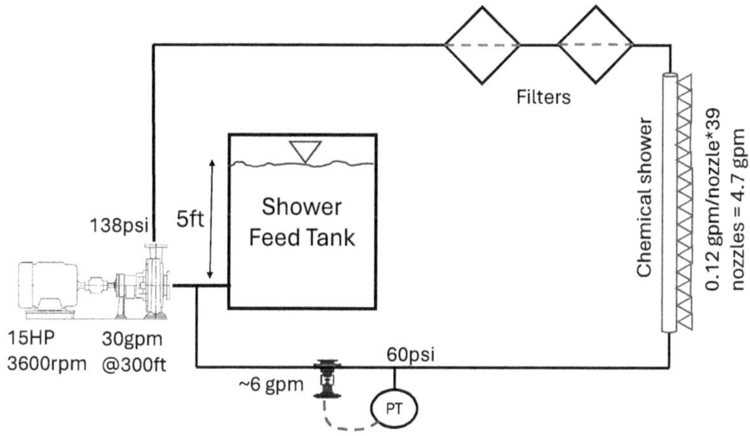

FIGURE 10.35 Process diagram for shower pump system.

Looking at historical pump maintenance records, the mean time between failure, MTBF, for the pumps was ~9 months. Most of the time there were seal failures or changing the pump for various issues. Sometimes it was a proactive move since the pump was apart due to unbalance and cleaning the impeller from buildup. The fluid would tend to buildup unbalanced mass on the impeller on occasion which required cleaning. Since there was an installed spare pump, each pump only operated half the year. The MTBF considering run time is more like 4.5 months.

Some of the short-term issues that were addressed were alignment errors, cleaning the impeller, correcting loose baseplate, and seal water hookup. Impeller buildup and loose baseplate were obvious issues found on initial inspection after seeing elevated vibration. The alignment error was found during pump repair after further questioning of precision maintenance plans. The pump laser alignment was completed, but the wrong reference speed was input into the laser – 1,800 rpm was entered instead of 3,600 rpm. The acceptable alignment tolerance was widened, resulting in a final alignment that was not as good as it should have been. The seal water was hooked up to the pump discharge as a product flush but there was no valve or pressure gauge to regulate the seal water pressure to the seal chamber.

The actual pump operating point was checked to assess the risk and potential effect on mechanical seal reliability. The pump TDH was measured to be ~316 ft and the motor load was also verified at ~10 HP. A mass balance was also verified using the shower nozzles and flow for the shower at ~5 gpm and the recirculation flow of ~6 gpm. The estimated operating point from three sources put the pump operating point all in the same general area on the pump curve as shown in Figure 10.36. The actual pump operating point was right at the low flow AOR limit for this pump and well off its original design target operating point.

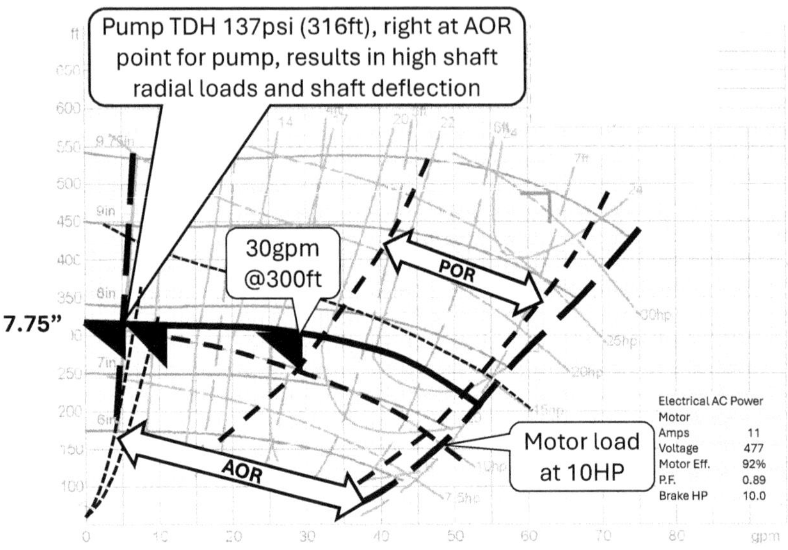

FIGURE 10.36 Pump and system curves with operating points.

At this operating point, there would be significant shaft forces and shaft deflection which would decrease mechanical seal reliability especially with this small pump which had a low shaft stiffness. A couple of options for improving the pump to improve mechanical seal reliability were available. First is to improve the pump shaft stiffness. The pump originally had a shaft sleeve (1.125″ shaft diameter) with a resulting shaft stiffness of ~144, this is much higher than desired especially running at the low flow AOR point. The shaft stiffness is based on the shaft diameter not the sleeve diameter. Converting the pump to a solid shaft (1.375″ shaft diameter) will improve the shaft stiffness to ~64.

Another potential easy system modification is to lower the header pressure setpoint slightly and recirculate more flow to get the flow near 30 gpm well inside the AOR and near the POR. This may require a larger control valve. Probably the main drawback on this option is the high flow added back to the pump suction. Changing back to the tank might be a better option at that high recirculation flow. The curve is very flat so just 3–4 psi lower pressure will accomplish the improved system curve. While this small amount should not greatly impact the shower operation, it will greatly affect the pump operation.

The last option was to install a different pump – 1 × 1.5–8 which will operate inside the current low flow operating point within the curve AOR, as shown in Figure 10.37. Also, if the flow is increased, the operating point will now be inside the POR for that pump.

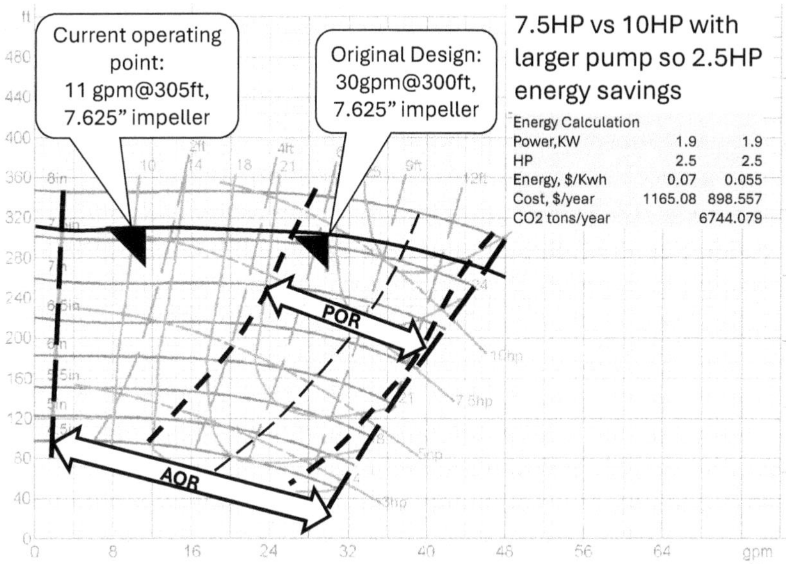

FIGURE 10.37 Proposed pump curve with new operating point.

A couple of other project-type improvements to this pump system were to relocate pumps to a new location to improve pump maintainability. The original pumps were located in a very tight access position which made service and alignment on the pumps very difficult. Also, the original pump skid did not have the foundation mass that it needed to properly dampen and ground the vibration.

Many of the pump reliability issues originated from design issues. The pump operating just at the edge of the AOR resulting in higher shaft loads coupled with using a shaft sleeve with lower shaft stiffness. Poor baseplate design and poor pump maintainability has also greatly affected the overall reliability of this pump system.

Case Study 58 Learning – Pump systems can have multiple issues that contribute to overall system reliability. Take a holistic approach to solving pump system problems.

One last case study showing the impact of operating the pump outside the AOR. This case is for a double suction pump instead of an end suction pump like the previous case studies. Double suction pumps can have some unique reactions to this operating point.

CASE STUDY 59 DOUBLE SUCTION PUMP – CHRONIC MECHANICAL SEAL FAILURE

A 14 × 16 double suction, single stage, horizontal split case pump was supplying dilution water to a headbox for consistency profile control. The pump was operated by a VFD on flow control and fed a screen before supplying dilution to the headbox on a paper machine. The process diagram is shown in Figure 10.38. The pump had a 15.5″ diameter impeller with a design point of 4,905 gpm at 275 ft. The pump was driven by a 500 HP, 1,800 rpm VFD motor. The pump had a NSS = 12,578.

The pump was originally supplied with packing for shaft sealing. Routine packing adjustments and packing replacement had gone on for ~10 years before an upgrade was executed to convert the pump sealing to a split mechanical seal to eliminate water to the floor. After converting to a split mechanical seal, the seal had leak issues in less than a month. A split seal was used for ease of maintenance and access not having to change bearings and limit disassembly of the pump for regular seal maintenance or replacement. The seal continued to have issues over the next nine years with a MTBF of ~1.1 years for having seal leaks in various forms.

The double suction pump had two to eight staggered vane impellers. The vibration signature showed the 8X and 16X vane pass vibration peaks. It also showed a high 4X peak as shown in Figure 10.39. It was not exactly clear the

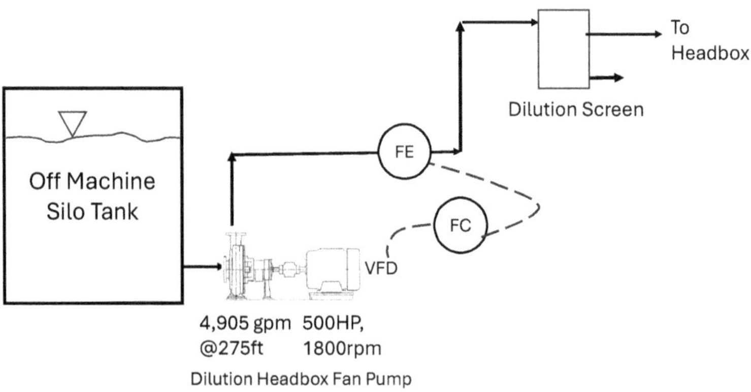

FIGURE 10.38 Process diagram for dilution pump system.

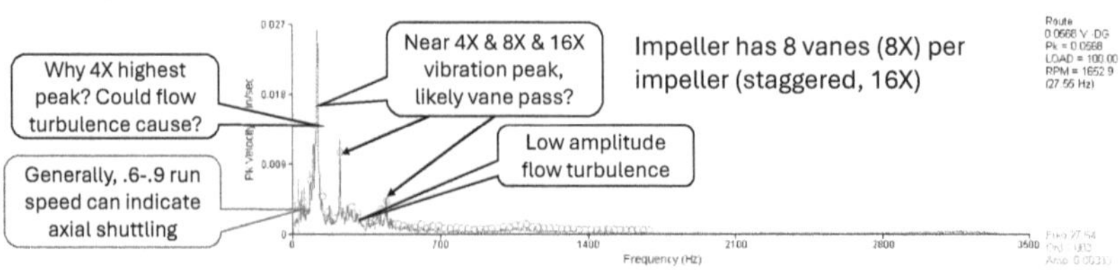

FIGURE 10.39 Typical vibration signature velocity spectrum.

reason for the high 4X peak versus the regular vane pass peaks. Raised floor indicating some flow turbulence was also present which covered any other small defect indications on the spectrum.

The pump had shown various types of mechanical seal failure such as springs were broken, scoring on the seal faces, leaking at the split joint area and several undocumented seal failure modes. One more important observation of the seal revealed that there had been some axial thrust that had marked one of the seal areas where it should not have had any markings. Figure 10.40 shows where the thrust showed on the rotating element which would open up the seal faces to leak. On a following outage, the drive end bearing was verified to have the retaining ring to make it the correct held end so that was ruled out as a possible root cause.

Several obvious things were looked at to determine hydraulic stability such as suction piping and air entrainment. The pump had an elbow ~4 foot from the pump suction where it should have had ~8.5 feet of straight run pipe, so this was not ideal but did have some straight run and the elbow was in vertical plane not horizontal plane. Submergence on the supply tank was sufficient so no issues with air vortex entering the pump. The casing also had three high point casing vents, and all were piped open for continued pump operation so any initial air would be quickly dispelled.

The seal water supply system was evaluated, and the seal water chamber pressure should be ~48 psi but the actual pressure on supply header pressure gauge was only 20–24 psi. Even though there had only been a few failures

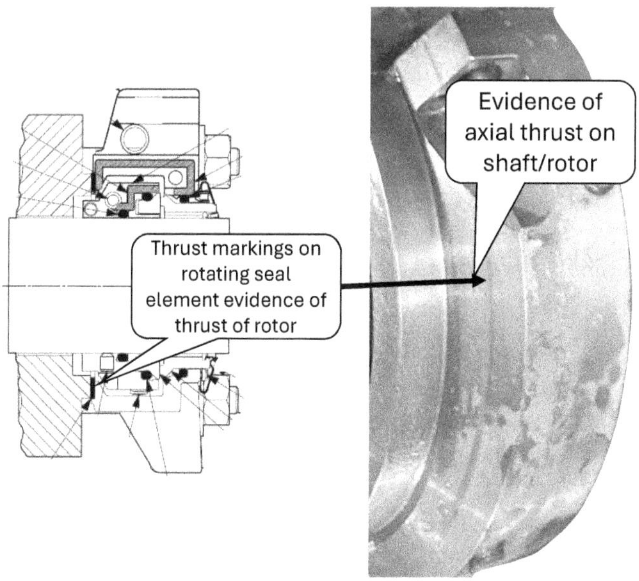

FIGURE 10.40 Seal face thrust indications on rotating element.

where some low consistency stock had gotten through the seal, low seal water supply pressure was an issue that needed to be corrected.

While not part of the pump failure mode, while inspecting the pump installation and operation, it was observed that the oil level for the bearings was low by about 3/16″ so oil was added to get to the center of the sightglass.

Looking back at the last maintenance records for the pump when the pump was opened up and the rotating assembly was changed (~9 years earlier), an interesting observation was seen from some pictures. The pump had some pitting damage on the cutwater indicating some recirculation-type cavitation damage, so an operating issue was potentially occurring even back when the pump was operating with packing.

Since the pump rotating assembly had operated for nine years (casing was ~20 years), a quick hydraulic efficiency check was completed. There was a flowmeter in the system which gave a good flow reading. The pump had a curve efficiency at the operating point of ~75%. The WHP calculated to be ~187 HP and the actual motor BHP calculated to be ~255 HP so the actual efficiency was ~73%. This was only 2% off the curve hydraulic efficiency so it appears that wear ring clearances were still in pretty good condition.

Looking at the pump operating point and hydraulics revealed much of the remaining mystery on the pump seal reliability issues. With a pump suction specific speed near 12,000, the pump operating point should > 50% of BEP flow to avoid suction recirculation. The actual pump operating point was 43% of BEP so suction recirculation is likely occurring. Now there are two pieces of evidence of suction recirculation cavitation potential.

Since the pump had a flowmeter, operating points were easy to determine. The pump was operating in a flow range of 2,460–2,900 gpm which was below the pump AOR as shown in Figure 10.41 pump curve. This was the third piece of collaborative evidence of suction and discharge recirculation flow turbulence and cavitation potential to cause issue for mechanical seal unreliability.

Operating in this region of the pump curve all the time increased pump shaft radial loads and shaft deflection which negatively affects mechanical seal reliability. With the recirculation flows, the pump shaft rotor becomes unstable and allowed the pump shaft to move axially as it shuttled back and forth between bearing clearances. A summary of the effects of the pump hydraulics is shown on the pump cross-section in Figure 10.42.

The evidence for rotor shuttling (axially moving) had all aligned together in the failure investigation.

- Recirculation cavitation from operating at low flow (outside AOR).
- Higher radial loads from outside AOR operation causing shaft deflection.
- Cavitation pitting around cutwater on pump casing.
- Mechanical seal rotating element thrust markings indicating thrusting had occurred.

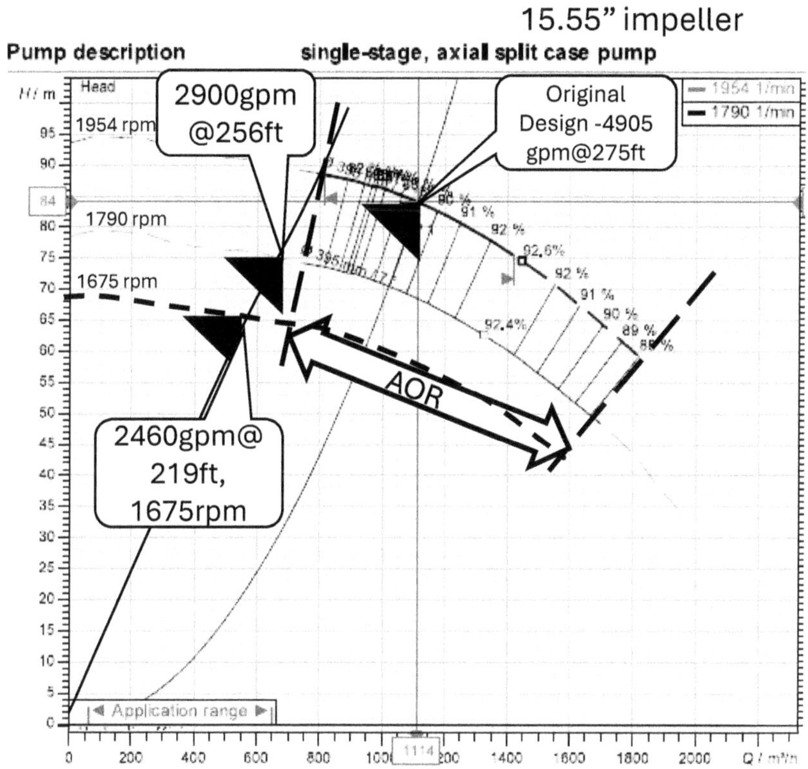

FIGURE 10.41 Pump curve and system curve with operating points.

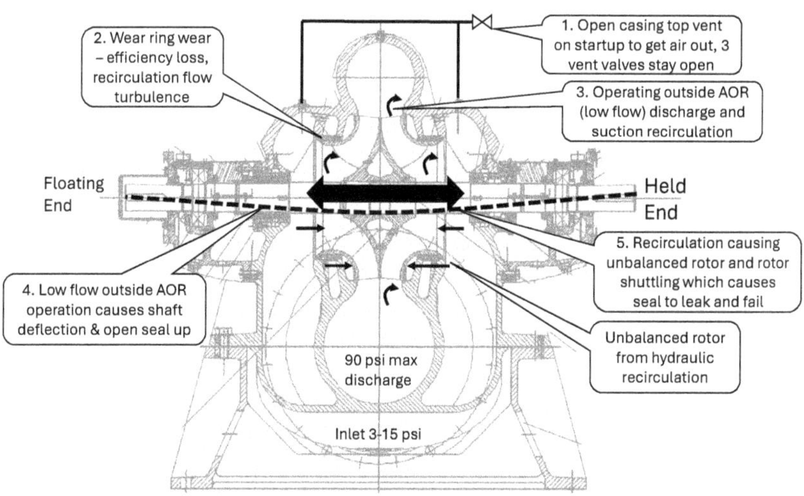

FIGURE 10.42 Pump cross-section showing pump hydraulic effects on pump shaft/rotor.

The root cause of failure was either a design issue (oversized pump for the process) or a process (pump system) change from the original process design that pushed the pump into operating outside its AOR. There is not an easy fix for the pump issue but there were many improvements that could be made to the pump and pump system. Below are some potential action items to address the deficiencies of this pump system reliability issues.

- Investigate the low-pressure issue on the seal water supply. The supply system has 80 psi on the header so appears to be a potential blockage in delivery pipe system to the pump seal.
- Install seal chamber bushing to give additional rotor stiffness and radial support to improve seal life.
- Convert sealing back to packing and remove the split mechanical seal.
- Consider potential option of a smaller pump for the lower current process conditions so the pump operates within the AOR and closer to POR.

Case Study 59 Learning – Ensure a double suction pump is operating within the AOR when using a mechanical seal to achieve good seal reliability.

With all the previous case studies involving pump failures and poor pump reliability, this last case study is highlighting very good pump reliability and what the pump operating conditions look like. There seem to be many more pump systems that operate with some deficiency that reduces pump reliability that it may seem hard to find the opposite condition where everything in the pump system from design, installation to maintenance and operation is right. The result of doing all that right is a pump that operates for a long time and with very high reliability.

CASE STUDY 60 PUMP OPERATING IN POR (AT BEP) – BEST RELIABILITY

This last short case study is one of success not failure. What does good pump operation look like with respect to pump reliability? As much of the focus is on root cause failure analysis, one could call this root cause success analysis of a pump system where great success was achieved without experiencing chronic failures. The final learnings may be the same, but in looking at healthy systems, the plant doesn't have to experience the pain and loss that failures give as the price for life experience.

The pump system is a simple system with a pump taking white water and stock from an atmospheric tank and supplying a diffuser in the floatation cell system, as shown in Figure 10.43. There is a small static head and pressure

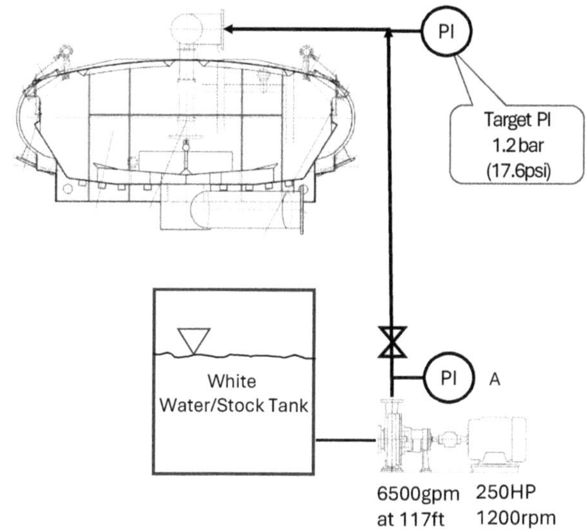

FIGURE 10.43 Process diagram for primary floatation cell feed pump.

head in the system as 17.6 psi is desired on the diffuser feed at the floatation cell inlet. All the head in the system is pretty much fixed and does not change so that does make the system more predictable and reliable. The pump is a 12 × 14 × 19 end suction stock pump with an 18.63″ × 17″ diameter impeller and driven by a 250 HP, 1,200 rpm fixed speed motor. The pump has an original design point of 6,500 gpm at 117 ft. The pump had a NSS = 10,056.

The pump and system have operated for ~15 years with little to no maintenance. The pump rotating assembly had never been changed. The mechanical seal had been changed twice and both times there were issues with the seal water flowmeter not providing enough seal water to the pump seal (flowmeter plugged with contamination). So, what is the root cause of success for this pump system? Why is it operating with such a high level of reliability?

One answer is that the pump operates in a steady state with basically one operating point. Unlike some complex systems where it is constantly changing what it is required to do, this one has a single operating condition. If that condition were not in a good spot for the pump, then poor reliability would result. The pump also has very simple controls and no automatic controls to manipulate. A pump system with multiple operating points or complex controls will likely have some operating points that are either not in POR or even near an AOR limit. It is also easy to have operator error with respect to operating the pump.

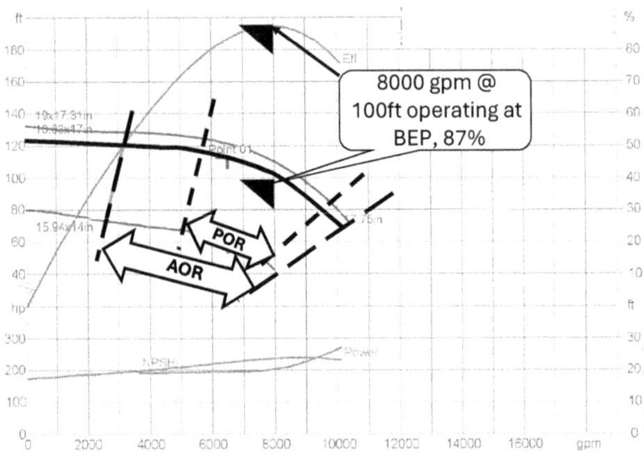

FIGURE 10.44 Pump curve with operating point at BEP.

This pump was operating at 8,000 gpm at 100 ft and 87% efficiency which is right at the BEP in the middle of the POR (even better than the original design point). Operating at the BEP, the pump has hydraulic conditions which are perfectly aligned for its hydraulic design so many of the typical issues are eliminated. Shaft loads are low; cavitation potential and other suction issues are non-existent. Even though the pump may have a design L10 bearing life of three years, under good conditions, it has operated over 15 years with no issues.

The pump operating point is shown on the pump curve in Figure 10.44, and it is right in the middle of the POR and at maximum efficiency at BEP. The pump suction-specific speed is ~11,000 so right at that edge of acceptable. The suction energy is $\sim 160 \times 10^6$ so NPSHmargin ~1.5 yielding an NPSHR ~30 ft, while the NPSHA is ~38 ft so cavitation is not a concern. The tank level typically runs ~12 ft level and steady state. The minimum submergence is ~8.5 ft so submergence and air entrainment are not an issue. All the other mechanical installation and design elements were also executed without any major issues, so the result is a reliable pump system with very little issues. The only issue that had occurred twice by maintenance records was around the mechanical seal and both times issues with the flowmeter function that caused the issue.

The reason for outstanding pump reliability results was very good design, good operation, and good maintenance. All three legs had to be executed in precision fashion.

Case Study 60 Learning – The better the pump operates in the POR, the better pump reliability will result.

From ten case studies in this book, the average suction-specific speed (NSS) was 10,241. Most of these pump reliability issues were resolved through some sort of pump or pump system modification and was restored to good pump reliability results. The last case study of a pump system which had very good pump reliability and one other pump case study after a pump change had an average NSS = 11,262. A general look at NSS would have a pump above NSS = 10,000–11,000 then poor reliability might be expected but that was not the case on any of these. One pump case study had a NSS = 553 and still had multiple reliability issues and short life. The takeaway would be that other factors, such as where the pump is operating within the AOR and POR, are a greater predictor of pump reliability than just looking at NSS as shown by the case studies in this chapter.

There are many details to pay attention in pump systems if a high reliability is desired as a result. As a summary, I have listed some key things to execute or evaluate to optimize your pump system reliability for the life cycle of your pump system.

1. Follow all pump safety guidelines, procedures, and regulations. This includes installation, maintenance, and operation. From rigging to energy isolation during maintenance to proper steps for startup.
2. Avoid dead head (no flow) pump operation for extended periods of time.
3. Size the right pump for the application to meet the process needs (flow and head).
4. Select the correct style pump for the application (end suction, double suction, open/closed/recessed impeller, etc.).
5. When evaluating an existing pump system, use multiple methods to determine pump operating points and conditions.
6. Select a pump at the lowest speed to meet process conditions when feasible. Lower speed means better reliability from mechanical to hydraulic.
7. Choose pump with a driver with full curve runout for impeller size if possible.
8. Select pump with good shaft stiffness ($L3/D4 < 55$) if available for the design point and application.
9. For optimum seal life, ensure that shaft deflection < 0.002″ over the operating range of the pump.
10. Use oil lubrication if feasible, available, or appropriate for selected pump design.
11. If pump shafts move (oscillate) or impeller clearance adjustment is often, or shaft deflection is likely then utilize packing for shaft sealing method if the application allows.
12. Provide proper seal water supply – quality (cleanliness) – filter with 30–50-micron filtration.
13. Provide proper seal water supply – pressure 10–15 psi above stuffing box/ seal chamber pressure.
14. Provide proper seal water supply – flow 0.3–1 gpm.
15. Use the right lubricant – correct viscosity (46, 68, or 100)
16. Use the right lubricant – amount/volume (level or grease dose)
17. Use the right lubricant – quality/cleanliness.

18. Change pump oil either annually for mineral oil or every two years with synthetic oil unless oil analysis or other known contamination has occurred.
19. Verify pump materials are appropriate for fluid conditions – impeller, wear plate, casing, sealing, shaft sleeve, etc.
20. Install pump and driver mechanical power transmission components correctly – motor, couplings, drive belts.
21. Make purchasing decisions considering LCC not just initial purchase price for pump system or pump components.
22. Get pump suction piping design and installation correct – pipe diameter sizing and fluid velocity for suction piping.
23. Get pump suction piping design and installation correct – straight run pipe into pump suction (10D).
24. Get pump suction piping design and installation correct – eccentric reducer installed correctly (flat side on top).
25. Get pump discharge piping design and installation correct – pipe diameter sizing and fluid velocity for discharge piping.
26. Get pump discharge piping design and installation correct – check valve and isolation valve in correct order and seating correct.
27. Install suction and discharge pressure gauges for pump performance evaluation.
28. Clean out sump pits annually to remove sediment buildup blocking pump suction intakes.
29. Inspect all process tanks annually for visual damage, cracks, leaks, etc.
30. Install proper pump system piping supports (type hangers, location, size) to eliminate pipe stress to pump nozzles.
31. Install solid foundation and baseplate for pump and driver – enough mass (3–5X pump mass) and size (30 angle from shaft centerline to floor).
32. Install solid foundation and baseplate for pump and driver – choose correct grout (epoxy grout for chemical, Cementous grout for temperature).
33. Install solid foundation and baseplate for pump and driver – sleeve anchor bolts and shim when grouting.
34. Install solid foundation and baseplate for pump and driver – install alignment jackbolts to aid alignment and motor movement.
35. Install solid foundation and baseplate for pump and driver – baseplate design to accept minimum 0.125″ shims at motor feet.
36. Align pump and driver within ISO standards.
37. Align pump and driver adjusting for thermal growth, when necessary, on first alignment or perform hot alignment on more critical applications.
38. Remove pipe strain from piping system when connecting to pump nozzles $<0.002''$ shaft impact.
39. Make sure suction and discharge pipes are aligned with pump nozzles.
40. Eliminate soft foot on pump and motor alignment $< 0.002''$.
41. Ensure shaft TIR $< 0.002''$ at coupling.
42. Satisfy minimum pump suction submergence to avoid vortex formation and air entrainment. Utilize vortex breaker if necessary.
43. Locate inlet piping to tanks away from pump suction.

44. Locate inlet piping into tanks below the fluid level in tanks to minimize air entrainment.
45. Set tank levels at proper level above agitators to eliminate air entrainment.
46. If recirculation line is used, route recirculation line back to tank away from pump suction.
47. Ensure NPSHmargin to avoid harmful cavitation (1.2–2.5 pending suction energy).
48. Operate suction tank level as high as possible to eliminate pump suction issues and improve fluid transfer efficiency.
49. Do not operate a pump in a drooping region of the pump curve.
50. Best to operate similar (or exactly the same) size pumps in parallel so each are hydraulically loaded evenly.
51. Optimize pump impeller size so the pump system control valves operate within the POR (40–60% open).
52. Utilize VFD in high friction head pump systems (> 50% friction head)
53. Verify motor torque at low speeds for VFD pump systems.
54. Always verify pump shaft rotation is correct.
55. Before pump startup, verify oil level is correct.
56. Before pump startup, verify that pump system valves are in the correct positions (open, shut or throttled).
57. Never apply open flame heat to an impeller to remove.
58. Inspect and measure pump impeller diameter during pump maintenance.
59. Inspect condition of wear plate wear during pump maintenance.
60. Do not operate pumps with excessive clearance between the impeller and stuffing box.
61. Verify the correct impeller is installed on the pump during pump maintenance – correct pattern #, # vanes, diameter.
62. Always replace impeller, impeller nut, and shaft sleeve orings on pump rebuilds.
63. Torque impeller nut to pump manufacturer specifications.
64. Set the impeller clearance to the pump manufacturer recommendations.
65. When using packing make sure the correct arrangement is installed for the specific pump design (std. 2L3).
66. Verify that the packing latern ring lines up with the seal water port on stuffing box.
67. Install packing correctly – correct length and OD for proper fit in the stuffing box.
68. Install packing correctly – seat packing in box.
69. Install packing correctly – stagger joints.
70. Set the seal water pressure to the correct setting for the application.
71. Use a dedicated and filtered seal water system supply when possible.
72. Consider a mud leg at the end of seal water supply lines.
73. Verify the correct oil level when the pump is down. Note running level if checking while operating (may be different than when pump is idle).
74. Install pumps level on installation so pump oil level is accurate.
75. If using a constant level oiler, make sure it is installed on the correct side of the pump rotation.

76. If using a constant level oiler, provide an additional visual sightglass to confirm oil level.
77. Ensure oil is clean to at least 17/15/12 ISO cleanliness level.
78. Avoid mixing lubricating greases if possible. If it becomes necessary, always verify the grease compatibility by consulting lubricant suppliers or performing grease compatibility testing.
79. Measure shafts, bearings and calculate bearing fits when rebuilding pump rotating assemblies.
80. Verify the correct bearings are used on pump rebuilds – size, arrangement, RIC, etc.
81. Avoid bearing housing cooling, if possible, to prevent bearing issues from excessive bearing RIC reduction.
82. For roller bearings, measure bearing RIC. Install bearings within the target RIC reduction targets by the bearing manufacturer.
83. Install all pump fasteners to the specific bolt torque per pump manufacturer.
84. Use appropriate bolt torque procedures for torque – dry or lubricated conditions, proper torque wrench procedure.
85. Use appropriate bolt torque procedures for torque – star pattern and ramped torque amplitudes on bolt circle.
86. Use appropriate bolt torque procedures for torque – verify surfaces are flat and clean.
87. Utilize predictive maintenance tools to prevent, predict, and monitor pump system health – vibration analysis, oil analysis, infrared, and ultrasonic.
88. Balance pump rotors within G1.0 to G6.3 balance grade most of the time depending on the pump type and speed. Consult pump manufacturers recommendations.
89. Utilize vibration analysis to identify basic pump defects – bearing defects (BPFO, BPFI, cage, ball spin), vane pass, flow turbulence, cavitation, misalignment, looseness, unbalance, resonance, etc.
90. During pump assembly install with correct keylength and key phase as pump is connected to driver.
91. Match the pump and pump system so that pump never operates outside AOR.
92. Design and operate pumps within the POR when possible.
93. Design and operate pumps as close to BEP as possible.
94. For on/off pump system controls, do not exceed NEMA starts/hr for electric motors.
95. If pump controls are run in manual, make sure that the pump operates within the AOR.
96. Install pumps for optimum maintainability where maintenance has proper access to perform routine precision maintenance.
97. Conduct proper training for pump maintenance technicians and operators on pump systems.
98. Establish basic care rounds to inspect pump systems for minor defects – leaks, noise, high temperature, etc.
99. Audit large or critical pump systems annually to evaluate pump performance and efficiency.
100. Perform root cause failure analysis on pump failures to determine corrective actions to prevent future failures.

CLOSING THOUGHTS

In the pursuit of pump and system reliability, there are many distractions that can blur the pathway to success. Some of these are clues or symptoms of a bigger issue. It can be very tempting to treat the symptom instead of solving the root cause of the symptom. Sometimes just trying to differentiate between the two is difficult.

Part of solving the riddle of pump failures and pump system issues is to first understand the fundamentals of how the pump and system works or is supposed to work. While there are many typical pump system elements which are the same, every pump system can have its own custom differences to evaluate. Solving pump reliability issues works in reverse much of the time. It is hard to solve a problem on a system that you don't understand. So first try and understand the pump system as it is designed.

Next define the pump failure mode and how it is failing or acting. Analyze the parts that are damaged or failed and identify the physics of failure for the pump components. Working this through all the possible causes of what can cause the damage on the components. This will require either personal knowledge on the pump and system or seeking the advice or assistance from a subject matter experts to find answers.

There are lots of mechanical design and operating conditions of the pump that affect pump reliability. However, based on many case studies and over 35 years in the pursuit of pump reliability, the system is most of the time the source of the issues in pump systems. It has been said that the pump is a dumb machine and only operates as the system allows it. While there are at times reliability issues within the pump design itself, the system is the driving factor to achieve pump reliability. When there are pump issues it is many times the result of improver maintenance where mistakes were made. The system curve and pump operating point being inside the AOR and POR are critical to address the hydraulic issues that drive pump reliability.

The solutions for pump reliability improvements are also many and encompass pump improvement and pump system elements. These case studies have shown the varied effect of pump operating points with respect to AOR and POR will be a major factor in how reliable the pump and system will ultimately achieve. Most of these case studies show how poor reliability resulted by operating the pump outside the POR and many times outside the AOR. Look back through each case study where the pump performance curve was shown and note the operating condition with respect to the AOR and POR. Poor reliability results every time the pump operates outside of AOR. On the opposite side, Case Study 60 showed that operating a pump inside the AOR and POR resulted in the best reliability.

I hope some of the case studies and experiences I've shared to improve pump system reliability over my career have helped young pump and maintenance reliability specialists achieve even greater success than what my experience has been and that others will push pump reliability to even greater levels of performance.

Index

Pages in *italics* refer to figures and pages in **bold**-*italics* refer to boxes.

R

S

T

For Product Safety Concerns and Information please contact our EU representative GPSR@taylorandfrancis.com
Taylor & Francis Verlag GmbH, Kaufingerstraße 24, 80331 München, Germany

www.ingramcontent.com/pod-product-compliance
Ingram Content Group UK Ltd.
Pitfield, Milton Keynes, MK11 3LW, UK
UKHW020022200926
475702UK00014B/2017
9781041217015